Polymer-Based Flexible Materials, 2nd Edition

Polymer-Based Flexible Materials, 2nd Edition

Guest Editors

Jiangtao Xu
Sihang Zhang

MDPI

Basel • Beijing • Wuhan • Barcelona • Belgrade • Novi Sad • Cluj • Manchester

Guest Editors

Jiangtao Xu
College of Materials
and Energy
South China Agricultural
University
Guangzhou
China

Sihang Zhang
School of Food Science and
Engineering
Hainan University
Haikou
China

Editorial Office
MDPI AG
Grossspeteranlage 5
4052 Basel, Switzerland

This is a reprint of the Special Issue, published open access by the journal *Polymers* (ISSN 2073-4360), freely accessible at: www.mdpi.com/journal/polymers/special_issues/9IY4FWHRRR.

For citation purposes, cite each article independently as indicated on the article page online and using the guide below:

Lastname, A.A.; Lastname, B.B. Article Title. *Journal Name* **Year**, *Volume Number*, Page Range.

ISBN 978-3-7258-3998-8 (Hbk)
ISBN 978-3-7258-3997-1 (PDF)
https://doi.org/10.3390/books978-3-7258-3997-1

Cover image courtesy of Sihang Zhang

Contents

About the Editors

Jiangtao Xu

Dr. Jiangtao Xu obtained his Ph.D. degree from Hong Kong Polytechnic University in 2021 and subsequently worked as a Postdoctoral Researcher there from 2021 to 2023. He is currently an Associate Professor and Master's Supervisor at South China Agricultural University. His research primarily focuses on biomass composite materials, the functional modification of biomass materials, and the comprehensive utilization of biomass resources. Specifically, he works with nanocellulose, wood, and other biomass-derived materials to develop high-performance SERS substrates, flexible stress sensors, and electromagnetic shielding materials. Dr. Xu is committed to innovating and developing high-value-added biomass materials for applications in food safety, environmental security, and healthcare. As a first author or corresponding author, he has published multiple SCI papers in prestigious journals such as *Materials Today*, *Nano Research*, *Composites Part B*, and *ACS Applied Materials & Interfaces*. He has also secured several competitive research grants, including projects funded by the National Natural Science Foundation of China (NSFC) and the Natural Science Foundation of Guangdong Province.

Sihang Zhang

Prof. Zhang Sihang is an Associate Professor at Hainan university. In June 2018, he obtained a Master's degree in Engineering from Sichuan University, and in June 2021, he earned a PhD in Materials Science from the State Key Laboratory of Polymer Materials Engineering at Sichuan University. Subsequently, from September 2021 to November 2023, he conducted postdoctoral research in the field of food safety detection at Hong Kong Polytechnic University. In 2023, he joined Hainan University as a Distinguished Associate Professor and a supervisor for Master's and PhD students. His research areas include food quality and safety detection, the preparation of flexible SERS substrates, and the design of organic and inorganic nanomaterials. He has published over 30 SCI papers in journals such as *J. Energy Chem.*, *J. Power Sources*, *ACS Appl. Mater. Interfaces*, *ACS Sustain. Chem. Eng.*, *Anal. Chem.*, *J. Hazard. Mater.*, and *Food Chem.* He serves as a Guest Editor for the SCI journal *Polymers*. He has been the principal investigator for projects such as the National Natural Science Foundation of China (Youth Program), Hainan Province Natural Science Foundation General Program, Hainan University High-level Talent Startup Fund, Hong Kong Special Administrative Region Government Research Talent Pool Postdoctoral Project (RTH-ITF), and the Sichuan Provincial Science and Technology Department Miaozi Project.

Preface

"Polymer-Based Flexible Materials, 2nd Edition" continues to explore this dynamic and rapidly evolving field, focusing on the diverse applications and innovations that have emerged since the publication of the first edition. This revised edition aims to provide a comprehensive overview and in-depth analysis of polymer-based materials, particularly those with key properties such as flexibility, elasticity, and stretchability—attributes that are critical for a wide range of modern applications.

The reprint covers the latest advancements in the synthesis, processing, and characterization of flexible polymer materials, with special emphasis on their potential in wearable technology, soft robotics, and flexible electronics. As the demand for lightweight, cost-effective, and multifunctional materials continues to grow, this work highlights the pivotal role of polymers in shaping the future of materials science and engineering. Designed as an authoritative resource for researchers, engineers, and students in materials science, polymer chemistry, and engineering, the reprint presents research on polymer-based flexible materials in an accessible yet scientifically rigorous manner—spanning fundamental principles to cutting-edge breakthroughs. It brings together the collective expertise of an interdisciplinary team of contributors, each specializing in various aspects of polymer-based materials. We extend our deepest gratitude to all the authors and contributors who shared their valuable insights and knowledge. Their dedication and hard work made this second edition possible.

We also express our sincere appreciation to the academic institutions and colleagues who provided support during the compilation of this book. Without their guidance and assistance, this project would not have been completed. It is our hope that "Polymer-Based Flexible Materials, 2nd Edition" will serve as a comprehensive and inspiring reference for researchers and practitioners in the field, while contributing to the continued advancement of this important discipline.

Jiangtao Xu and Sihang Zhang
Guest Editors

polymers

MDPI

Article

Preparation of Polymerized High Internal Phase Emulsion Membranes with High Open-Cellular Extent and High Toughness via RAFT Polymerization

Yulan Wu [1], Jie Huang [1], Zanru Guo [1], Qian Yang [1], Chunmiao Xia [2] and Zhenan Zheng [1,*]

1 Fujian Provincial Key Laboratory of Modern Analytical Science and Separation Technology, College of Chemistry, Chemical Engineering and Environment, Minnan Normal University, Zhangzhou 363000, China
2 Anhui Laboratory of Clean Energy Materials and Chemistry for Sustainable Conversion of Natural Resources, School of Chemical and Environmental Engineering, Anhui Polytechnic University, Wuhu 241000, China
* Correspondence: zzn1963@mnnu.edu.cn

Abstract: Porous polymer membranes with highly interconnected open-cellular structure and high toughness are crucial for various application fields. Polymerized high internal phase emulsions (polyHIPEs), which usually exist as monoliths, possess the advantages of high porosity and good connectivity. However, it is difficult to prepare membranes due to brittleness and easy pulverization. Copolymerizing acrylate soft monomers can effectively improve the toughness of polyHIPEs, but it is easy to cause emulsion instability and pore collapse. In this paper, stable HIPEs with a high content of butyl acrylate (41.7 mol% to 75 mol% based on monomers) can be obtained by using a composite emulsifier (30 wt.% based on monomers) consisting of Span80/DDBSS (9/2 in molar ratio) and adding 0.12 mol·L^{-1} $CaCl_2$ according to aqueous phase concentration. On this basis, polyHIPE membranes with high open-cellular extent and high toughness are firstly prepared via reversible addition–fragmentation chain transfer (RAFT) polymerization. The addition of the RAFT agent significantly improves the mechanical properties of polyHIPE membranes without affecting open-cellular structure. The toughness of polyHIPE membranes prepared by RAFT polymerization is significantly enhanced compared with conventional free radical polymerization. When the molar ratio of butyl acrylate/styrene/divinylbenzene is 7/1/1, the polyHIPE membrane prepared by RAFT polymerization presents plastic deformation during the tensile test. The toughness modulus reaches 93.04 ± 12.28 kJ·m^{-3} while the open-cellular extent reaches 92.35%, and it also has excellent thermal stability.

Keywords: polyHIPE membranes; RAFT polymerization; open-cellular extent; toughness

Academic Editors: Mauro Carraro and Francesco Galiano

Received: 26 December 2024
Revised: 24 January 2025
Accepted: 14 February 2025
Published: 17 February 2025

Citation: Wu, Y.; Huang, J.; Guo, Z.; Yang, Q.; Xia, C.; Zheng, Z. Preparation of Polymerized High Internal Phase Emulsion Membranes with High Open-Cellular Extent and High Toughness via RAFT Polymerization. *Polymers* **2025**, *17*, 515. https://doi.org/10.3390/polym17040515

1. Introduction

Porous polymer membranes play a key role in various application fields, such as separation analysis [1–3], sensors [4,5], energy storage and conversion, etc. [6–9]. An ideal porous polymer membrane should possess a highly interconnected open-cellular structure and high toughness. The highly interconnected open-cellular structure is beneficial to achieving high mass transfer flux with low resistance, while high toughness helps to maintain structure integrality when subjected to external force. The preparation method of porous polymer materials usually includes foaming, phase separation, pore-forming agent, etc. [10–12]. However, these methods lack the regulation of the pore structure of materials.

By contrast, polymerized high internal phase emulsions (polyHIPEs) provide a facile method to build open-cellular structures with high porosity and good pore connectivity.

Typical polyHIPEs are monoliths with a high glass transition temperature (T_g), which are produced by free radical crosslinking copolymerization of monomers such as styrene (St) and divinylbenzene (DVB). Although having a high modulus and strength, typical polyHIPEs are extremely brittle, thus easily leading to fragmentation during the preparation of membranes. The mechanical properties of typical polyHIPEs can only be measured by a compression test instead of a tensile test [13]. Therefore, improving the toughness of polyHIPEs is critical in order to expand their application fields [13,14]. In order to address the above problems, Kovačič et al. [13,15–17] prepared polyHIPE membranes with improved toughness by ring-opening metathesis polymerization (ROMP) of dicyclopentadiene (DCPD). However, ROMP of DCPD requires strict control of reaction conditions, such as temperature, pressure, and solvent selection, to avoid the occurrence of side reactions and the instability of the product structure [15,16].

Free radical polymerization is a more convenient process due to the wide range of applied monomers and mild reaction conditions. But to prepare polyHIPE membranes with high open-cellular extent and high toughness via free radical polymerization is still challenging. By copolymerizing acrylate soft monomers, the T_g of the polymer can be significantly reduced, thereby improving the toughness of polyHIPEs [18]. However, due to the high polarity of the acrylate monomer, Ostwald ripening is overwhelming when its dosage is large, which is detrimental to the formation of a stable emulsion [19]. Furthermore, copolymerizing too many soft monomers will lead to the collapse and closure of pores, affecting the formation of an open-cellular structure [20]. To date, the concerned research is limited, and there is a lack of discussion on the intrinsic structure and mechanical properties of the materials.

Previous studies have shown that controlled/"living" free radical polymerization (CLRP) realizes the controllability of the polymerization process by changing the kinetic characteristics, thereby obtaining a more uniform crosslinking structure and further improving the mechanical properties of the material [21–26]. This provides a new direction for the preparation of polyHIPE membranes with high open-cellular extent and high toughness. In this paper, polyHIPE membranes composed of monomers, including styrene (St), butyl acrylate (BA), and divinylbenzene (DVB), composite emulsifiers including sorbitol monooleate (Span 80) and sodium dodecyl benzene sulfonate (DDBSS), are firstly prepared via reversible addition–fragmentation chain transfer (RAFT) polymerization, one of the controlled/"living" free radical polymerizations. The effects of RAFT polymerization and monomer composition on the intrinsic structure and mechanical properties of polyHIPE membranes are systematically investigated in order to achieve high open-cellular extent and high toughness.

2. Experimental Section

2.1. Materials

In order to remove the inhibitor, styrene (St, M_n = 104.15 g·mol^{-1}, Xilong Scientific Co., Ltd., Shantou, Guangdong, China) was distilled under reduced pressure, and butyl acrylate (BA, M_n = 128.19 g·mol^{-1}, Xilong Scientific Co., Ltd., Shantou, Guangdong, China) was washed with a 10 wt.% sodium hydroxide solution for more than three times. Divinylbenzene (DVB, M_n = 130.19 g·mol^{-1}, 80%, Sigma-Aldrich, St. Louis, MO, USA), 2,2′-azobis(2-methylpropionitrile) (AIBN, M_n = 164.21 g·mol^{-1}, 99%, Shanghai Aladdin Biochemical Technology Co., Ltd., Shanghai, China), sorbitan oleate (Span 80, M_n = 428.61 g·mol^{-1}, Xilong Scientific Co., Ltd., Shantou, Guangdong, China), sodium dodecyl benzene sulfonate (DDBSS, M_n = 348.48 g·mol^{-1}, 90%, Macklin, Shanghai, China), and calcium chloride ($CaCl_2$, M_n = 110.98 g·mol^{-1}, anhydrous, Xilong Scientific Co., Ltd., Shantou, Guangdong, China) were used as purchased without further purification. The

2-{[(Dodecyl-sulfanyl) carbonothioyl] sulfanyl} propanoic acid (M_n = 350 g·mol^{-1}) was synthesized and purified as described in reference [27], which was used as the RAFT agent. The molecular formula of the RAFT agent is shown in Figure 1.

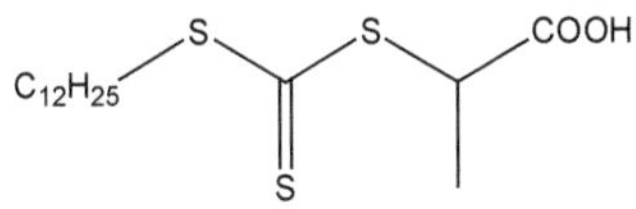

Figure 1. The molecular formula of RAFT agent.

2.2. Preparation of polyHIPE Membranes

Take RAFT-2 as an example: the organic phase was firstly prepared by mixing 4.480 g BA (3.5 × 10^{-2} mol), 0.809 g St (7.77 × 10^{-3} mol), 0.506 g DVB (3.89 × 10^{-3} mol), and 0.0652 g the RAFT agent (1.86 × 10^{-4} mol), 1.472 g Span 80 (3.43 × 10^{-3} mol) and 0.0513 g AIBN (3.12 × 10^{-4} mol). Then, it was poured into a three-necked round-bottom flask. An amount of 0.266 g $CaCl_2$ (2.40 × 10^{-3} mol), 0.266 g DDBSS (7.63 × 10^{-4} mol) was dissolved in 20 g of deionized water to form the aqueous phase. The aqueous phase was then dropwise added to the organic phase in the flask while stirring at 300 rpm. Once the addition was completed, the stirring speed was adjusted to 1600 rpm and continued for 30 min to obtain a viscous, homogeneous emulsion. The emulsion was poured into a mold and transferred to a 70 °C drying oven for polymerization. After 72 h, the samples were removed from the mold and placed in a Soxhlet extractor. They were first extracted with deionized water for 24 h, followed by anhydrous ethanol for another 24 h. After extraction, the samples were vacuum dried at 70 °C for 10 h, and polyHIPE membranes were finally obtained.

2.3. Characterization

Pore structure. The scanning electron microscope (JSM-6010LA, JEOL Ltd., Tokyo, Japan) was applied to observe the internal structure of polyHIPE membranes. The membrane surface was sputtered for 120 s in a vacuum before observation, and the test voltage was 10 kV.

Specific surface area. The surface area and pore size analyzer (Tristar II 3020, Micromeritics, Norcross, GA, USA) was applied to measure the specific surface area of polyHIPE membranes. The temperature of nitrogen adsorption was set as −196 °C, and the degassing temperature was 100 °C.

Porosity. The mercury porosimeter (Auto Pore IV 9510, Micromeritics, Norcross, GA, USA) was applied to measure the porosity of polyHIPE membranes.

Tensile properties. The tensile properties of polyHIPE membranes were measured using a Zwick/Roell Z020 (ZwickRoell, Ulm, Germany). The shape of the test bar conformed to GB/T 1040.2-2006 [28]. The testing was conducted at 15 °C with a tensile speed of 1 mm·min^{-1}. The tensile test was repeated at least three times. The modulus of toughness (M_T) was calculated using Equation (1) as follows:

$$M_T = \int_0^{\varepsilon_f} \sigma d\varepsilon \quad (1)$$

where σ is tensile strength, ε is strain, and ε_f is elongation at break.

Heat stability. The heat stability of polyHIPE membranes was measured using a Pyris 1 TGA (Perkin-Elmer, Ulm, Germany) in a nitrogen atmosphere. The range of test temperature was 30 °C to 700 °C with a heat speed of 10 °C·min^{-1}.

Glass transition temperature. A TA Q200 differential scanning calorimeter (TA Instruments, New Castle, DE, USA) was applied to measure the T_g of polyHIPE membrane. The test was conducted in a nitrogen atmosphere, and the test temperature range was −40 °C to 100 °C with a heat speed of 10 °C·min^{-1}.

3. Results and Discussion

The introduction of polar monomers into the HIPE is conducive to Ostwald ripening, resulting in emulsion instability and thus affects the structure and properties of polyHIPEs [29]. Previous studies have shown that Ostwald ripening can be effectively inhibited by introducing electrolyte [30] and a composite emulsifier [31] composed of nonionic and anionic emulsifiers. In this paper, a composite emulsifier (Span80/DDBSS equal to 9/2 in molar ratio) and an electrolyte (0.12 mol·L^{-1} $CaCl_2$) were applied to enhance the emulsion stability. The amount of composite emulsifier was 30 wt.% of the total monomers. The synthesized recipes and the corresponding emulsion stability are summarized in Table 1. It can be seen that emulsions with a BA content less than 75 mol% remain stable at 25 °C for more than 15 days, indicating superior stability. However, when the BA content reaches 92 mol%, the emulsion becomes unstable after one day, which is attributed to serious Ostwald ripening.

Table 1. The recipe and stability of HIPEs.

Sample	BA/St/DVB [a]	RAFT [b] (mol%)	Emulsifier (30 wt.%)	Electrolyte (mol·L^{-1})	Initiator [b] (mol%)	O/W [c]	Emulsion Stability (days)
			Span80/DDBSS	$CaCl_2$	AIBN		
RAFT-1	11/0/1	0.4	9/2	0.12	0.67	1/3	>1
FRP-2	9/2/1	—					>15
RAFT-2	9/2/1	0.4					>15
RAFT-3	7/4/1	0.4					>15
RAFT-4	5/6/1	0.4					>15
RAFT-5	3/8/1	0.4					>15
RAFT-6	7/2/3	0.4					>15

[a] molar ratio, [b] molar ratio of monomers, [c] volume ratio.

The stable HIPEs were poured into the mold followed by thermally initiated polymerization to prepare polyHIPE membranes. The effect of conventional free radical polymerization (FRP) and RAFT polymerization on the structure and properties of polyHIPE membranes are compared while maintaining the molar ratio of BA/St/DVB equal to 9/2/1. As shown in Figure 2, the size of the samples is basically the same and the appearance is white, indicating that there is an obvious phase separation structure inside the samples. As presented in Figure 3, two kinds of polyHIPE membranes both exhibit an interconnecting open-cellular structure. Herein, the pore connectivity rate (P_C) is defined to characterize the open-cellular extent inside polyHIPE membranes, which can be calculated by Equation (2):

$$P_C = P_A/P_T \tag{2}$$

where P_A is the actual porosity measured by a mercury porosimeter and P_T is the theoretical porosity calculated by Equation (3):

$$P_T = \frac{V_{H_2O}}{V_{H_2O} + V_{St} + V_{DVB} + V_{BA}} = \frac{m_{H_2O}/\rho_{H_2O}}{m_{H_2O}/\rho_{H_2O} + m_{St}/\rho_{St} + m_{DVB}/\rho_{DVB} + m_{BA}/\rho_{BA}} \tag{3}$$

where the V_{H_2O} is the volume of water phase, V_{St}, V_{DVB}, V_{BA} are the volumes of different ingredients, m_{St}, m_{DVB}, m_{BA} are the actual weights of different ingredients, and

$\rho_{H_2O}, \rho_{St}, \rho_{DVB}, \rho_{BA}$ are the densities of H_2O, St, DVB and BA, which are 1, 0.902, 0.919 and 0.898 g·cm^{-3}, respectively [32].

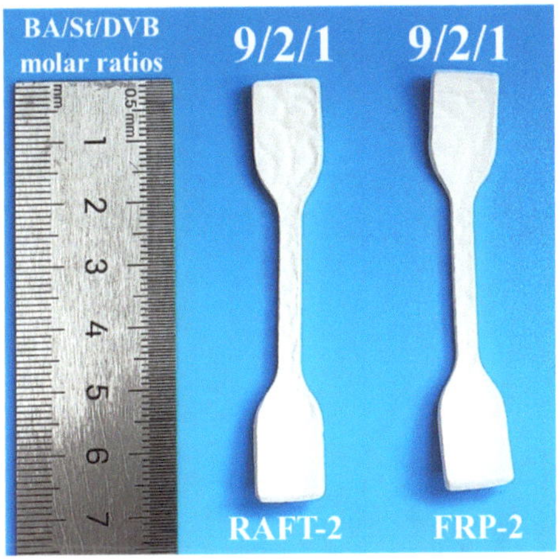

Figure 2. The optical images of polyHIPE membranes prepared by FRP and RAFT polymerization.

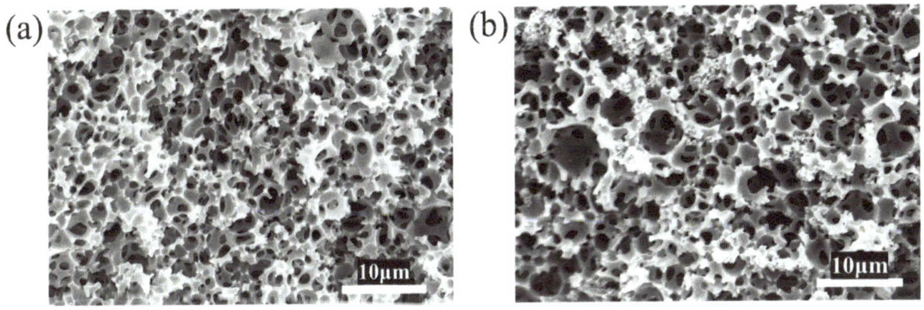

Figure 3. SEM images of polyHIPE membranes prepared by FRP and RAFT polymerization: (**a**) RAFT-2, (**b**) FRP-2.

The results of mercury porosimeter are summarized in Table 2. As can be seen, the specific surface area and pore connectivity of RAFT-2 and FRP-2 are basically consistent. However, the Pc of RAFT-2 and FRP-2 is merely around 90%. It is speculated that the high proportion of the soft monomer leads to the insufficient mechanical strength and thus results in partial pore collapse and closure [20,33], which will be elaborated in the following discussion. As shown in the pore size distribution curves (Figure 4), RAFT-2 contains a large amount of smaller mesopores with a size about 1–2 nm, which does not appear in FRP-2, indicating remarkable difference in microstructure for materials prepared by RAFT polymerization and conventional free radical polymerization [34]. The stress–strain curves of synthesized polyHIPE membranes are shown in Figure 5 with results summarized in Table 2. As can be seen, the tensile strength, elongation at break, and modulus of toughness of RAFT-2 are obviously higher than those of FRP-2. Specifically, the tensile strength enhances from 266.8 ± 7.05 kPa to 347 ± 4.31 kPa, and elongation at break increases from 17.36 ± 2.40% to 24.00 ± 3.26%. Correspondingly, the M_T of RAFT-2 reaches to 47.60 ± 7.88 kJ·m^{-3}, which is about twice that of FRP-2.

Table 2. The structure and mechanical properties of polyHIPE membranes prepared by FRP and RAFT polymerization.

Sample	P_A (%)	P_T (%)	P_C (%)	BET ($m^2 \cdot g^{-1}$)	E [a] (MPa)	σ [b] (kPa)	ε_f [c] (%)	M_T [d] ($kJ \cdot m^{-3}$)
RAFT-2	67.21	75.69	88.80	3.42	2.04 ± 0.34	347 ± 4.31	24.00 ± 3.26	47.60 ± 7.88
FRP-2	68.71	75.99	90.42	3.25	1.63 ± 0.76	266.8 ± 7.05	17.36 ± 2.40	23.86 ± 8.71

[a] Young's modulus, [b] ultimate tensile strength, [c] elongation at break, [d] modulus of toughness.

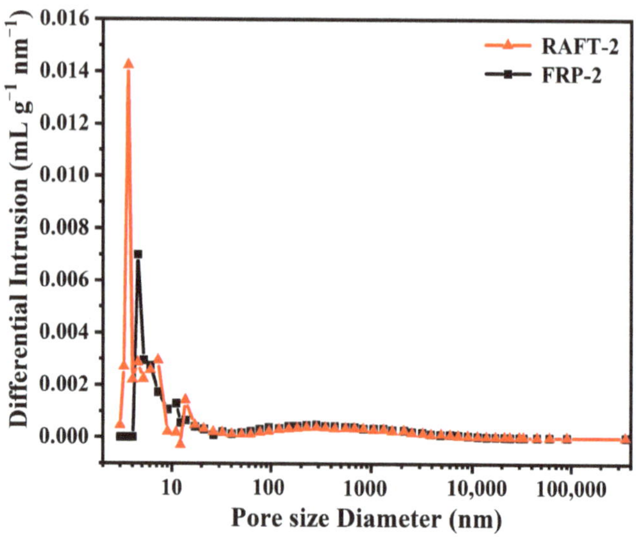

Figure 4. The pore size distribution curves of polyHIPE membranes prepared by FRP and RAFT polymerization.

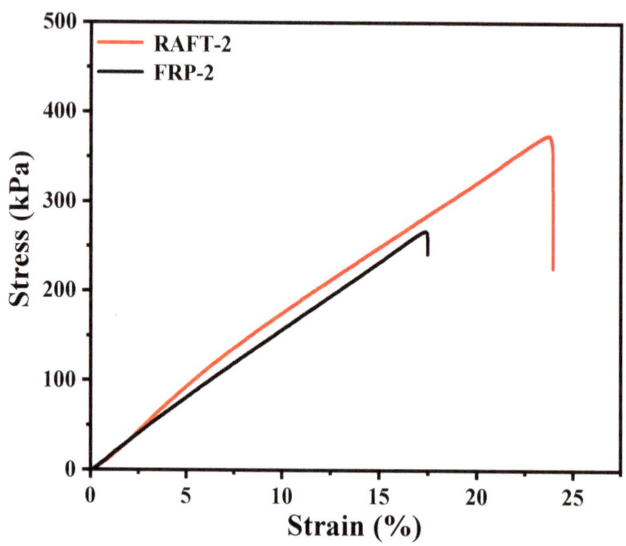

Figure 5. The stress–strain curves of polyHIPE membranes prepared by FRP and RAFT polymerization.

The difference in the structure and properties of polyHIPE membranes prepared by RAFT polymerization and conventional free radical polymerization are intimately related to the kinetic characteristics of the polymerization process. In conventional free radical polymerization, the radicals formed by the decomposition of the initiator rapidly grow into polymer chains with high molecular weight, and then terminate in a very short time. Due to the low concentration of polymer chains, cyclization and intramolecular crosslinking reactions are prone to form microgels. As the reaction proceeds, the microgels form a network structure through intermolecular crosslinking [35], as illustrated in Figure 6a. The crosslinking network formed by microgels is heterogeneous and contains a lot of structural defects, which is unfavorable to the mechanical properties. By contrast, all of the polymer chain are generated at the beginning and grow simultaneously during RAFT polymerization. Therefore, a large number of the oligomer living chains tend to form a uniform intermolecular crosslinking network with less structural defects, as illustrated in Figure 6b, which helps to improve the mechanical properties [36].

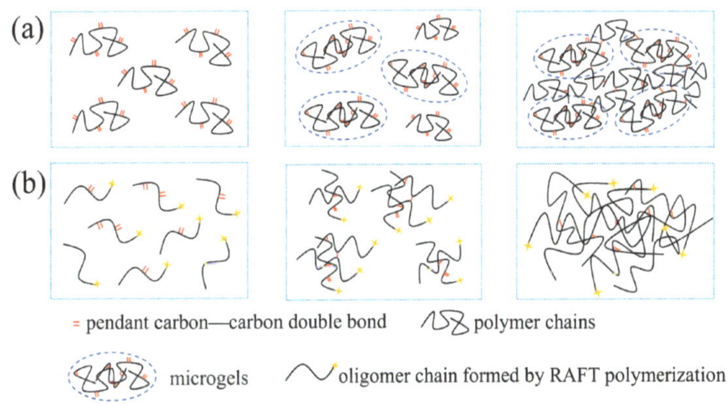

Figure 6. Illustration of the crosslinking process of (**a**) conventional free radical polymerization and (**b**) RAFT polymerization.

Although RAFT polymerization effectively improves the toughness of polyHIPE membranes, the DSC results show that the T_g of RAFT-2 is about −6.7 °C (Figure 7), which is lower than test temperature (i.e., 15 °C), indicating that RAFT-2 is in a rubbery state, so Young's modulus and the tensile strength of RAFT-2 are still low. The insufficient mechanical strength leads to incomplete open-cellular structures, as described above. Aiming at the above problems, the strength of the material is enhanced by increasing the molar ratio of DVB or St to obtain higher toughness and open-cellular extent. Five kinds of polyHIPE membranes with different monomer ratios based on stable emulsions are synthesized via RAFT polymerization, as listed in Table 1. When the molar ratio of BA/St/DVB changes from 9/2/1 to 7/2/3, RAFT-6 shows brittle fracture, as shown in Figure 8, indicating that the amount of DVB has a significant effect on the brittleness of the material. By changing the molar ratio of BA/St/DVB from 9/2/1 to 5/6/1, RAFT-3 and RAFT-4 remain intact and the degree of shrinkage is reduced. However, when the molar ratio of BA/St/DVB changes to 3/8/1 in RAFT-5, brittle fracture is observed again, as shown in Figure 8. The SEM image (Figure 9) shows that all polyHIPE membranes have an interconnected open-cellular structure when the molar ratio of BA/St/DVB changes from 9/2/1 to 5/6/1. The changes of molar ratios of BA/St/DVB also cause the P_C to increase from 88.80% to 100%, which is consistent with the change of sample size, indicating that the collapse and closure of pores can be avoided effectively by increasing the St content. The specific surface area increases from 3.42 $m^2 \cdot g^{-1}$ to 5.19 $m^2 \cdot g^{-1}$, as shown in Table 3.

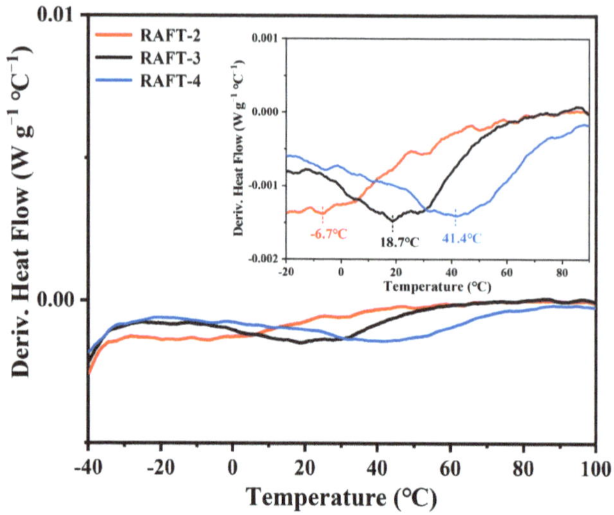

Figure 7. The derivative DSC curves of RAFT-2, RAFT-3, and RAFT-4.

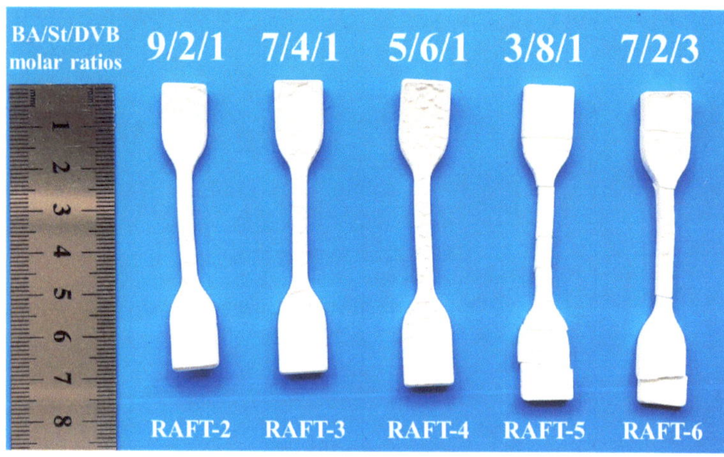

Figure 8. The optical images of polyHIPE membranes obtained by different molar ratios of BA/St/DVB prepared by RAFT polymerization.

A tensile test of polyHIPE membranes with different molar ratios of BA/St/DVB were further conducted. The test results are summarized in Table 3, while the stress–strain curves are shown in Figure 10. When the molar ratio of BA/St/DVB changes from 9/2/1 to 7/4/1, the T_g of RAFT-3 increases to 18.7 °C (Figure 7), and a distinct yield point is observed in the stress–strain curve. Because the test temperature is close to the T_g, the yield stress of RAFT-3 is lower than the fracture stress. The frozen chain segment of the glassy RAFT-3 begins to move under external force after the yield point, and the extension of the polymer chain provides significant deformation of the material. Therefore, RAFT-3 exhibits plastic deformation with ductile fracture. Specifically, although elongation at break decreases from 24.00 ± 3.26% to 17.50 ± 3.90%, Young's modulus and the tensile strength increase to 10.09 ± 1.82 MPa and 819.62 ± 33.31 kPa, respectively. Accordingly, the M_T increases from 47.60 ± 7.88 kJ·m^{-3} to 93.04 ± 12.28 kJ·m^{-3}. When the molar ratio of BA/St/DVB changes to 5/6/1, the T_g of RAFT-4 increases to 41.7 °C (Figure 7) and

Young's modulus of RAFT-4 increases sharply, indicating that the strength of the polymer skeleton is greatly enhanced. Simultaneously, a large number of macropores with a size of several hundred nanometers to microns appear inside the material, while the mesopores with a size below 10 nm basically disappear, as shown in Figure 11, which is consistent with the research of Luo et al. [36]. The results indicate that the existence of mesopores inside RAFT-2 and RAFT-3 with high BA content may be attributed to the collapse of pores caused by insufficient polymer skeleton strength, rather than structural defects caused by uneven crosslinking. However, because the test temperature is much lower than the T_g, the yield stress of RAFT-4 is higher than the fracture stress. Thus, elongation at break and toughness of RAFT-4 is sharply reduced, and RAFT-4 responds in an almost linear–elastic manner and fails by brittle fracture. Therefore, the adjustment range of the molar ratios of BA/St/DVB is quite narrow in order to achieve high open-cellular extent and high toughness simultaneously in polyHIPE membranes.

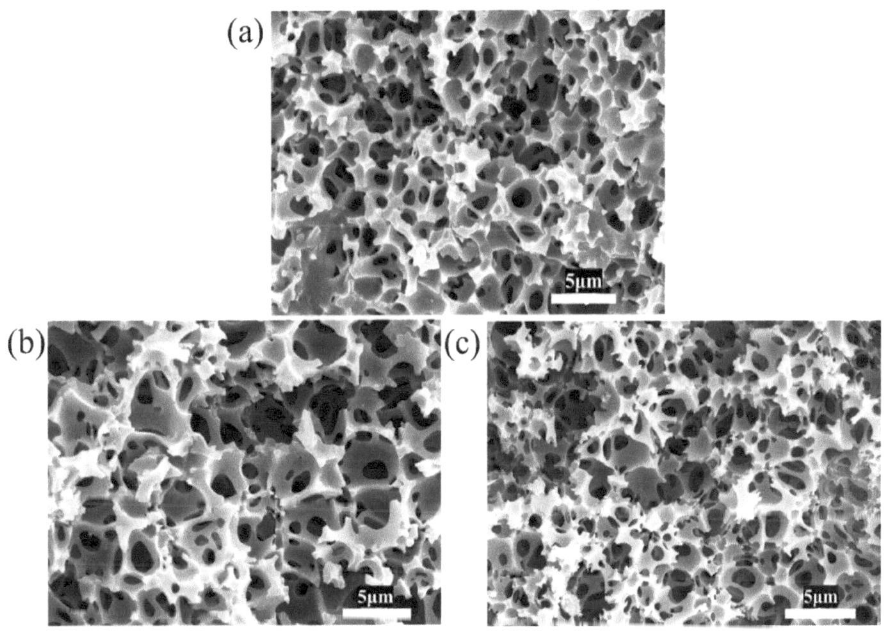

Figure 9. SEM images of polyHIPE membranes with different molar ratios of BA/St/DVB prepared by RAFT polymerization: (**a**) 9/2/1, (**b**) 7/4/1, and (**c**) 5/6/1.

Table 3. The structure and mechanical properties of polyHIPE membranes with different molar ratios.

Sample	BA/St/DVB	E [a] (MPa)	σ [b] (kPa)	ε [c] (%)	M_T [d] ($kJ \cdot m^{-3}$)	BET ($m^2 \cdot g^{-1}$)	P_A (%)	P_T (%)	P_C (%)
RAFT-2	9/2/1	2.04 ± 0.34	347 ± 4.31	24.00 ± 3.26	47.60 ± 7.88	3.42	67.21	75.69	88.80
RAFT-3	7/4/1	10.09 ± 1.82	819.62 ± 33.31	17.50 ± 3.90	93.04 ± 12.28	3.85	69.79	75.57	92.35
RAFT-4	5/6/1	51.10 ± 2.68	664.4 ± 41.37	1.60 ± 0.66	6.24 ± 3.45	5.19	76.65	75.71	100 [e]

[a] Young's modulus, [b] ultimate tensile strength, [c] elongation at break, [d] modulus of toughness, [e] Deviation in the theoretical calculation, P_C is recorded as 100% when P_A is greater than P_T.

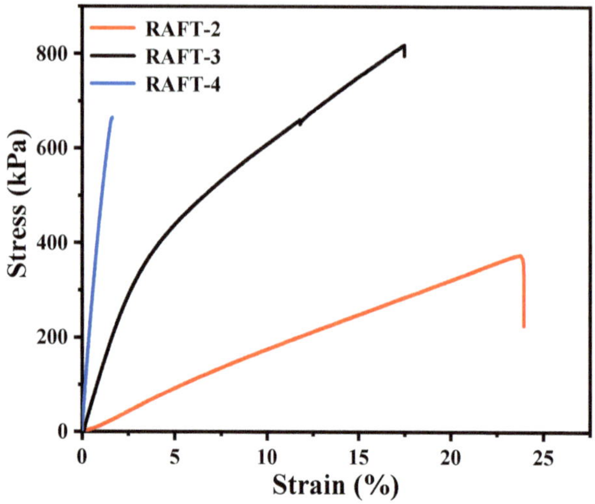

Figure 10. The stress–strain curves of polyHIPE membranes with different molar ratios of BA/St/DVB prepared by RAFT polymerization.

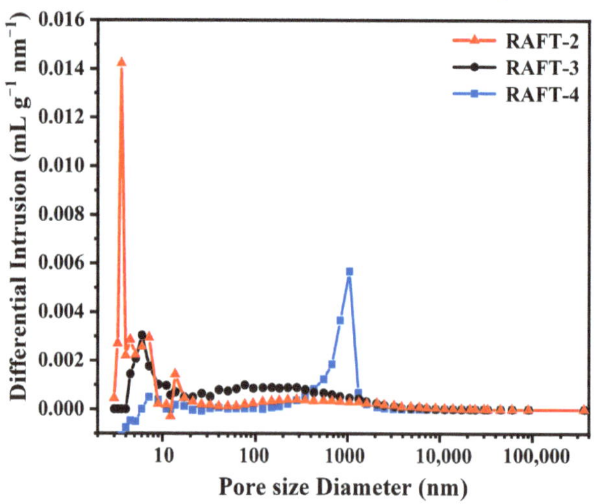

Figure 11. Pore size distribution curves of polyHIPE membranes with different molar ratios of BA/St/DVB prepared by RAFT polymerization.

The heat stability of RAFT-2, RAFT-3, and RAFT-4 was also analyzed. The TGA results show that the initial decomposition temperatures of the samples all exceed 300 °C, and the heat resistance is improved by increasing the St content, as shown in Figure 12. Moreover, no shrinkage or yellowing phenomenon is observed after 200 °C treatment in air for 30 min, as shown in Figure 13, proving excellent thermal stability of all the samples. The synthesized polyHIPE membranes have wide application prospect in the field of lithium ion batteries as separators due to their high open-cellular structures and high toughness. Since the traditional polyolefin separators possesses poor heat stability, the superior heat stability of the polyHIPE membrane can effectively delay the thermal runaway and thus improve the safety of lithium ion batteries.

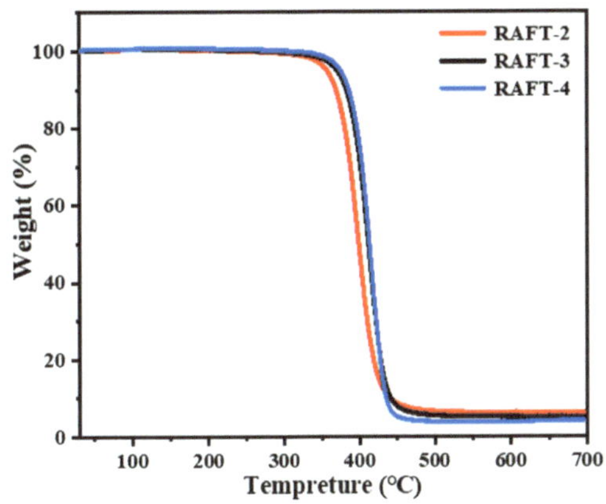

Figure 12. Thermogravimetry analysis curves of RAFT-2, RAFT-3, and RAFT-4.

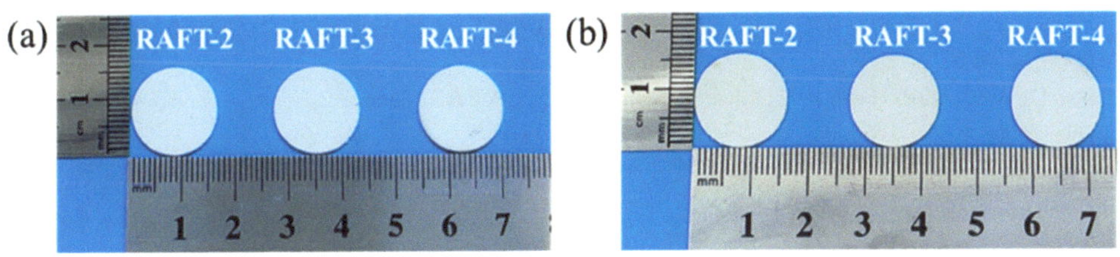

Figure 13. Optical photographs of RAFT-2, RAFT-3 and RAFT-4: (**a**) before and (**b**) after exposure at 200 °C for 30 min.

4. Conclusions

In this paper, stable HIPEs with a high butyl acrylate content (41.7 mol% to 75 mol% based on monomers) can be obtained by using a composite emulsifier (30 wt.% based on monomer) consisting of Span80/DDBSS (9/2 in molar ratio) and adding 0.12 mol·L^{-1} $CaCl_2$ according to aqueous phase concentration. On this basis, polyHIPE membranes with high open-cellular extent and high toughness are firstly prepared by RAFT polymerization. RAFT polymerization can significantly improve the toughness of the material. When the molar ratio of BA/St/DVB equals 9/2/1, the M_T of the polyHIPE membrane prepared by RAFT polymerization is twofold that of traditional free radical polymerization, reaching 47.6 ± 7.88 kJ·m^{-3}, while the P_C almost remains the same. When the molar ratio of BA/St/DVB equals 7/4/1, the polyHIPE membrane prepared by RAFT polymerization shows plastic deformation during the tensile test, and the M_T is further improved to 93.04 ± 12.28 kJ·m^{-3} with a P_C of 92.35%, and it also exhibits excellent thermal stability. This work provides a new idea to prepare polyHIPE membranes with high open-cellular structures and high toughness, and thus expands the application field of polyHIPEs.

Author Contributions: Conceptualization, Z.Z.; Methodology, J.H.; Validation, Z.G. and C.X.; Formal analysis, Y.W. and J.H.; Investigation, Y.W.; Data curation, J.H. and Q.Y.; Writing—original draft, Y.W.; Writing—review & editing, J.H., Z.G. and Z.Z.; Supervision, Z.Z.; Funding acquisition, Z.G., C.X. and Z.Z. All authors have read and agreed to the published version of the manuscript.

Funding: This research was funded by Fujian Provincial Natural Science Foundation grant number 2022J05172 and 2022J01900 and Foundation of Anhui Provincial Higher Education Institutions grant number 2022AH050959. The APC was funded by Research Fund of Minnan Normal University.

Institutional Review Board Statement: Not applicable.

Data Availability Statement: The original contributions presented in this study are included in the article. Further inquiries can be directed to the corresponding author.

Conflicts of Interest: The authors declare no conflict of interest.

References

1. Favre, E. The Future of Membrane Separation Processes: A Prospective Analysis. *Front. Chem. Eng.* **2022**, *4*, 916054. [CrossRef]
2. DuChanois, R.M.; Porter, C.J.; Violet, C.; Verduzco, R.; Elimelech, M. Membrane Materials for Selective Ion Separations at the Water–Energy Nexus. *Adv. Mater.* **2021**, *33*, 2101312. [CrossRef]
3. Aristizábal, S.L.; Lively, R.P.; Nunes, S.P. Solvent and Thermally Stable Polymeric Membranes for Liquid Molecular Separations: Recent Advances, Challenges, and Perspectives. *J. Membr. Sci.* **2023**, *685*, 121972. [CrossRef]
4. Song, Z.; Ma, Y.; Morrin, A.; Ding, C.; Luo, X. Preparation and Electrochemical Sensing Application of Porous Conducting Polymers. *TrAC Trends Anal. Chem.* **2021**, *135*, 116155. [CrossRef]
5. Li, J.; Li, N.; Zheng, Y.; Lou, D.; Jiang, Y.; Jiang, J.; Xu, Q.; Yang, J.; Sun, Y.; Pan, C.; et al. Interfacially Locked Metal Aerogel inside Porous Polymer Composite for Sensitive and Durable Flexible Piezoresistive Sensors. *Adv. Sci.* **2022**, *9*, 2201912. [CrossRef] [PubMed]
6. Zheng, B.; Lin, X.; Zhang, X.; Wu, D.; Matyjaszewski, K. Emerging Functional Porous Polymeric and Carbonaceous Materials for Environmental Treatment and Energy Storage. *Adv. Funct. Mater.* **2020**, *30*, 1907006. [CrossRef]
7. Shi, J.; Qin, M.; Aftab, W.; Zou, R. Flexible Phase Change Materials for Thermal Energy Storage. *Energy Storage Mater.* **2021**, *41*, 321–342. [CrossRef]
8. Düerkop, D.; Widdecke, H.; Schilde, C.; Kunz, U.; Schmiemann, A. Polymer Membranes for All-Vanadium Redox Flow Batteries: A Review. *Membranes* **2021**, *11*, 214. [CrossRef]
9. Chen, W.; Xiang, Y.; Kong, X.-Y.; Wen, L. Polymer-Based Membranes for Promoting Osmotic Energy Conversion. *Giant* **2022**, *10*, 100094. [CrossRef]
10. Yue, H.; Yuan, L.; Zhang, W.; Zhang, S.; Wei, W.; Ma, G. Macrophage Responses to the Physical Burden of Cell-Sized Particles. *J. Mater. Chem. B* **2018**, *6*, 393–400. [CrossRef]
11. Stucki, M.; Loepfe, M.; Stark, W.J. Porous Polymer Membranes by Hard Templating—A Review. *Adv. Eng. Mater.* **2018**, *20*, 1700611. [CrossRef]
12. Liu, Q.; Xiong, J.; Lin, W.; Liu, J.; Wan, Y.; Guo, C.F.; Wang, Q.; Liu, Z. Porous Polymers: Structure, Fabrication and Application. *Mater. Horiz.* **2025**. [CrossRef]
13. Kovačič, S.; Žagar, E.; Slugovc, C. Strength versus Toughness of Emulsion Templated Poly(Dicyclopentadiene) Foams. *Polymer* **2019**, *169*, 58–65. [CrossRef]
14. Ronan, W.; Deshpande, V.S.; Fleck, N.A. The Tensile Ductility of Cellular Solids: The Role of Imperfections. *Int. J. Solids Struct.* **2016**, *102–103*, 200–213. [CrossRef]
15. Kovačič, S.; Krajnc, P.; Slugovc, C. Inherently Reactive polyHIPE Material from Dicyclopentadiene. *Chem. Commun.* **2010**, *46*, 7504. [CrossRef]
16. Kovačič, S.; Preishuber-Pflügl, F.; Slugovc, C. Macroporous Polyolefin Membranes from Dicyclopentadiene High Internal Phase Emulsions: Preparation and Morphology Tuning. *Macromol. Mater. Eng.* **2014**, *299*, 843–850. [CrossRef]
17. Kovačič, S.; Jeřábek, K.; Krajnc, P.; Slugovc, C. Ring Opening Metathesispolymerisation of Emulsion Templated Dicyclopentadiene Giving Open Porous Materials with Excellent Mechanical Properties. *Polym. Chem.* **2012**, *3*, 325–328. [CrossRef]
18. Cameron, N.R.; Sherrington, D.C. Preparation and Glass Transition Temperatures of Elastomeric PolyHIPE Materials. *J. Mater. Chem.* **1997**, *7*, 2209–2212. [CrossRef]
19. Carnachan, R.J.; Bokhari, M.; Przyborski, S.A.; Cameron, N.R. Tailoring the Morphology of Emulsion-Templated Porous Polymers. *Soft Matter* **2006**, *2*, 608. [CrossRef]
20. Pulko, I.; Smrekar, V.; Podgornik, A.; Krajnc, P. Emulsion Templated Open Porous Membranes for Protein Purification. *J. Chromatogr. A* **2011**, *1218*, 2396–2401. [CrossRef] [PubMed]
21. Patten, T.E.; Xia, J.; Abernathy, T.; Matyjaszewski, K. Polymers with Very Low Polydispersities from Atom Transfer Radical Polymerization. *Science* **1996**, *272*, 866–868. [CrossRef] [PubMed]
22. Chiefari, J.; Chong, Y.K.; Ercole, F.; Krstina, J.; Jeffery, J.; Le, T.P.T.; Mayadunne, R.T.A.; Meijs, G.F.; Moad, C.L.; Moad, G.; et al. Living Free-Radical Polymerization by Reversible Addition−Fragmentation Chain Transfer: The RAFT Process. *Macromolecules* **1998**, *31*, 5559–5562. [CrossRef]
23. Nothling, M.D.; Fu, Q.; Reyhani, A.; Allison-Logan, S.; Jung, K.; Zhu, J.; Kamigaito, M.; Boyer, C.; Qiao, G.G. Progress and Perspectives Beyond Traditional RAFT Polymerization. *Adv. Sci.* **2020**, *7*, 2001656. [CrossRef]
24. Silverstein, M.S. PolyHIPEs: Recent Advances in Emulsion-Templated Porous Polymers. *Prog. Polym. Sci.* **2014**, *39*, 199–234. [CrossRef]

25. Lubomirsky, E.; Khodabandeh, A.; Hofe, T.; Susewind, M.; Preis, J.; Hilder, E.F.; Arrua, R.D. Single-Step Synthesis of Methacrylate Monoliths with Well-Defined Mesopores. *Polymer* **2024**, *299*, 126921. [CrossRef]
26. Li, K.; Zhang, Q.; Li, J.; Dai, H.; Zhou, D. Dual-Templating Synthesis of Superhydrophobic Fe3O4/PSt Porous Composites for Highly Efficient Oil-Water Separation. *J. Mol. Liq.* **2024**, *396*, 123922. [CrossRef]
27. Ferguson, C.J.; Hughes, R.J.; Nguyen, D.; Pham, B.T.T.; Gilbert, R.G.; Serelis, A.K.; Such, C.H.; Hawkett, B.S. Ab Initio Emulsion Polymerization by RAFT-Controlled Self-Assembly. *Macromolecules* **2005**, *38*, 2191–2204. [CrossRef]
28. *GB/T 1040.2-2006;* Determination of Tensile Properties of Plastics Part 2: Test Conditions for Molded and Extruded Plastics. National Standardization Administration of the People's Republic of China: Beijing, China, 2006.
29. Yadav, A.; Ghosh, S.; Samanta, A.; Pal, J.; Srivastava, R.K. Emulsion Templated Scaffolds of Poly(ε-Caprolactone)—A Review. *Chem. Commun.* **2022**, *58*, 1468–1480. [CrossRef]
30. Aronson, M.P.; Petko, M.F. Highly Concentrated Water-in-Oil Emulsions: Influence of Electrolyte on Their Properties and Stability. *J. Colloid Interface Sci.* **1993**, *159*, 134–149. [CrossRef]
31. Foudazi, R. HIPEs to PolyHIPEs. *React. Funct. Polym.* **2021**, *164*, 104917. [CrossRef]
32. van Krevelen, D.W.; Te Nienhuis, K. *Properties of Polymers;* Science Press: Beijing, China, 2010.
33. Zhang, F.; Wang, C.; Huang, X.; Dong, X.; Chi, H.; Xu, K.; Bai, Y.; Wang, P. New Approach for Preparation of Porous Polymers with Reversible Pore Structures for a Highly Safe Smart Supercapacitor. *ACS Appl. Mater. Interfaces* **2024**, *16*, 19442–19452. [CrossRef]
34. Koler, A.; Brus, J.; Krajnc, P. RAFT Polymerisation and Hypercrosslinking Improve Crosslink Homogeneity and Surface Area of Styrene Based PolyHIPEs. *Polymers* **2023**, *15*, 2255. [CrossRef] [PubMed]
35. Yu, Q.; Xu, S.; Zhang, H.; Ding, Y.; Zhu, S. Comparison of Reaction Kinetics and Gelation Behaviors in Atom Transfer, Reversible Addition–Fragmentation Chain Transfer and Conventional Free Radical Copolymerization of Oligo(Ethylene Glycol) Methyl Ether Methacrylate and Oligo(Ethylene Glycol) Dimethacrylate. *Polymer* **2009**, *50*, 3488–3494. [CrossRef]
36. Luo, Y.; Wang, A.-N.; Gao, X. Pushing the Mechanical Strength of PolyHIPEs up to the Theoretical Limit through Living Radical Polymerization. *Soft Matter* **2012**, *8*, 1824–1830. [CrossRef]

polymers

Article

Numerical Simulation of Airflow Organization in Vulcanization Tanks for Waste Tires

Tianxi Su [1], Yongzhi Ma [1,*], Baolin Wang [2], Xiaowen Luan [2], Hui Li [2] and Xuelong Zhang [1]

[1] College of Mechanical and Electrical Engineering, Qingdao University, Qingdao 266071, China; sutianxi@qdu.edu.cn (T.S.); zhangxuelong@qdu.edu.cn (X.Z.)
[2] Qingdao Wanfang Recycling Technology Co., Ltd., Qingdao 266071, China; wangbl@qdecolan.com (B.W.); luanxw@qdecolan.com (X.L.); lih@qdecolan.com (H.L.)
* Correspondence: hiking@126.com; Tel.: +86-138-542-83816

Abstract: Currently, in the domestic practice of retreading tires using vulcanization tanks, some tanks exhibit uneven temperature distributions leading to low retreading success rates. To address that, this paper simulated the temperature and velocity fields during the heating process of vulcanization tanks for waste tire retreading. The results indicated that a higher heating power reduces the time required for the vulcanizing agent to reach the vulcanization condition, but it also increases the difference in tire temperature in the tank, with a severely uneven distribution of the temperature field. Subsequently, to improve the uniformity of temperature distribution and enhance the retreading rate of waste tires, this paper proposed two types of orifice plates to adjust the airflow organization. The results show that both the plain orifice plate and the frustum cone orifice plate can enhance the uniformity of the temperature field within the vulcanization tank and reduce the temperature difference between tires. Moreover, at the same heating power, the presence of the orifice plates increases the rate of temperature increase in the tires and the vulcanizing agent compared to the original vulcanization tank, improving the thermal efficiency of the vulcanization tank heater.

Keywords: waste tire retreading; vulcanization tank; airflow organization; orifice plate; numerical simulation

Academic Editor: Changwoon Nah

Received: 20 December 2024
Revised: 14 January 2025
Accepted: 16 January 2025
Published: 17 January 2025

Citation: Su, T.; Ma, Y.; Wang, B.; Luan, X.; Li, H.; Zhang, X. Numerical Simulation of Airflow Organization in Vulcanization Tanks for Waste Tires. *Polymers* **2025**, *17*, 232. https://doi.org/10.3390/polym17020232

1. Introduction

With rapid development and global industrialization, the growing world population's demand for automobile production is increasing [1], and the disposal of waste tires has become a global environmental protection challenge. The stable growth of the automotive industry inevitably leads to a significant increase in the number of end-of-life tires (ELT) [2], most of which come from waste passenger cars and truck tires [3,4]. Waste tires, known as "black pollution", pose significant challenges in recycling and disposal technology. According to statistics, over the next thirty years, the global waste output is expected to increase by 70%, reaching 3.4 billion tons [1]. Thanks to the continuously growing automotive industry, China alone discards a staggering 16 million tons of waste tires annually, while the global volume has reached an astonishing 1.5 billion tons [5–7]. Against this backdrop, the recycling of waste tires is particularly important [8].

The treatment of waste tires mainly includes renovation and reuse, that is, the utilization of reclaimed rubber from waste tires, pyrolysis, and direct incineration. The reclaimed rubber from waste tires uses physical means to crush, separate, and sort waste car tires, allowing the production of rubber particles and rubber powder with minimal pollution at room temperature. Although the production of reclaimed rubber powder from waste car tires can achieve the harmless treatment of this "black pollution", it ultimately results in additives and degradation materials derived from the waste tire rubber that may leach into the environment [9].

Pyrolysis involves the controlled combustion of rubber waste without oxygen, leading to the decomposition of rubber into smaller components such as fuel oil, gas, carbon black, sulfur, and metals [10,11]. The main objective of pyrolysis is to extract energy from waste tire rubber. The fuel produced through this process is purified to remove sulfur, char, and ash, ensuring the production of high-quality fuel, suitable for enhancing engine performance. The gas produced during the pyrolysis process is used as for heat and electricity generation in power plants [12]. Additionally, the carbon black produced by pyrolysis can be mixed with plastics or EVA foam (Ethylene-Vinyl Acetate copolymer foam), or it can be further processed into activated carbon [13]. In summary, pyrolysis products have a wide range of applications with considerable utilization value, aligning with the trends of the times, saving energy, being environmentally friendly, and being one of the best ways to create a sustainable development-oriented society. However, this solution for waste tire management requires large-scale pyrolysis equipment, with high construction and operation costs (high temperature and low pressure), limiting its large-scale industrial application [14].

Incineration is a self-sustaining exothermic process that occurs above 400 °C, and since the calorific value of waste tires is higher than that of coal (18.6–27.9 MJ/kg), they are used for energy recovery. The calorific value of waste tires is 32.6 MJ/kg, which can be used as a fuel source for producing steam, electricity, pulp, paper, lime, and steel. Additionally, Oriaku et al. [15] reported the recovery of carbon black (CB) through the combustion of tires in limited air supply through incineration. The recovered material can be used in small-scale industries for the production of printing inks and paints. The main advantages of incineration are the low cost of energy production and the maximum recovery of heat. However, the atmospheric pollution caused by emissions of flue gas and particulate matter is a serious air pollution source that needs to be addressed [16].

Tire retreading, as the primary and most effective method used for recycling, has characteristics such as multiple retreading, low material consumption, low cost, and long service life. Retreading is the process of replacing the worn tread of discarded tires with new tread so that the tires can be reused [17,18]. In the tire retreading process, the first step involves a thorough inspection of the discarded tire body to assess its suitability for reuse. After this evaluation, the tire crown is separated from the tire body through grinding and subsequent repair. After the necessary repairs, a cushion rubber sheet is applied, and a pre-vulcanizing agent is used to vulcanize the tread rubber. As shown in Figure 1, the process involves applying a vulcanizing agent to the outer tire and then attaching it to the inner tire. Among them, vulcanizing agents play a role in significantly improving the physical and chemical properties of rubber, such as elasticity, strength, heat resistance, medium resistance, and durability, even if rubber is transformed from thermoplastic rubber to thermosetting rubber through vulcanization reactions, which play a crucial role in enhancing the adhesion, physical properties, and production efficiency of refurbished tires, while reducing energy consumption and costs. The final stage includes the vulcanization of the tread rubber and the final inspection of the product's quality before it is ready for the market [19]. Studies have shown that each retreaded tire requires only 30% of the energy and 25% of the raw materials needed to produce a new tire [20,21]. Each retreading can regain 60–90% of the service life of a new tire, with an average driving mileage of 50,000–70,000 km. In 2019, the United States produced over 4.05 million metric tons of waste tires, while the European Union produced 3.56 million metric tons. In recent years, in the United States, given the increasing number of waste tires, more tires are being recycled or used as energy [22]. Currently, in China, there is a large output of waste tires [23], a low volume of retreading with a retreading rate of about 4%, and generally low retreading rates [24], which are somewhat behind those of developed countries; however, significant progress has been made compared to the past. In China, with the rapid growth in the number of vehicles, the generation of waste tires is also increasing at a double-digit rate, posing severe challenges for environmental protection and resource conservation. Therefore, promoting the circular utilization of waste tires not only helps alleviate the issue of rubber resource shortage in China but also reduces environmental pollution, promoting the construction of a circular economy and a conservation-oriented society.

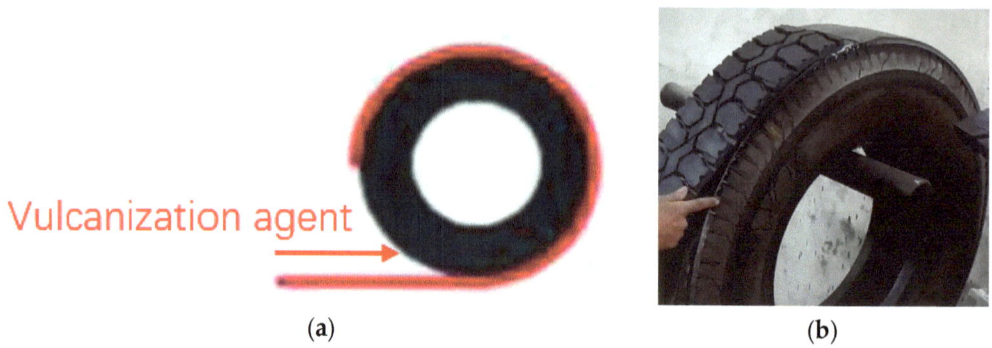

Figure 1. Operation diagram of bonding rubber sheets from waste tire vulcanizing agents. (**a**) Position of vulcanizing agent; (**b**) tire retread operation diagram.

There are mainly two methods of retreading, hot vulcanization and cold vulcanization. Hot retreading of waste tires refers to the traditional retreading method, with vulcanization temperatures generally around 145~155 °C, with the hot retreading process temperature being far above the critical temperature, causing significant damage to the tire body. The use of rigid molds in the hot retreading process can easily lead to tire deformation, thus generating internal stress, causing layering, shoulder voids, and a high probability of tire blowouts, affecting the service life of the tire which can only be used for passenger car tires [17]. In this method, a new rubber layer is formed on the tire, and the entire tire is vulcanized at 150 °C to 180 °C to mold the tread pattern. Hot vulcanization uses mature technology, with a long history, cheap equipment costs, lower investment costs. Some tires, like airplane tires, can only be retreaded with hot vulcanization. The cold vulcanization method, also known as the pre-vulcanization method, has a general vulcanization temperature below 120 °C and is suitable for commercial vehicle tires. Since the critical temperature for the denaturation of tire rubber is 120 °C, this means that above 120 °C, the physical performance indicators of the tire rubber significantly decrease. For example, the adhesion strength between the rubber and the framework material decrease in such cases, leading to tire body layer separation, voids, or even blowouts. Therefore, this method, compared to traditional hot retreading, does not damage the retreaded tire body and does not affect the service life of the tire body. The optimal vulcanization temperature for cold retreading is generally around 100 °C. It uses pre-molded vulcanized strips or rings that are applied to the polished old tire body, which is vulcanized at low temperatures in the cylinder, potentially saving energy and reducing the aging phenomenon caused by secondary vulcanization, thus protecting tire body quality and extending service life. Under normal conditions (the wear resistance of the pre-vulcanized tread rubber reaches 80,000 km or more), the service life of hot retreaded tires should be 60~80% of that of new tires, while tires retreaded using the pre-vulcanization method can approach 100%. However, the current domestic cold retreading technology is not mature, and the general uneven temperature distribution in cold retreading vulcanization tanks leads to a low tire retreading rate.

To explore and solve the mentioned issues, this paper studies the temperature and velocity fields in the vulcanization tank and the development process of tire temperature within the tank through numerical simulations in ANSYS Fluent 2022R1. The investigation scope of tire temperature includes temperature differences, average tire temperature, and the temperature of vulcanizing agent which plays multiple roles in tire retreading, including restoring tire performance, providing adhesion, improving durability, maintaining tread shape, and promoting environmental protection. Subsequently, based on the simulation results, two types of orifice plate structures are proposed, the flat orifice plate and the frustum cone orifice plate, which significantly improve the uneven temperature distribution in the vulcanization tank's temperature field, laying the foundation for the advancement of the tire retreading industry in China.

2. Materials and Methods

The vulcanization tank, as shown in Figure 2a,b, has two heating air ducts in the heating part of the tank body. The specifications of the screw-type armored thermocouple temperature meter are M27·1.5, which means the nominal diameter is 27 mm and the pitch is 1.5 mm. It is worth mentioning that the length of the thermometer is 150 mm. The temperature error is ±0.35 °C. The air flow direction and the position of the thermometer are shown in Figure 3. The schematic diagram of the energy transfer is shown in Figure 4. The heated air circulates clockwise, absorbs heat through an electric heating tank, and then releases heat through waste tires, repeating the cycle. Schematic diagrams and simplified geometric structures of the vulcanization tank are shown in Figure 2b,c. The detail can be seen in Table 1. The vulcanization tank consists of a tank, air ducts, and a driving fan, all made of 304 stainless steel [20,21]. The temperature of the laboratory environment is 7 °C. Each heating flue is equipped with three U-shaped heating tubes, the model of the waste tires are 1200r20 tires, and the operation procedure of the vulcanization tank includes the following steps:

1. Close the vulcanization tank with 21 waste tires;
2. Add compressed air at a pressure of 6 atmospheres to the tank;
3. Turn on the driving fan and heating power to cause the air inside the tank to circulate and heat up; the heating power is 88 kw and the heating duration is 1600 s;
4. Evenly heat the waste tires in the tank to facilitate the vulcanization of the vulcanizing agent.

Table 1. Parameters of the vulcanization tank used in the experiment.

Parameters	Values
L_1	8500 mm
L_2	8400 mm
L_3	4063 mm
L_4	400 mm
L_5	200 mm
D_1	1600 mm
D_2	487.2 mm
R_1	800 mm
R_2	700 mm
r_1	30°
r_2	33.4°

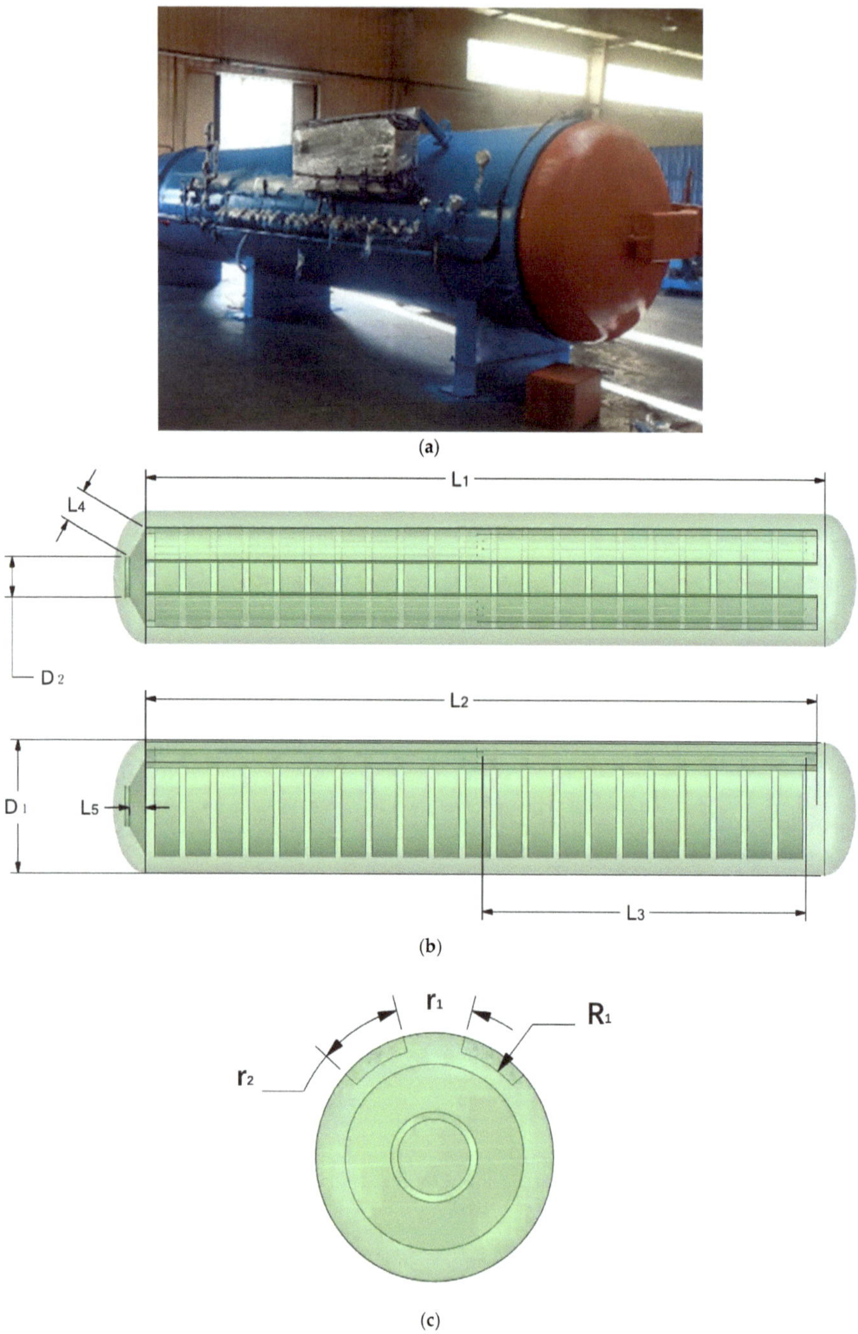

Figure 2. Experimental equipment. Vulcanization tank: (**a**) physical drawing; (**b**) top view and front view; (**c**) side view.

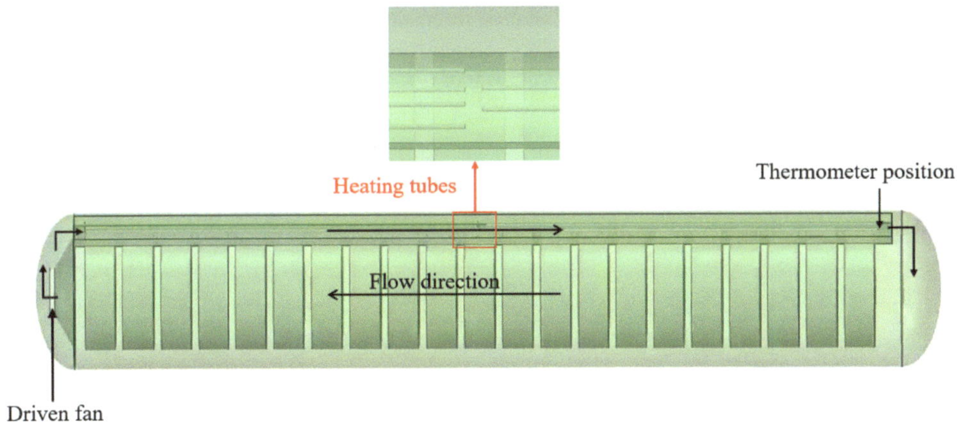

Figure 3. Flow direction and thermometer position.

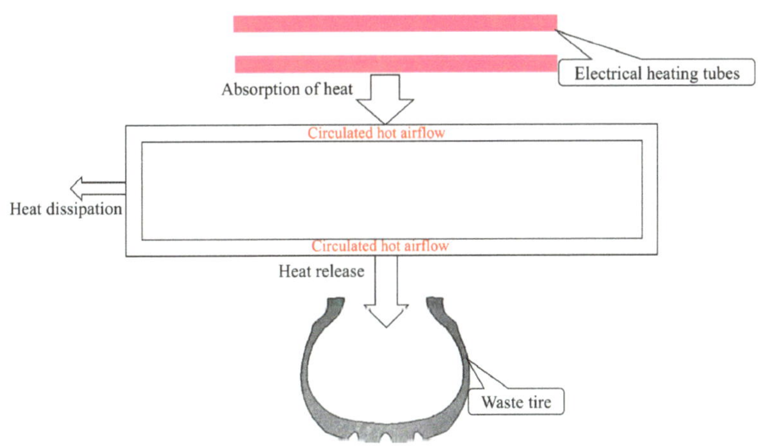

Figure 4. Schematic diagram of energy transfer.

3. Geometric Model

All experimental models are simulated through ANSYS Fluent 2022R1. To simplify the model, each U-shaped heating tube is modeled as two longitudinal heating tubes, ignoring the influence of the electronic control system. The relevant physical parameters of the material and the boundary conditions for simulation are shown in Table 2. It is worth noting that the power of the heat source is set as P = 45, 60, 75, 90 kW. Among them, the heating conditions are that the thermometer temperature < 100 °C and the tire temperature <80 °C. The driving fan operates with a flow rate of 4500 m^3/h, with the minimum Reynolds number being 17,663, which is far greater than 4000, thus, the model is a turbulent flow model. Additionally, there are two other types of airflow organization orifice plate structures, which are the flat orifice plate and the frustum cone orifice plate, as shown in Figures 5 and 6. The 5 cm aperture diameter and the 2 mm thickness of the orifice plate can be ignored. The diameter of the flat orifice plate is 1460 mm. The bottom diameter of the frustum cone orifice plate is 560 mm, and the diameter of the upper circle is 230 mm.

From left to right, there are 28 temperature monitoring points set for the vulcanizing agent, with their specific locations shown in Figure 7.

A simulation model of the waste tire curing tank is presented in the text, which includes fluid flow physical problems and fluid–solid heat transfer physical problems. In the computational domain, air is assumed to be incompressible and with constant properties. The flow is considered to be three-dimensional and steady. This study applies three governing equations.

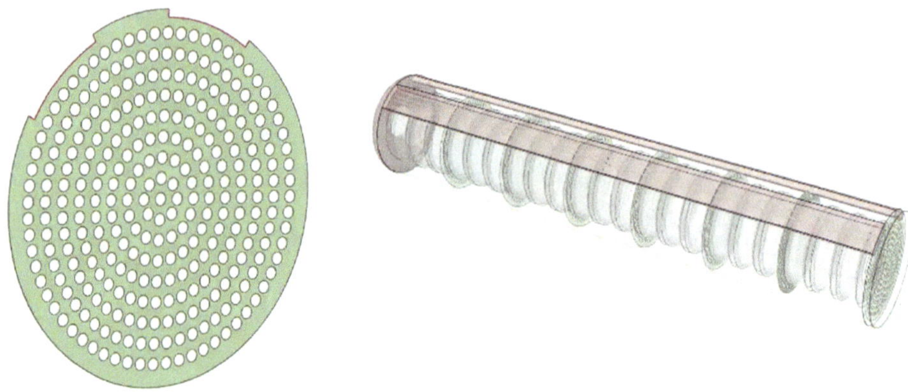

(**a**) (**b**)

Figure 5. Schematic diagram of the plain orifice plate: (**a**) flat orifice plate; (**b**) vulcanization tank with flat orifice plates.

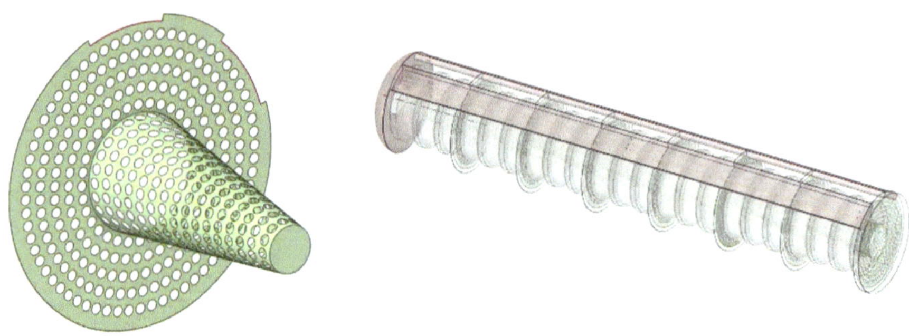

(**a**) (**b**)

Figure 6. Schematic diagram of the plain orifice plate: (**a**) frustum cone orifice plate; (**b**) vulcanization tank with frustum cone orifice plates.

The continuity equation and turbulence momentum equation are expressed as follows:

$$\rho\frac{\partial u_i}{\partial x_i} = 0 \tag{1}$$

$$\rho\frac{\partial(u_i u_j)}{\partial x_i} = -\frac{\partial P}{\partial x_j} + \frac{\partial}{\partial x_i}\left(u\left(\frac{\partial U_i}{\partial x_j} + \frac{\partial u_j}{\partial x_i}\right)\right) \tag{2}$$

The turbulence kinetic energy equation in the fluid region and turbulence kinetic energy dissipation rate equation are expressed as follows:

$$\rho u_i\frac{\partial k}{\partial x_i} = \frac{\partial}{\partial x_i}\left[\left(\mu + \frac{\mu_t}{\sigma_k}\right)\frac{\partial k}{\partial x_i}\right] + \frac{\mu_t}{2}\left(\frac{\partial u_i}{\partial x_j} + \frac{\partial u_j}{\partial x_i}\right) - \rho\varepsilon \tag{3}$$

$$\rho u_i \frac{\partial \varepsilon}{\partial x_i} = \frac{\partial}{\partial x_i}\left[\left(\mu + \frac{\mu_t}{\sigma_k}\right)\frac{\partial \varepsilon}{\partial x_i}\right] + C_1 \frac{\mu_t}{2} \tag{4}$$

where u_i and u_j represent the mean velocity components. P and ε represent the mean pressure and dissipation rate of TKE, respectively. ρ, u, and μ_t are the air density, air molecular dynamic viscosity coefficient, and air turbulence dynamic viscosity, respectively.

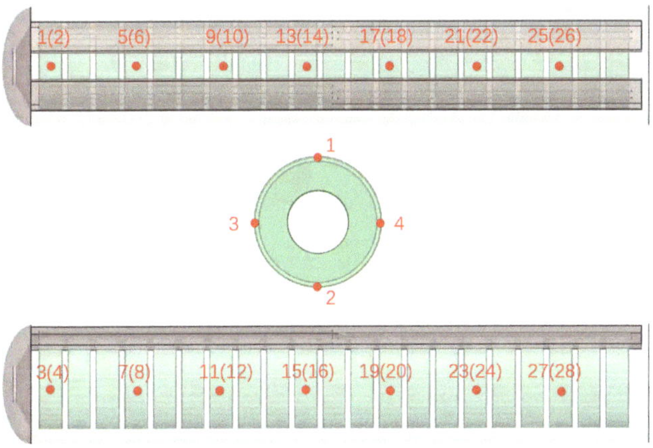

Figure 7. Location diagram of the temperature monitoring positions for the vulcanizing agent.

In this model, the turbulence model is a typical turbulence model in ANSYS Fluent. Previous studies [25,26] have all used typical turbulence models to simulate heat exchanges. Therefore, the typical parameters [23] in this turbulence model are set as follows:

$Pr_{TKE} = 1$; $Pr_{TDR} = 1.3$; $Pr_{wall} = 0.85$; $Pr_e = 0.85$; $C_1 = 1.44$; $C_2 = 1.92$; $C_\mu = 0.09$.

The energy transport equation in the airflow is expressed as follows:

$$\rho_{air} C_{p,air} \frac{\partial T_{air}}{\partial_t} + \rho_{air} C_{p,air} u_i \frac{\partial T_{air}}{\partial x_i} = \frac{\partial}{\partial x_i}\left[\left(\lambda_a + \frac{\mu_t}{\sigma_T}\right)\frac{\partial T_{air}}{\partial x_i}\right] \tag{5}$$

where T_{air}, λ_a, and $C_{p,air}$ respectively represent the air temperature, thermal conductivity of air, and heat capacity of air. The difference in tire temperature is as follows:

$$\Delta T = T_{\max} - T_{\min} \tag{6}$$

where $T_{\max}$ and $T_{\min}$ are the maximum and minimum temperatures of the tire, respectively.

$$T_{\max} = Max\{T_1, T_2 \ldots T_{28}\} \tag{7}$$

$$T_{\min} = Min\{T_1, T_2 \ldots T_{28}\} \tag{8}$$

The standard deviation was calculated as follows:

$$s = \sqrt{\frac{1}{28-1}\sum_{i=1}^{28}(T_i - \overline{T})} \tag{9}$$

In order to save computational resources, this paper adopts a standard turbulence model to simulate the heating process of the vulcanizing tank for retreading waste tires, following the example of some similar studies [27–29]. In this simulation, Patankar's [30] SIMPLE algorithm is used to solve the pressure–velocity field. The finite volume method is employed for the discretization of the governing equations. To reduce computational resources and improve calculation speed, the momentum equation terms, turbulence kinetic energy terms, specific dissipation rate terms, and energy terms are discretized using a first-order upwind scheme, and the pressure terms are discretized using a first-order scheme. Compared with second-order discretization schemes, the first-order discretization scheme has better convergence. Although the use of this scheme helps increase discretization errors, it may lead to shorter simulation times compared to second-order schemes.

The expected criterion for solver convergence is based on the absolute residual parameter. Parameters for the continuity equation, momentum equation, kinetic energy equation, turbulence equation, and specific dissipation rate equation adopt convergence criterion 10^{-3}. For the energy equation, the convergence criterion is more strictly set to 10^{-5}. This means that the solver's goal is to achieve residuals below these thresholds to ensure the convergence and accuracy of the solution to the governing equations.

Table 2. Thermophysical properties of 304 stainless steel and air.

Domain	Density ρ	Thermal Conductivity Coefficient k	Specific Heat Capacity c	Viscosity μ
Air calculation area	7.1587 kg/m^3	0.02516 W/(m·K)	1.015 kJ/(kg·K)	1.78247 × 10^{-5} kg/(m·s)
Heating tube (304 SS [31])	8002 kg/m^3	0.014 t + 14.63 W/(m·K)	0.1467 t + 495 kJ/(kg·K)	-
Retreaded tire [32–34]	950 kg/m^3	−0.00048 t + 0.355 W/(m·K)	1300-0.0025 t kJ/(kg·K)	-

4. Grid Independence and Numerical Model Validation

To validate the numerical simulation of the vulcanization tank, this study compares the simulation results with experimental data. Figure 8 shows the comparison of the exit temperature of the heating flue gas between the experimental results and the simulation results. The mean relative error (*RE*) is expressed by Equation (10) as follows:

$$RE = \frac{1}{N}\sum \frac{|T_{exp} - T_{si}|}{T_{exp}} \times 100 \quad (10)$$

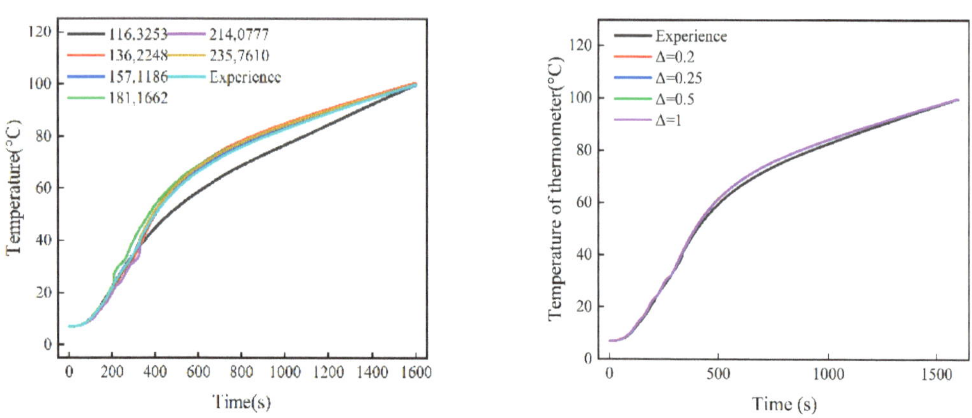

(**a**) (**b**)

Figure 8. Grid and time independence verification: (**a**) grid independence verification; (**b**) time independence verification.

In summary, the relative error between the simulation data and the experimental data are within 2.2%, further proving the accuracy of the computational method for the vulcanization tank.

To demonstrate the independence of the grid, Figure 8 shows the outlet temperatures of the heating pipe obtained using five different grid sizes. For grid sizes of 1,163,253, 1,362,248, 1,571,186, 1,811,662, and 2,140,777 cells, the relative errors between simulation and experiment are approximately 5.69%, 2.73%, 2.63%, 3.12%, 2.74%, and 1.43%, respectively. The results indicate that when the number of grids is 1,362,248 or more, the effect of grid size changes on the increase in temperature can be neglected. Therefore, to ensure more reliable results and shorter simulation times, a grid size of 1,362,248 was finally chosen for the subsequent simulations, with the simulation time unit step being consistent with the unit step of the experimental data recording set to 1 s.

In Figure 8b, a grid consisting of 1,362,248 cells was used to monitor air temperature, with different time steps of 0.02 s, 0.25 s, 0.5 s, and 1 s. The results indicate that the differences between the various time steps are minimal, with the average temperature change between the smallest and largest time steps (0.025 s and 0.5 s) being within 2.04%. Since the experimental data were recorded with a consistent time step of 1 s, we considered setting the time step to 1 s in order to ensure numerical stability while conforming to the experimental recording step size.

5. Results and Discussion

In this section, for the simulation of heating in the vulcanization tank under different heating powers, this paper compares the heating rate at fixed points, the temperature of the vulcanizing agent, the overall heating rate of the tire, the temperature difference analysis of the tire, as well as the temperature and velocity distribution diagrams. Subsequently, in order to improve the uniformity of temperature distribution to enhance the recycling rate of waste tires, this paper conducts a simulation analysis for the airflow organization of two types of orifice plate structures and compares the simulation results with those of the original model, proving that the orifice plates play a significant role in improving the uniformity of temperature distribution.

5.1. Model Validation

Figure 9a,b shows the temperature rise process of the thermometer temperature and tire temperature under different heating powers. It is worth mentioning that the label "Origin-45" refers to the original experimental simulation model with a power of 45 kW. The results indicated that there is an inverse relationship between the heating power and the time required to reach the target temperature increase. In other words, the higher the heating power, the shorter the time needed for both the thermometer temperature and tire temperature needed to reach the target. The time required for the tire temperature to reach 80 °C at heating powers of 90 kW, 75 kW, 60 kW, and 45 kW is 7330 s, 8760 s, 10,910 s, and 14,490 s, respectively.

The reason for this phenomenon is that the higher the power of the heater, the higher the temperature of the heating element. Due to the increased temperature difference between the heated air and the dry-burn tube, the amount of heat absorbed by the air per unit of time increases, and thus, the thermometer and the tire absorb more heat per unit of time, leading to a faster rate of increase in temperature. Combining Figure 9a,b shows that when the heating power was 90 kW, the thermometer's temperature had risen to 100 °C but heating had yet to stopped; instead, it continued increasing for a while before stopping because the tire's temperature had yet to reach the target temperature (80 °C).

Similarly, Figure 9c presents the temperature profile of the vulcanizing agent during the heating process of the vulcanizing tank. The results demonstrate that increasing the heating power helps the vulcanizing agent reach the vulcanization temperature sooner, thereby shortening the retreading time. Specifically, the time required to reach the vulcanization temperature with a heating power of 90 kW is almost 40% of that required with a heating power of 45 kW.

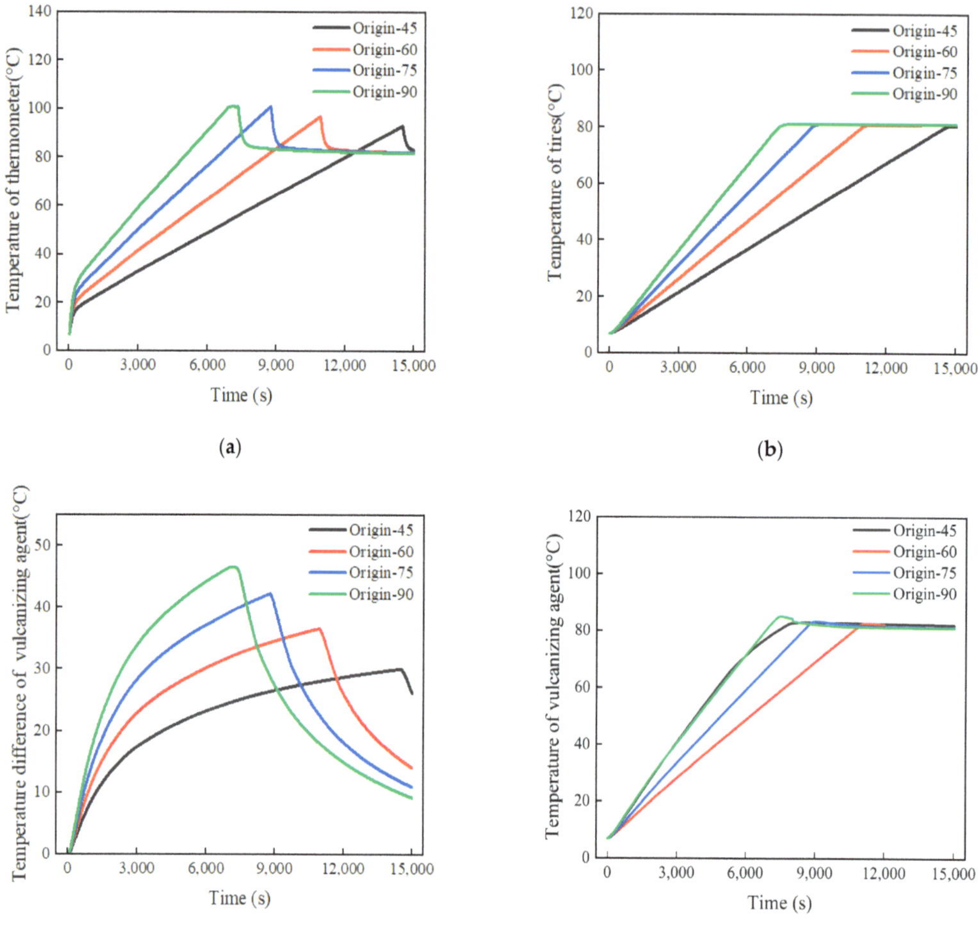

Figure 9. Temperature history: (**a**) temperature history of the thermometer; (**b**) temperature history of the tire; (**c**) temperature difference history of the vulcanizing agent; (**d**) temperature history of the vulcanizing agent.

Figure 9d shows the temperature difference diagram of the vulcanizing agent during the heating process. It can be observed from Figure 9d that in the early stage of heating, an increase in heating power at the same time point leads to an increase in the maximum temperature difference, which may be due to the existence of a heat transfer blind zone with air flow inside the tire and excessive heat concentration. However, it can be noted that in the later stage of heating, at the same time point, the higher the heating power, the smaller the temperature difference. This is because the higher the heating power, the easier it is for the vulcanization tank to reach the conditions required to stop heating, and then the tire, under the combined effects of forced convection and natural convection, carries away the heat from the higher temperature parts of the tire, causing the temperature difference to drop rapidly.

Figures 10 and 11 show the xy slice temperature distribution and velocity distribution during the heating process of the vulcanization tank with a heating power of 60 kW. It can be observed from the figures that at times of 3000 s, 6000 s, and 9000 s, both the highest and lowest temperatures increased significantly. The temperature difference across the slice increased, and the general temperature distribution of the tire was from the bottom to the top and from right to left. The temperature then gradually decreased, with the right-bottom tread being higher in temperature and the left-top tread being lower in temperature. By analyzing the velocity distribution, it can be found that the right side of the vulcanization tank under the tire receives significant concentrated heating, while the upper side of the left tire lacks heating. In addition, the internal air temperature on the left side of the tire is significantly lower than that in other areas, indicating an uneven distribution of air temperature.

For a time period of 12,000 s, the maximum temperature and the section temperature decreased significantly. This is because the heating stop condition was reached at t = 10,900 s, but the fan continued to work, and forced convection combined with natural convection promoted heat transfer, reducing the temperature difference. It is worth noting that at that time, the minimum temperature on the left tire was 65 °C, which had yet to reach the vulcanization temperature. The temperature on the right side reached 89 °C.

In summary, the increase in heating power helps reduce the heating time required for the vulcanizing agent, the vulcanization tank thermometer temperature, and the tire temperature needed to reach the target. However, this also increases the temperature difference on the tread, leading to concentrated heating on the lower part of the right tire, which may cause over-vulcanization or under-vulcanization in some areas of the tread, thus affecting the quality of the retreaded tires.

5.2. Analysis of the Optimization Results for Airflow Organization

In order to adjust the heat accumulation caused by airflow and ensure the uniform heating of waste tires, this paper proposed two orifice plate structures to adjust the airflow organization flow, as shown in Figures 4 and 5. Figure 12 shows the temperature heating curves of two types of vulcanizers with and without orifice plates. "Origin" refers to the untreated vulcanizer, while "Flat" and "Frustum" are the vulcanizers with flat orifice plates and frustum cone orifice plates.

From the graph, we can observe that the thermometer temperature of the vulcanization tank with the orifice plate treatment first rises, then levels off, and finally decreases. This is because the thermometer had reached the target temperature, but the overall temperature of the tire had yet to reach the vulcanization temperature, and the heating temperature had to be maintained at 100 °C so that the tire temperature reached 80 °C before heating stopped, causing the thermometer temperature to decrease.

Compared to the untreated vulcanization tank, the vulcanization tank with orifice plates reaches the target temperature earlier, and the time required to heat to the target temperature decreases as the heating temperature increases. This is because the presence of the orifice plate slows down the flow velocity of the fluid near the thermometer, leading to the rapid accumulation of heat and a quick increase in temperature, ultimately reaching the target temperature. Additionally, in the vulcanization tank with orifice plates, the time required to reach the target temperature is similar for frustrum cone orifice plates and flat orifice plates, but the frustrum cone orifice plate reaches the point of stopping heating sooner than the flat orifice plate, and the distance increases with the rise in heating power. This means that when the heating power is 45 kW, 60 kW, 75 kW, and 90 kW, the presence of the orifice plate can approximately save at least 90 kWh, 68 kWh, 50 kWh, and 36 kWh, respectively.

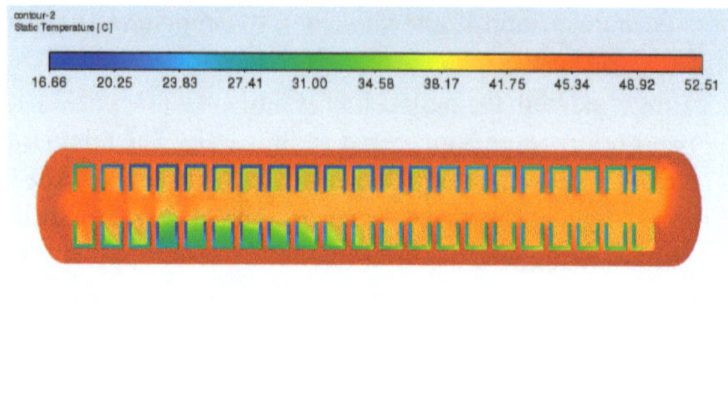

(a)

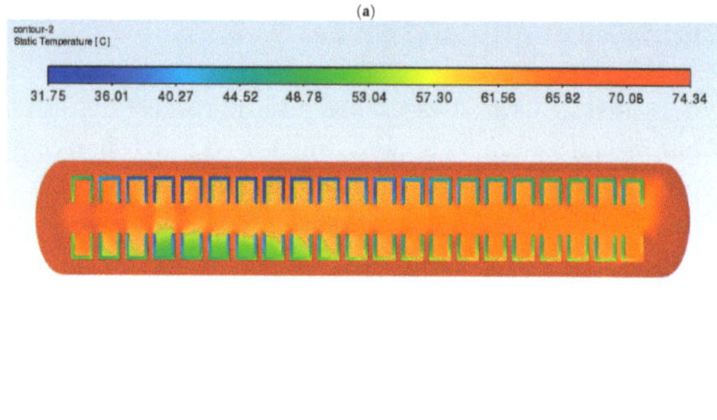

(b)

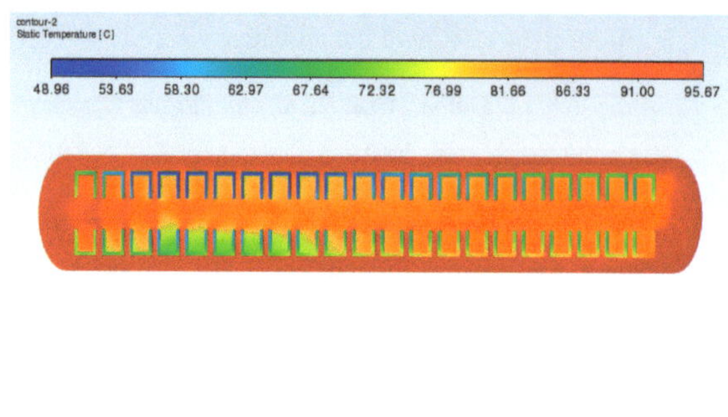

(c)

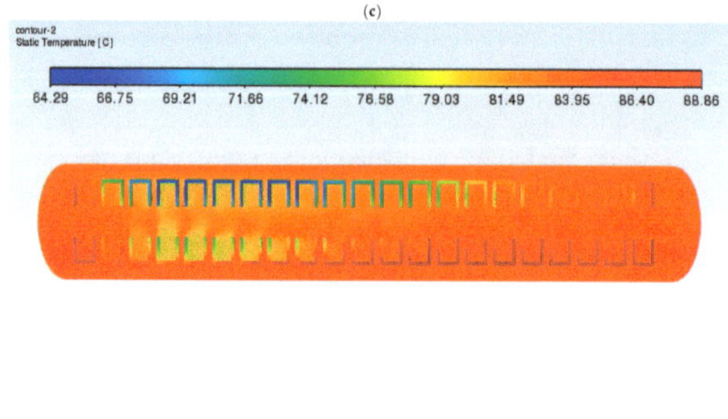

(d)

Figure 10. Cross-sectional xy slice temperature distribution of the vulcanization tank (60 kW): (**a**) time = 3000 s; (**b**) time = 6000 s; (**c**) time = 9000 s; (**d**) time = 12,000 s.

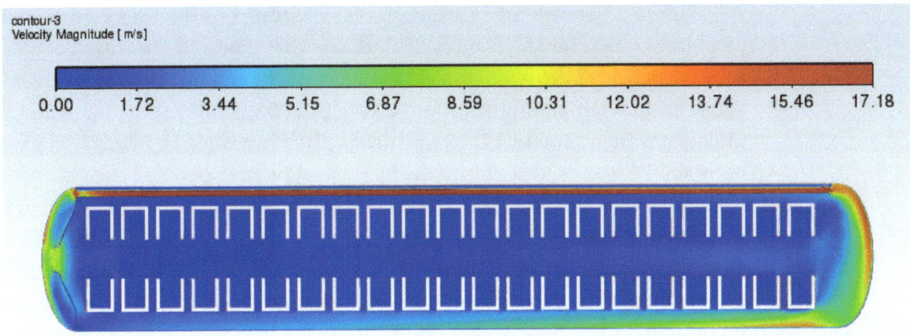

Figure 11. Cross-sectional xy slice velocity distribution of the vulcanization tank (60 kW, 3000–12,000 s).

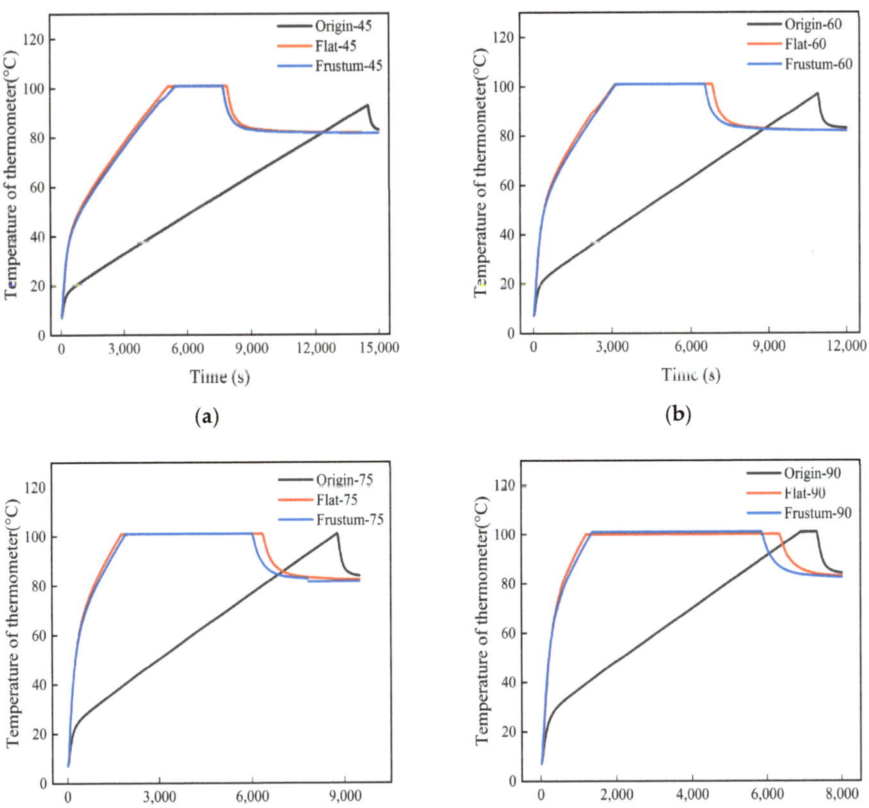

Figure 12. Temperature history diagrams of thermometers under different power levels: (**a**) 45 kW; (**b**) 60 kW; (**c**) 75 kW; (**d**) 90 kW.

Figure 13 shows the heating process diagrams of the tires in three different types of vulcanization tanks. The results indicate that when the heating power is 45 kW, the time required for the waste tires in the vulcanization tank with orifice plates to reach the vulcanization temperature is approximately 7800 s, while for the untreated vulcanization tank waste tires, it takes 14,500 s, which is directly reduced by 46.2%. The time difference between the two decreases with the increase in heating power. When the heating power is 90 kW, the time required for the waste tires in the vulcanization tank with orifice plates to reach the vulcanization temperature is about 5900 s, and for the untreated vulcanization tank waste tires, it takes 7300 s, which is reduced by 19.2%. The results show that the orifice plate adjusts the flow of the heating air, making the heating air more concentrated towards the tires, thus making it easier for the tires to reach the target temperature. In the comparison of the time required for the two types of orifice plate tires to reach the vulcanization temperature, the frustum cone orifice plate has a slight advantage over the flat orifice plate, with a time difference ranging from 200 to 480 s.

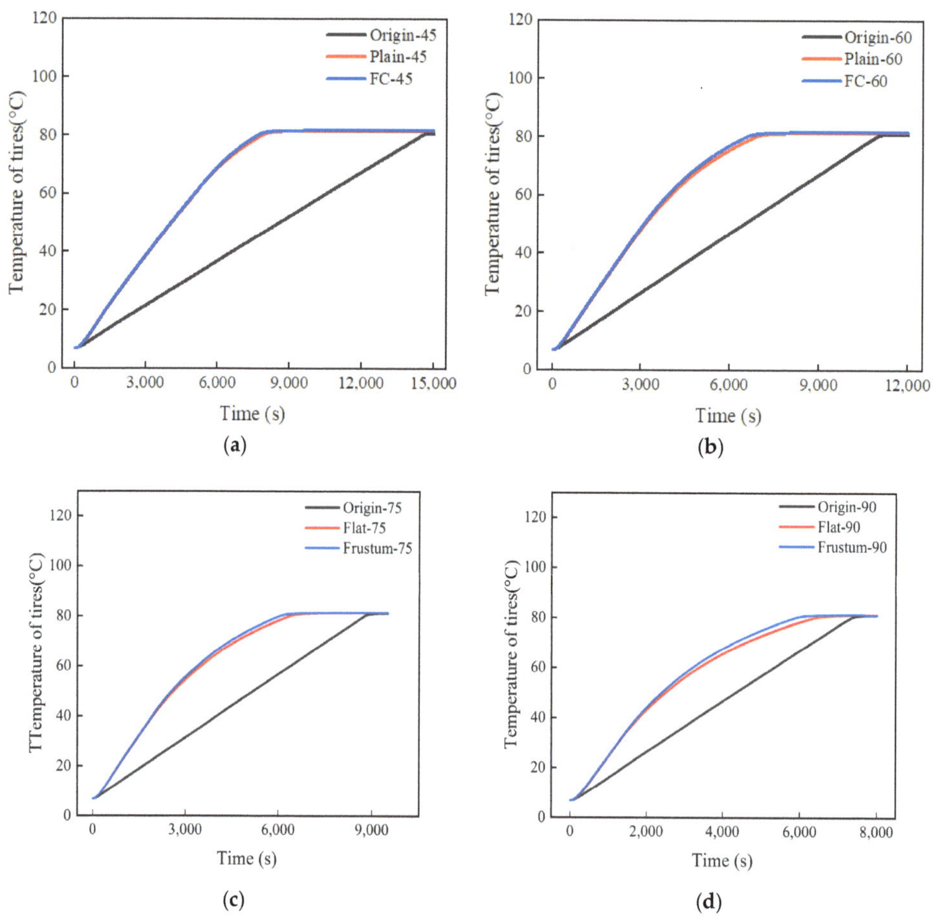

Figure 13. Temperature history diagrams of tire under different power levels: (**a**) 45 kW; (**b**) 60 kW; (**c**) 75 kW; (**d**) 90 kW.

Figure 14 shows the temperature heating history of the vulcanizing agent under different power levels. The process of retreading old tires actually involves heating the vulcanizing agent to the vulcanization temperature to bond the old tire with a new tread, creating a new tire. Under different heating powers, the effects of the frustrum cone orifice plates and flat orifice plates on increasing the temperature of the vulcanizing agent are almost the same, which may be due to the similarity in their structures. The time required for the vulcanizing agent with orifice plates to reach the vulcanization temperature is approximately 7300 s, while the time required for the untreated vulcanizing agent to reach the vulcanization temperature is about 14,200 s, which means that at a heating power of 45 kW, the presence of the orifice plates can help reduce the time required to reach vulcanization temperature by 48.6%. Similarly, this time difference decreases with the increase in heating power, reaching a reduction of 20.7% at a heating power of 45 kW. The results indicate that the presence of the orifice plates regulates the flow of the heating air, making it more concentrated towards the vulcanizing agent, thus facilitating the vulcanizing agent to reach the vulcanization temperature more easily.

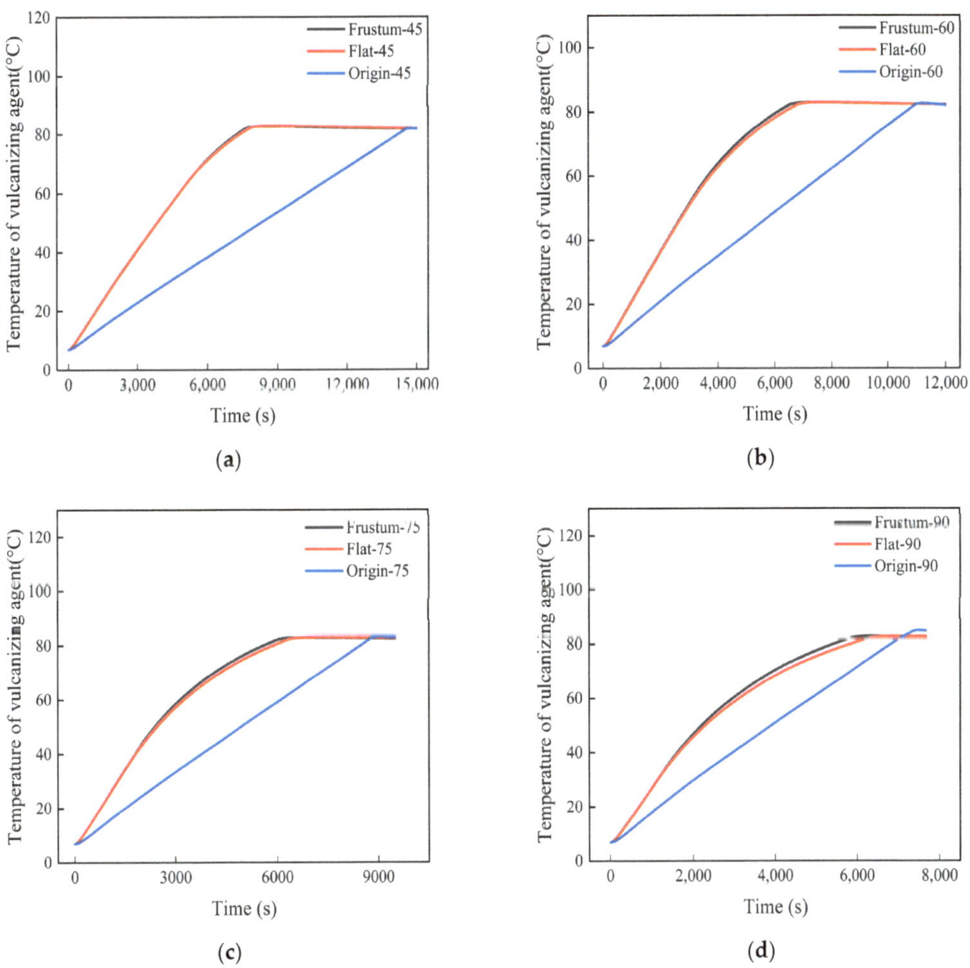

Figure 14. Temperature history diagrams of the vulcanizing agent under different power levels: (**a**) 45 kW; (**b**) 60 kW; (**c**) 75 kW; (**d**) 90 kW.

A large number of studies [35–38] have shown that the retreading results of tires are related to the temperature of the vulcanizing agent, and an excessively large temperature difference can easily lead to the over-vulcanization reaction of the vulcanizing agent in some positions, which affects the quality of retreaded tires. Figure 15 shows the heating process of the vulcanizing agent temperature difference at different powers. The results show that the difference in the temperature of the vulcanizing agent treated with the orifice plate can greatly reduce the maximum temperature difference. At the same time, with an increase in heating power, the difference between the temperature difference of the vulcanizing agent treated with the orifice plate and the difference in the temperature of the vulcanizing agent without an orifice plate will increase. This means that as the heating power increases, the temperature regulation of the orifice plate becomes more obvious. In the comparison of orifice vulcanization tanks, the temperature regulation of the frustrum cone orifice plate is superior to that of the flat orifice plate, resulting in the better quality of the retread tires.

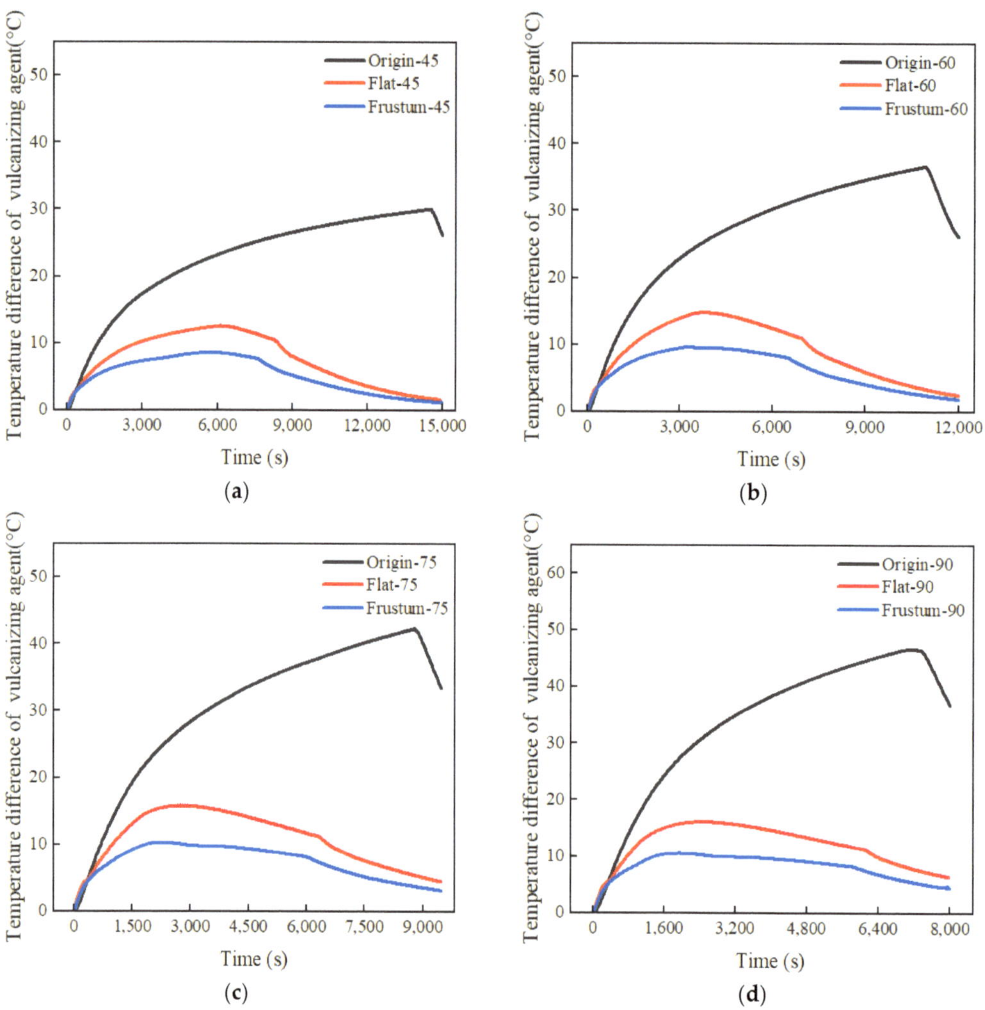

Figure 15. Temperature difference heating process diagram of the vulcanizing agent under different powers: (**a**) 45 kW; (**b**) 60 kW; (**c**) 75 kW; (**d**) 90 kW.

Figure 16 shows the heating process diagram of the standard deviation of the vulcanizing agent temperature under different powers. When the heating power is 45 kW, the standard deviation of the flat plate orifice plate and the frustum cone orifice plate is 7.01, which is 7.2 lower than that of the untreated vulcanization tank. When the heating power is 90 kW, the standard deviation of the flat plate orifice plate and the frustum cone orifice plate is 11.1, which is 11.75 lower than that of the untreated vulcanization tank. The results indicate that the presence of the orifice plates can greatly improve the uneven temperature distribution of the vulcanizing agent, thereby enhancing the quality of retreaded tires, and the regulation effect becomes more pronounced with the increase in heating power.

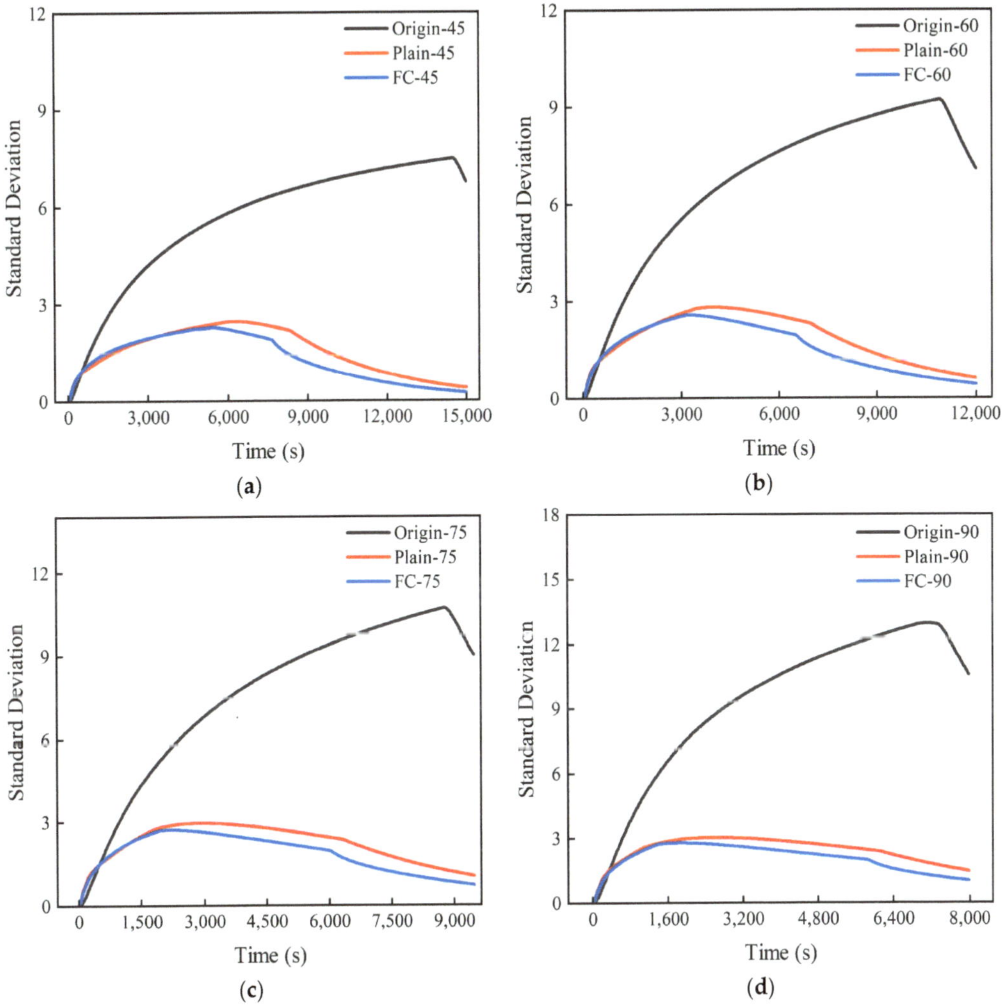

Figure 16. Standard deviation heating process diagram of the vulcanizing agent under different powers: (**a**) 45 kW; (**b**) 60 kW; (**c**) 75 kW; (**d**) 90 kW.

Figures 17 and 18 show the cross-sectional xy slice temperature distribution of a vulcanizer with orifice plates during the heating process at a heating power of 60 kW. As can be seen from the diagram, when the time is 3000 s, there is still a slight trend of the temperature inside the vulcanizing tank being lower than outside. However, at

6000 s, this phenomenon improves, and the temperature difference between the inside and outside of the retreaded tire is significantly reduced. By 9000 s, this phenomenon is further reduced. At 12,000 s, the temperatures inside and outside of the retreaded tire are essentially consistent. The results indicate that the presence of the orifice plate and the frustrum cone orifice plate can effectively improve the uneven temperature distribution issue in the retreaded tire vulcanization process, thereby enhancing the quality of the retreaded tires.

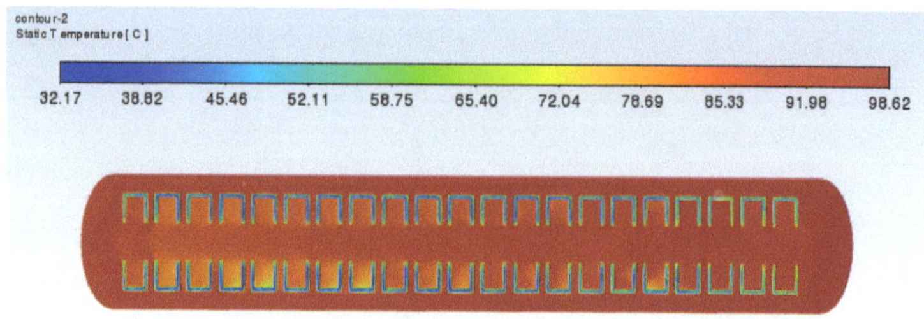

(a)

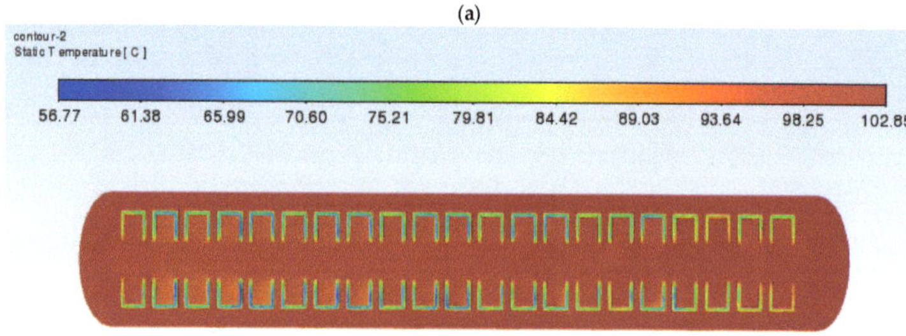

(b)

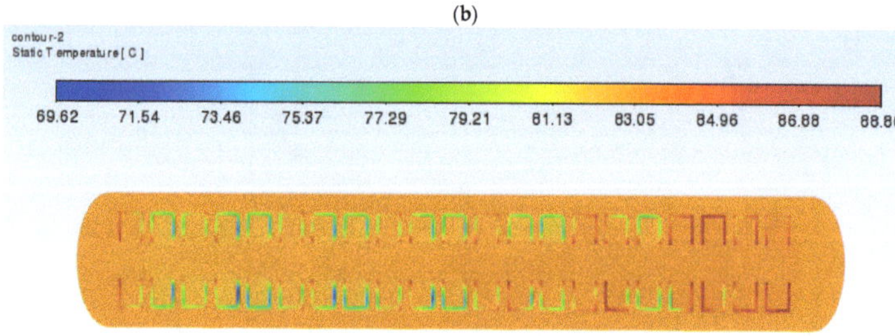

(c)

Figure 17. *Cont.*

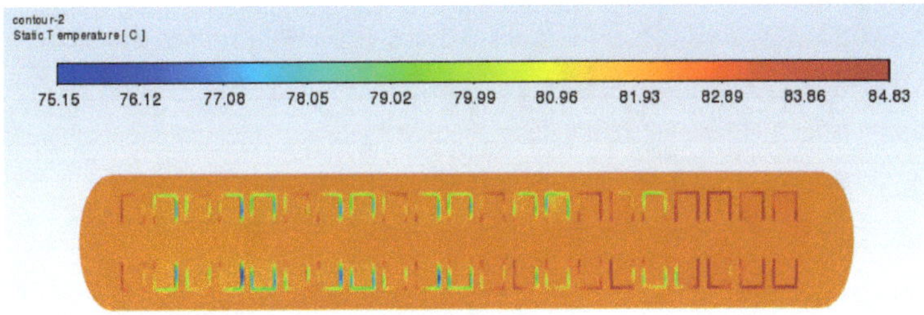

(d)

Figure 17. Cross-sectional xy slice temperature distribution of the vulcanization tank with flat plates (60 kW): (**a**) time = 3000 s; (**b**) time = 6000 s; (**c**) time = 9000 s; (**d**) time = 12,000 s.

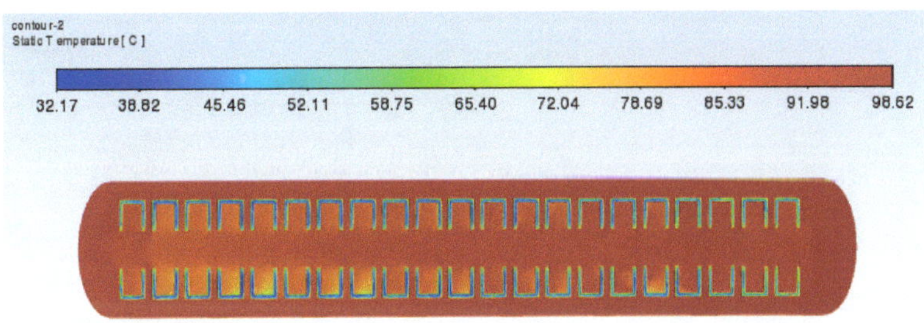

(a)

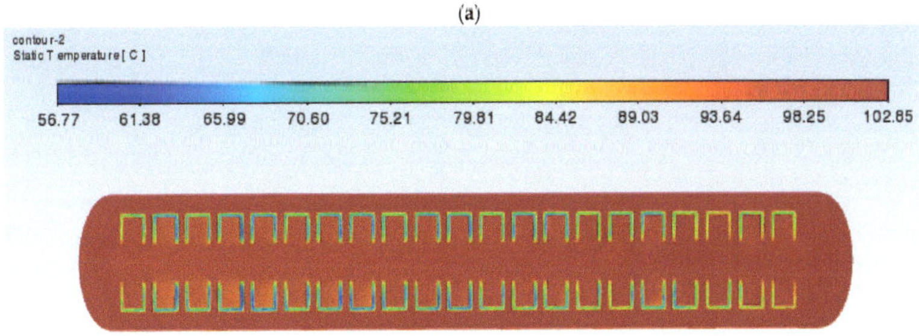

(b)

Figure 18. *Cont.*.

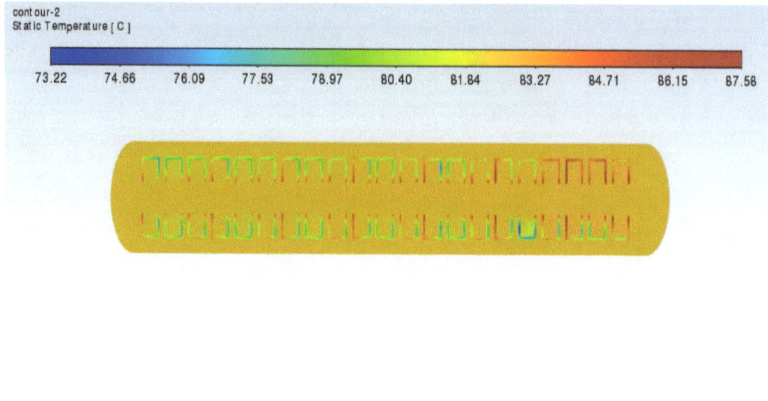

(c)

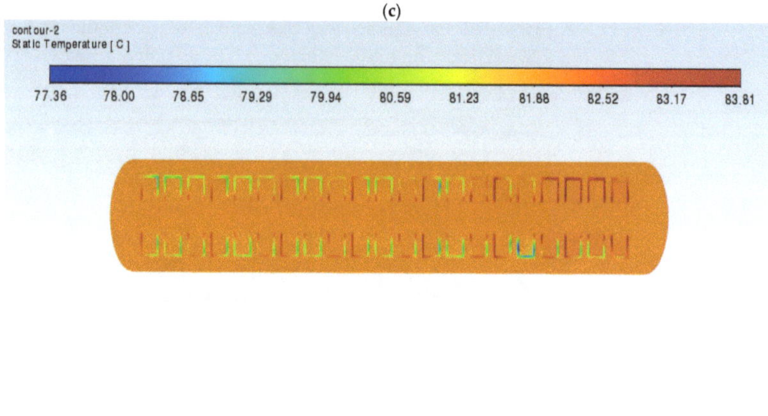

(d)

Figure 18. Cross-sectional xy slice temperature distribution of the vulcanization tank with frustum cone plates (60 kW): (**a**) time = 3000 s; (**b**) time = 6000 s; (**c**) time = 9000 s; (**d**) time = 12,000 s.

Additionally, Figures 19 and 20 show the velocity distribution diagrams of the cross-sections of the xy slice during the heating process of the vulcanization tank with frustum cone orifice plates and flat orifice plates at a heating power of 60 kW. The results indicate that the orifice plates can mitigate the issue of excessive heat concentration.

Although the improvement effect of the frustum cone orifice plate is better than that of the flat orifice plate, the area of the frustum cone orifice plate is 76% higher than that of the flat orifice plate, which means that the manufacturing cost of the frustum cone orifice plate is 76% higher than that of the flat orifice plate. Moreover, the shape of the frustum cone orifice plate is more complex than that of the flat orifice plate, which implies higher costs.

The results indicated that the presence of the orifice plate prevents heat accumulation and addresses the issue of excessive temperature difference during the heating of used tires, which can cause over-vulcanization or under-vulcanization, thereby enhancing the quality of the retreaded tires. In the future, the structure of the orifice plate can be optimized to concentrate heat more effectively on the vulcanizing agent, reducing the heating time and retreading duration of the tires, ultimately achieving energy savings and efficiency improvements.

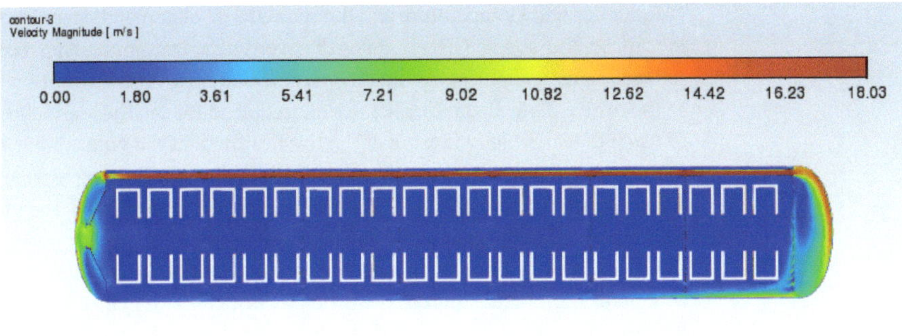

Figure 19. Cross-sectional xy slice velocity distribution of the vulcanization tank with flat plates (60 kW).

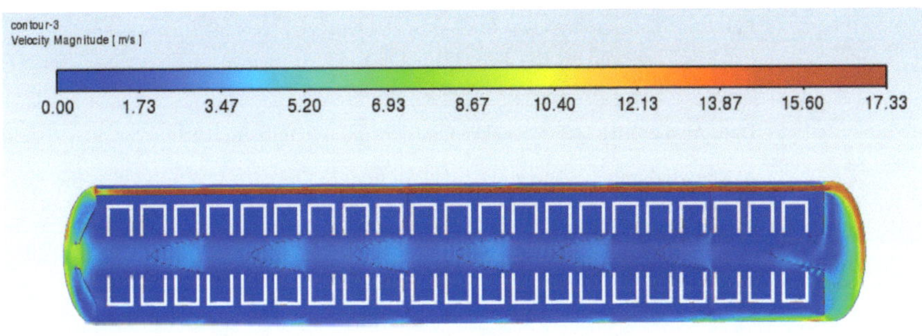

Figure 20. Cross-sectional xy slice temperature distribution of the vulcanization tank with frustrum cone orifice plates (60 kW).

6. Conclusions

The main conclusions of this study are as follows:

(1) During the heating process of a vulcanization tank without orifice plate, an increase in heating power leads to a decrease in the time required for the thermometer temperature and the tire to reach the target temperature, as well as for the vulcanizing agent to reach the vulcanization temperature. This helps improve the renovation rate of retreaded tires. However, the increase in heating power can also lead to the accumulation of heat, resulting in an increase in the temperature difference within the vulcanizing agent and an enlargement of the standard deviation at the monitoring points. This can cause the over-vulcanization or under-vulcanization of the retreaded tire, ultimately affecting its quality.

(2) During the heating process, placing orifice plates in the vulcanization tank can significantly improve the uneven temperature distribution issue that occurs during the original vulcanization tank heating process, greatly reducing the difference in the

vulcanizing agent temperature and the standard deviation of the monitoring points for the vulcanizing agent. This mitigates the over-vulcanization or under-vulcanization of retreaded tires, enhancing the quality of retreaded tires. At the same time, the presence of the orifice plate also reduces the time required for the thermometer temperature to reach the target temperature of the tire heating and the vulcanizing agent heating to the vulcanization temperature, thereby reducing the heating time and ultimately achieving energy savings. When the heating power is 45~90 kW, the presence of the orifice plate can approximately save at least 90~36 kW*h, respectively.

(3) In the comparison of orifice plates, the performance of the frustum cone orifice plate is shown to be better than that of the flat orifice plate. However, the area of the frustum cone orifice plate is 76% higher than that of the flat orifice plate, and it is more complex to manufacture, increasing costs. Therefore, considering all factors, this paper recommends using the flat orifice plate for improving the uneven temperature distribution issue during the vulcanization tank heating process.

Author Contributions: Methodology, T.S.; Validation, X.L. and H.L.; Data curation, B.W.; Writing—original draft, T.S.; Writing—review and editing, T.S.; Supervision, Y.M., B.W. and X.Z.; Funding acquisition, B.W. All authors have read and agreed to the published version of the manuscript.

Funding: This research was funded by the National Natural Science Foundation of China (Grant No. 51575286), and the Key R&D Plan of Shandong Province (public welfare science and technology research) (Grant No. 2018GGX 105007).

Data Availability Statement: Data are contained within the article.

Acknowledgments: The authors are grateful to Qingdao University and the company Qingdao Wanfang Recycling Technology Co., Ltd., China for their financial and moral support. Moreover, the authors are also indebted to the reviewers for their positive suggestions that helped improve the content of this study.

Conflicts of Interest: Authors Baolin Wang, Hui Li, and Xiaowen Luan were employed by the company Qingdao Wanfang Recycling Technology Co., Ltd. The remaining authors declare that the research was conducted in the absence of any commercial or financial relationships that could be construed as a potential conflict of interest.

References

1. Nuzaimah, M.; Sapuan, S.M.; Nadlene, R.; Jawaid, M. Recycling of Waste Rubber as Fillers: A Review. *IOP Conf. Ser. Mater. Sci. Eng.* **2018**, *368*, 012016. [CrossRef]
2. Danon, B.; van der Gryp, P.; Schwarz, C.E.; Görgens, J.F. A Review of Dipentene (Dl-Limonene) Production from Waste Tire Pyrolysis. *J. Anal. Appl. Pyrolysis* **2015**, *112*, 1–13. [CrossRef]
3. Fazli, A.; Rodrigue, D. Recycling Waste Tires into Ground Tire Rubber (GTR)/Rubber Compounds: A Review. *J. Compos. Sci.* **2020**, *4*, 103. [CrossRef]
4. Fiksel, J.; Bakshi, B.R.; Baral, A.; Guerra, E.; De Quervain, B. Comparative Life Cycle Assessment of Beneficial Applications for Scrap Tires. *Clean Technol. Environ. Policy* **2011**, *13*, 19–35. [CrossRef]
5. Alkadi, F.; Lee, J.; Yeo, J.-S.; Hwang, S.-H.; Choi, J.-W. 3D Printing of Ground Tire Rubber Composites. *Int. J. Precis. Eng. Manuf. -Green Technol.* **2019**, *6*, 211–222. [CrossRef]
6. Toncheva, A.; Brison, L.; Dubois, P.; Laoutid, F. Recycled Tire Rubber in Additive Manufacturing: Selective Laser Sintering for Polymer-Ground Rubber Composites. *Appl. Sci.* **2021**, *11*, 8778. [CrossRef]
7. Thai, Q.B.; Chong, R.; Nguyen, P.T.T.; Le, D.K.; Le, P.T.K.; Phan-Thien, N.; Duong, H.M. Recycling of Waste Tire Fibers into Advanced Aerogels for Thermal Insulation and Sound Absorption Applications. *J. Environ. Chem. Eng.* **2020**, *8*, 104279. [CrossRef]
8. Stevenson, K.; Stallwood, B.; Hart, A.G. Tire Rubber Recycling and Bioremediation: A Review. *Bioremediation J.* **2008**, *12*, 1–11. [CrossRef]
9. Dabic-Miletic, S.; Simic, V.; Karagoz, S. End-of-Life Tire Management: A Critical Review. *Environ. Sci. Pollut. Res.* **2021**, *28*, 68053–68070. [CrossRef]

10. Deng, N.; Wang, W.; Chen, G.; Zhang, Y.; Zhang, Y.; Ma, H. Pyrolysis Characteristics of Rubber Compositions in Medical Waste. *J. Cent. South Univ.* **2013**, *20*, 2466–2471. [CrossRef]
11. Alsaleh, A.; Sattler, M.L. Waste Tire Pyrolysis: Influential Parameters and Product Properties. *Curr. Sustain. Renew. Energy Rep.* **2014**, *1*, 129–135. [CrossRef]
12. Čabalová, I.; Ház, A.; Krilek, J.; Bubeníková, T.; Melicherčík, J.; Kuvik, T. Recycling of Wastes Plastics and Tires from Automotive Industry. *Polymers* **2021**, *13*, 2210. [CrossRef] [PubMed]
13. Tsai, W.-T.; Chen, C.-C.; Lin, Y.-Q.; Hsiao, C.-F.; Tsai, C.-H.; Hsieh, M.-H. Status of Waste Tires' Recycling for Material and Energy Resources in Taiwan. *J. Mater. Cycles Waste Manag.* **2017**, *19*, 1288–1294. [CrossRef]
14. Sathiskumar, C.; Karthikeyan, S. Recycling of Waste Tires and Its Energy Storage Application of By-Products—A Review. *Sustain. Mater. Technol.* **2019**, *22*, e00125. [CrossRef]
15. Oriaku, E.C. Waste to Wealth through the Incineration of Waste Tyres and Recovery of Carbon Black. *Int. J. Multidiscip. Sci. Eng.* **2013**, *4*, 30–36.
16. Singh, S.; Nimmo, W.; Gibbs, B.M.; Williams, P.T. Waste Tyre Rubber as a Secondary Fuel for Power Plants. *Fuel* **2009**, *88*, 2473–2480. [CrossRef]
17. Dabić-Ostojić, S.; Miljuš, M.; Bojović, N.; Glišović, N.; Milenković, M. Applying a Mathematical Approach to Improve the Tire Retreading Process. *Resour. Conserv. Recycl.* **2014**, *86*, 107–117. [CrossRef]
18. Ounsaneha, W.; Buadit, T.; Rattanapan, C. Assessment of Human Health Impact Based on Life Cycle Assessment: A Case Study of Thai Retread Tire. *IOP Conf. Ser. Mater. Sci. Eng.* **2020**, *773*, 012038. [CrossRef]
19. Qiang, W.; Li, J.; Yunlong, W.; Xiaojie, Q.; Guotian, W. Discussion on Tire Retreading and Reuse Technology. *IOP Conf. Ser. Earth Environ. Sci.* **2020**, *512*, 012146. [CrossRef]
20. Gieré, R.; Smith, K.; Blackford, M. Chemical Composition of Fuels and Emissions from a Coal + Tire Combustion Experiment in a Power Station. *Fuel* **2006**, *85*, 2278–2285. [CrossRef]
21. Ferrão, P.; Ribeiro, P.; Silva, P. A Management System for End-of-Life Tyres: A Portuguese Case Study. *Waste Manag.* **2008**, *28*, 604–614. [CrossRef] [PubMed]
22. US EPA, O. Durable Goods: Product-Specific Data. Available online: https://www.epa.gov/facts-and-figures-about-materials-waste-and-recycling/durable-goods-product-specific-data (accessed on 19 November 2024).
23. Shih, T.-H.; Liou, W.W.; Shabbir, A.; Yang, Z.; Zhu, J. A New k-ϵ Eddy Viscosity Model for High Reynolds Number Turbulent Flows. *Comput. Fluids* **1995**, *24*, 227–238. [CrossRef]
24. Yu, S., Peng, Q.; Jun, X. Status and Prospect of General Industrial Solid Wastes in China's Different Regions. *Mod. Chem. Ind.* **2022**, *42*, 11–17. [CrossRef]
25. Pandya, M.; Patel, R.N.; Amarnath, S.K.P. Determination of Time Delay and Rate of Temperature Change during Tyre Curing (Vulcanizing) Cycle. *Procedia Eng.* **2013**, *51*, 828–833. [CrossRef]
26. Zhang, J.; Jiao, Z.; Liu, Y.; Zhang, Y.; Yang, W. Development of Inner Mould Direct-Pressure Tire Vulcanization Technology Equipment. *Key Eng. Mater.* **2013**, *561*, 351–356. [CrossRef]
27. Ti, S.; Chen, Z.; Kuang, M.; Li, Z.; Zhu, Q.; Zhang, H.; Wang, Z.; Xu, G. Numerical Simulation of the Combustion Characteristics and NOx Emission of a Swirl Burner: Influence of the Structure of the Burner Outlet. *Appl. Therm. Eng.* **2016**, *104*, 565–576. [CrossRef]
28. Yu, L.; Chang, Z.; Liu, H.; Yang, G.; Li, X. Research on the Numerical Simulation of Secondary Air Diffusion Angle to the Swirl Burners Combustion Process. *IOP Conf. Ser. Mater. Sci. Eng.* **2017**, *231*, 012034. [CrossRef]
29. Sharikov, Y.V.; Cabascango, V.E.Q. Mathematical Modeling of Mass, Heat and Fluid Flow in a Reverberatory Furnace for Melting Nickel-Containing Raw Materials. *J. Phys. Conf. Ser.* **2021**, *1753*, 012064. [CrossRef]
30. Alzwayi, A.; Paul, M.C. Heat Transfer Enhancement of a Lithium-Ion Battery Cell Using Vertical and Spiral Cooling Fins. *Therm. Sci. Eng. Prog.* **2024**, *47*, 102304. [CrossRef]
31. Roger, C.R.; Yen, S.H.; Ramanathan, K.G. Temperature Variation of Total Hemispherical Emissivity of Stainless Steel AISI 304. *J. Opt. Soc. Am.* **1979**, *69*, 1384–1390. [CrossRef]
32. Shida, Z.; Koishi, M.; Kogure, T.; Kabe, K. A Rolling Resistance Simulation of Tires Using Static Finite Element Analysis. *Tire Sci. Technol.* **1999**, *27*, 84–105. [CrossRef]
33. Mars, W.V.; Luchini, J.R. An Analytical Model for the Transient Rolling Resistance Behavior of Tires. *Tire Sci. Technol.* **1999**, *27*, 161–175. [CrossRef]
34. Lin, Y.-J.; Hwang, S.-J. Temperature Prediction of Rolling Tires by Computer Simulation. *Math. Comput. Simul.* **2004**, *67*, 235–249. [CrossRef]
35. Schlanger, H.P. A One-Dimensional Numerical Model of Heat Transfer in the Process of Tire Vulcanization. *Rubber Chem. Technol.* **1983**, *56*, 304–321. [CrossRef]
36. Toth, W.J.; Chang, J.P.; Zanichelli, C. Finite Element Evaluation of the State of Cure in a Tire. *Tire Sci. Technol.* **1991**, *19*, 178–212. [CrossRef]

37. Gehman, S.D. Heat Transfer in Processing and Use of Rubber. *Rubber Chem. Technol.* **1967**, *40*, 36–99. [CrossRef]
38. Li, Q.L.; Fan, L.; Li, T.; Chen, H.-L. Study on the Influence of Initial Temperature of Tire on Curing Temperature Field and Cure State. *Adv. Mater. Res.* **2011**, *221*, 528–532. [CrossRef]

polymers

MDPI

Article

Self-Healable, Antimicrobial and Conductive Hydrogels Based on Dynamic Covalent Bonding with Silver Nanoparticles for Flexible Sensor

Te Qi [1], Xuefeng Liu [1], Nan Zheng [1], Jie Huang [1], Wenlong Xiang [1], Yujin Nie [1], Zanru Guo [1,*] and Baixue Cai [2,*]

1 College of Chemistry, Chemical Engineering and Environmental Science, Minnan Normal University, Zhangzhou 363000, China; hj1745@mnnu.edu.cn (J.H.); wenlongx2290@mnnu.edu.cn (W.X.)
2 Chongqing Academy of Metrology and Quality Inspection, Chongqing 401120, China
* Correspondence: guozanru@mnnu.edu.cn (Z.G.); baicai1901@163.com (B.C.)

Abstract: Dynamic hydrogels have attracted considerable attention in the application of flexible electronics, as they possess injectable and self-healing abilities. However, it is still a challenge to combine high conductivity and antibacterial properties into dynamic hydrogels. In this work, we fabricated a type of dynamic hydrogel based on acylhydrazone bonds between thermo-responsive copolymer and silver nanoparticles (AgNPs) functionalized with hydrazide groups. The hybrid hydrogels exhibited sol–gel transition, self-healable, injectable and thermo-responsive abilities. The self-healing efficiency was over 92%. Moreover, the hydrogel displayed antimicrobial properties and high conductivity (6.85 S/m). Notably, the fabricated hydrogel-based sensors exhibited strain and temperature sensing (22.05%/°C) and could detect human motion and speech, and electrocardiographic (ECG) and electromyography (EMG) signals. Overall, this work provides a simple strategy to synthesize AgNPs-based dynamic hydrogels with multi-functions, and the hydrogels may find potential applications in antibacterial wearable electronics, health monitoring and speech recognition.

Keywords: hydrogel sensor; self-healing; silver nanoparticles; antimicrobial properties; acylhydrazone bonds

check for updates

Academic Editor: Tsuyoshi Michinobu

Received: 5 December 2024
Revised: 24 December 2024
Accepted: 27 December 2024
Published: 29 December 2024

Citation: Qi, T.; Liu, X.; Zheng, N.; Huang, J.; Xiang, W.; Nie, Y.; Guo, Z.; Cai, B. Self-Healable, Antimicrobial and Conductive Hydrogels Based on Dynamic Covalent Bonding with Silver Nanoparticles for Flexible Sensor. *Polymers* **2025**, *17*, 54. https://doi.org/10.3390/polym17010054

1. Introduction

As one of the soft materials with 3D network structure storing large amounts of water, hydrogels have tremendous applications ranging from drug delivery [1], wound dressings [2], soft robotics [3,4], flexible electronics [5,6] and so on. Among them, there is an increasing interest in hydrogel-based flexible electronics [6], because hydrogels possess softness comparable to biological tissues and can provide an environment similar to that of extracellular matrix [7,8]. Thus, hydrogel-based flexible sensors could contact with target tissue seamlessly in comparison with traditional rigid devices, which enable the sensors to record motion and electrophysiological (EP) or other physiological signals [9]. As for hydrogel sensors, conductive hydrogels were usually employed, as they converted the mechanical or physiological signals into electrical signals [10]. Conductive hydrogel contains a hydrogel matrix and conductive component [6]. However, it is hard for hydrogel sensors to avoid the mechanical damage by external pressure when they are used [11]. To extend the service life of hydrogel sensors and reduce waste, it is of great significance to develop self-healable hydrogel matrix capable of autonomously repairing damage without any external intervention.

Self-healable hydrogels would form by the introduction of dynamic covalent bonds (DCBs) or noncovalent bonds [12]. Compared with noncovalent bonds, dynamic covalent bonds can dissociate, recombine and exchange reversibly in hydrogel networks [13], which integrate the reversibility of noncovalent bonds and the stability of covalent bonds [12,13]. DCBs serving as crosslinking in hydrogel were concerned. To date, several DCBs such as phenylboronate esters [14], disulfide bonds [15] and acylhydrazone [16] have been used to prepare self-healable hydrogels [17,18]. Among them, acylhydrazone bonds, formed from the condensation of acylhydrazine and aldehyde/ketone, were concerned due to their easy designability and thermodynamic stability [19]. Recently, we and other researchers [12,13,19–22] have prepared hydrogels by using acylhydrazone bonds as crosslinking. The hydrogels exhibited self-healing ability without any external stimuli and possessed injectable and sol–gel transition properties. However, those hydrogels were mainly used as drug delivery vehicles and cell culture scaffolds in view of their biocompatibility [12,19]. Acylhydrazone-based hydrogels were rarely explored for the application of flexible sensors.

For hydrogel sensors, it is necessary to endow hydrogels with improved conductivity. Embedding conductive fillers, such as ionic liquids [23], semiconductors [24], carbon materials [25] and nano-metals [26–28], is considered a facile and feasible method to enhance the conductivity of hydrogels [29]. Among these conductive fillers, silver nanoparticles (AgNPs) could be regarded as a potential choice for the conductor due to its low toxicity, broad-spectrum antibacterial properties and high conductivity [30]. A hydrogel sensor with antibacterial ability will reduce bacterial infection when it is applied to the human body. Han et al. [31] introduced the lignin–silver hybrid nanoparticles into the polyvinyl alcohol matrix and found that the hydrogel has sensitivity for compression. Fan et al. [32] dispersed AgNPs-attached CNCs in the hydrogel matrix and obtained the hydrogels with monitoring for various human movements. Wang et al. [33] prepared hydrogels with AgNPs for antibacterial strain sensors. However, AgNPs in those hydrogels were mainly dispersed in the hydrogel network rather than participating in the formation of the network [30–33]. AgNPs might have dropped from the carrier and agglomerated in the network after repeated stretching and compression. In addition, the hydrogels with AgNPs were mainly used to detect the motion [34,35], and the monitoring for other physiological signals such as temperature were rarely examined. When AgNPs have functional groups and form acylhydrazone bonds with temperature-responsive polymers, it is anticipated that the AgNPs will immobilize in the network hydrogels, and the hydrogel not only possesses self-healable, antibacterial, injectable and sol–gel transition properties but also exhibits temperature sensing due to the thermo-responsive hydrogel network.

In this paper, we report a type of self-healable, injectable and antimicrobial conductive hydrogel possessing motion and temperature sensing based on a thermo-responsive copolymer and AgNPs functionalized with hydrazide groups, as illustrated in Scheme 1. The thermo-responsive copolymer poly[(*N*-isopropylacrylamide)-co-(diacetone acrylamide) -co-poly[(acrylamide)] (P(NIPAM-co-DAAM-co-AM)) was synthesized by reversible addition–fragmentation chain-transfer (RAFT) polymerization. AgNPs with hydrazide groups were obtained by surface functionalization and hydrazinolysis. The conductive hydrogels were generated by mixing the copolymer and AgNPs via the formation of acylhydrazone bonds between ketone groups and hydrazides (Scheme 1). Based on dynamical acylhydrazone crosslinking, the obtained hydrogels exhibited sol–gel transition, self-healable, injectable and thermo-responsive abilities. Moreover, the hydrogel displayed antimicrobial properties, and the fabricated hydrogel-based sensors exhibited strain and temperature sensing. The hydrogel can also be used as an electrode for ECG and EMG signals detection. With these attractive characteristics, the dynamic hydrogels may find potential applications in antibacterial wearable electronics, health monitoring and speech recognition.

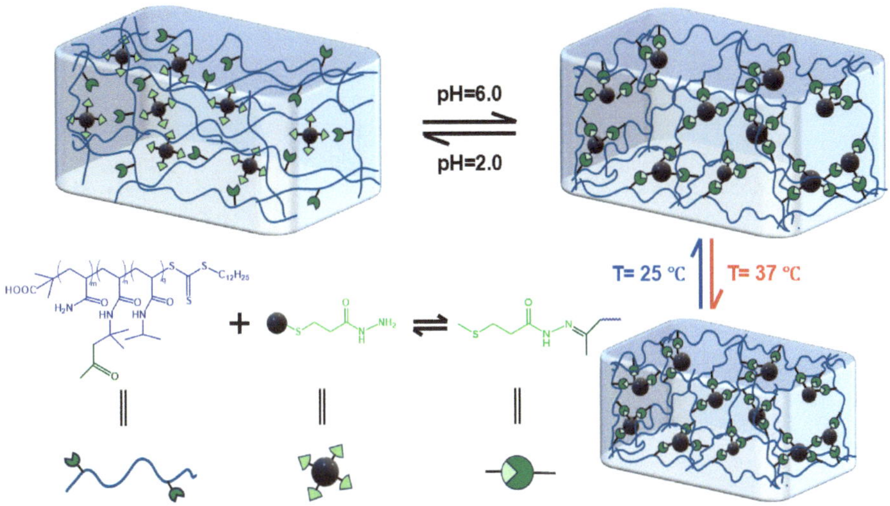

Scheme 1. Schematic of the fabrication of self-healable and sol–gel transition hydrogels based on AgNPs.

2. Materials and Methods

2.1. Materials

Silver nitrate ($AgNO_3$, 99.8%, Aladdin Co., Ltd., Shanghai, China), methyl 3-mercaptopropionate (98%, Aladdin) and acrylamide (AM, >98%, Aladdin Co., Ltd.), sodium borohydride ($NaBH_4$, 98%, Aladdin Co., Ltd.), hydrazine hydrate monohydrate (>98%, Aladdin Co., Ltd.), diacetone acrylamide (DAAM, 98%, TCI) and azodiisobutyronitrile (AIBN, 99%, Aladdin Co., Ltd.) were used as received. *N*-isopropyl acrylamide (NIPAM, 98%, TCI) was purified by recrystallization. A total of 4 M NaOH and HCl was used to adjust the pH of solutions. 2-(1-carboxy-1-methyl-ethylsulfanylthio- -carbonylsulfanyl)-2-methylpropionic acid was used as a chain transfer agent (CTA) and was synthesized following a previous report [36]. All other chemicals were analytical grade and used as received.

2.2. Preparation of Temperature-Sensitive Polymer Containing Ketone Groups

The copolymer was prepared according to our previous report [37]. NIPAM (4.0 g, 35.40 mmol), AM (0.718 g, 10.10 mmol) and DAAM (0.855 g, 5.06 mmol) were added to a two-necked vial with 30 mL dimethyl sulfoxide (DMSO), and then nitrogen was bubbled into the solution for 30 min to remove oxygen. The chain transfer agent CTA (922.0 mg, 0.25 mmol) and AIBN (2.1 mg, 0.013 mmol) were added with the protection of N_2. The solution was stirred continuously for another 5 min and placed in an oil bath at 70 °C for polymerization. After 24 h, a large amount of ether was added dropwise to obtain white precipitate. The product was collected and redissolved in DMSO and was reprecipitated into excess ether. This purification cycle was repeated twice. The resulting products were freeze-dried for 24 h to yield the objective copolymer, denoted as P_{70}, where "P" means the polymer and the subscript "$_{70}$" is the mole ratio of NIPAM to the total moles of the copolymer. The molar ratio of NIPAM, AM and DAAM in copolymer was 70:20:10.

2.3. Preparation of AgNPs Containing Hydrazide Groups

Three types of AgNPs containing hydrazide groups were prepared (Table S1). A typical experiment of AgNPs1 was as follows: methyl 3-mercaptopropionate (0.1200 g) dissolved in methanol (10 mL) was added dropwise into methanol (100 mL) with 0.85 g of $AgNO_3$ under high-speed stirring with a shading condition. After stirring for 4 h, $NaBH_4$ (0.5630 g) was slowly added to the solution (the solution immediately turned black). After the addition of $NaBH_4$, the mixture was stirred for another 4 h. Then, hydrazide hydrate (0.2300 g) was added and the solution refluxed at 70 °C for 12 h. Subsequently, AgNPs were obtained by centrifugation at 10,000 rpm for 5 min. The remaining solid was washed with anhydrous ethanol three times to remove any impurities. AgNPs containing hydrazides were dried in a vacuum oven at 50 °C for 24 h. Pure AgNPs were synthesized by adding $NaBH_4$ (0.5630 g) into $AgNO_3$ (0.85 g) and methanol (100 mL), and used as control.

2.4. Fabrication of Conductive Hydrogels

The conductive hydrogels were prepared by a simple mixing of the copolymer P_{70} solution and AgNPs solutions, where the number of hydrazide groups from AgNPs was equivalent to that of the ketone groups of the copolymer. The hydrazide groups per gram of AgNPs1, AgNPs2 and AgNPs3 were 1.14 mmol, 1.70 mmol and 2.24 mmol, respectively, which were determined by TGA (discussed in Section 3.2). The ketone groups per gram of P_{70} were 0.947 mmol, based on the unit's molar ratio (15.3:10.7:74.0) calculated by ^{1}H NMR spectra (discussed in Section 3.1). The solid content was kept at 10 wt%. The obtained hydrogels based on AgNPs1, AgNPs2 and AgNPs3 were labeled as HAg1, HAg2 and HAg3, respectively. The compositions of hydrogels are listed in Table S3. The procedure for the fabrication of HAg1 was as follows: 1.66 g AgNPs1 was dispersed into 32.94 g D. I. water by sonication for 30 min, then 2.00 g copolymer was dissolved in the dispersion by stirring. The pH of the mixture solution was tuned to ca. 6.0 by using 4 M HCl and 4 M NaOH. The viscous solutions became hydrogels after ca. 1 h at room temperature. The hydrogels were kept at room temperature for 24 h before testing. The preparation procedures for HAg2 and HAg3 were similar to that of AgNPs1.

2.5. Self-Healing and Injectable Properties of the Hydrogel

For the self-healing testing, one disk-shaped hydrogel was cut into two halves. Thereafter, two semicircular hydrogels rejoined for 15 min at room temperature. To calculate the self-healing efficiency, the tensile stress–strain curves of the rectangle-shaped hydrogels (40 mm × 10 mm × 10 mm) and the self-healed hydrogels were measured. The self-healing efficiency is defined as the ratio of the stress at fracture to the original stress. The mean values were calculated from at least three replicate experimental data.

The injectability experiment of the hydrogel was implemented by extruding HAg1 from a syringe.

2.6. Antibacterial Test of Hydrogels

Staphylococcus aureus (*S. aureus*, ATCC 25923) was used as the tested strain, and the trypticase soy agar (TSA) nutrient medium was used to cultivate it at 37 °C for 24 h. The cylindrical hydrogels (the diameter was 1.4 cm) were transferred to a Luria–Bertani (LB) agar plate covered with fresh *S. aureus* suspension. After incubation at 37 °C for 24 h, the diameter of the antibacterial circle was measured with vernier calipers, and the average value was obtained from at least three replicate experimental data.

The morphology changes of microorganisms on hydrogels were examined with SEM. A total of 1 mL of fresh *S. aureus* suspension (1×10^6 CFUs/mL) was incubated with the hydrogels at 37 °C for 6 h. The hydrogel samples were washed with PBS 3 times and then fixed with 2.5% glutaraldehyde solution for 4 h. The samples were dehydrated in a graded ethanol series from 50% to 100% each for 10 min, and the samples were dried in air at room temperature. They were observed with SEM for microbe morphology changes after coating with gold.

2.7. Characterization

^1H NMR analyses were conducted on a Brucker 300 MHz spectrometer (Brucker, Wissembourg, France) at room temperature, and D_2O was used as the solvent.

Fourier transform infrared (FT-IR) spectra were obtained on a Nicolet MX-1E FTIR spectrophotometer(Thermo Electron Corporation, Waltham, MA, USA). The dried samples were mixed with KBr and compressed into films.

Gel permeation chromatography (GPC) was performed using HCC-8320GPC (Higashima Co., Ltd., Tokyo, Japan) with equipped with a TSK GEL Super AWM-H column and refractive index detector. DMF was used as the eluent (the flow rate is 0.4 mL min^{-1}). Mono-dispersed polymethyl methacrylate was used as the standard.

SEM observation was conducted on a JSM-6510 (Japan Electronics Co., Ltd, Tokyo, Japan). The hydrogel samples were frozen by liquid nitrogen and then dried by vacuum freeze-drying.

X-ray diffraction (XRD) patterns were recorded using a RigakuDMAX 2200 (Rigaku corporation, Tokyo, Japan) with Ni-filtered CuKα radiation in the scanning range of 30° to 90°, at 40 kV and 40 mA X-ray power.

The conductivity of hydrogels was measured at room temperature by a four-probe instrument SB1201 (Shanghai Qianfeng Electronic Instrument Co., Ltd., Shanghai, China). The round hydrogels (5 mm thickness, 30 mm diameter) were used for the conductivity test. The mean values and errors were calculated from at least three independent samples for each type of hydrogel.

Thermogravimetry analysis (TGA) was conducted on a Diamond TG/DTA thermal analysis system (PerkinElmer, Hopkinton, MA, USA). Samples were heated in a flow of N_2 (20 mL min^{-1}) from room temperature to 800 °C at a heating rate of 10 °C min^{-1}.

Rheological characterization was performed on a Physica MCR 302 (Anton Paar, Graz, Austria) rheometer by the flat plate mode with a diameter of 25 mm. The experiments were conducted by the stress- or strain-controlled mode. For the modulus changing upon temperature, the temperature increasing rate is 0.2 °C min^{-1}.

The tensile strain–stress tests of the hydrogels (40 mm × 10 mm × 10 mm) were conducted on a universal testing machine BJ-SPLZ (Guangzhou Biaoji Packaging Equipment Co., Ltd., Guangzhou, China) fitted with a 20 N load cell at room temperature. The tensile speed was 50 mm/min.

Sensor tests were recorded on a digital multimeter TH2829A LCR meter (Changzhou Tonghui Electronics Co., Ltd., Changzhou, China). The gauge factor (GF) used to evaluate the sensitivity of sensors was calculated by the following equation: $GF = \Delta R/R_0/\varepsilon$, where ΔR and R_0 are the resistance difference after stretching and the initial resistance, and ε is the tensile strain of hydrogels. For the fabrication of the hydrogel-based sensor, both ends of the hydrogel (20 × 10 × 2 mm^3) were winded with copper wires serving as the current collectors. The other ends of copper wires were connected to the LCR meter for recording the resistance. The test was carried out by stretching with a universal testing machine or attaching to the subjects' bodies, such as the wrist and throat.

Electrocardiographic (ECG) and electromyography (EMG) signals were recorded by a commercial ECG and EMG monitor with three electrodes. Round hydrogels (1.5 mm thickness, 20 mm diameter) were used as electrodes and were directly adhered to the skin.

3. Results and Discussion

3.1. Fabrication and Structural Characterization of Temperature-Responsive Polymer

To obtain the polymer with a desirable molecular weight, RAFT polymerization was employed to prepare the copolymer [38]. According to our previous report [37], the cloudy point of the temperature-responsive polymer containing ketones could be well tuned by variation of the components of NIPAM, DAAM and AM units. NIPAM, DAAM and AM three monomers with a molar ratio of 70: 10: 20 were chosen to form the copolymer P_{70}, as its cloudy point was close to the human body temperature (37 °C) [19,37]. The structure of P_{70} was confirmed by ^1H NMR spectra (Figure S1a). The signals at 0.9~1.2 ppm and 3.7~4.0 ppm were assigned to the protons from NIPAM units [39]. The chemical shift peaks of $-(CH_3)_2C-$ on DAAM units and the protons on the main chain of P_{70} occurred at 1.2~1.8 ppm. The peaks of 1.8~2.3 ppm were ascribed to the protons from $-CH-$ of the main chain and $-CH_3$ on DAAM units. The peak at 2.8–3.3 ppm was assigned to $-CH_2-$ protons on DAAM units [40]. The signals of hydrogen protons were in agreement with our previous reports [37,39]. According to our previously reported equation [37,39], the unit ratios of P_{70} were calculated based on the integration of the proton peaks in the corresponding units. The molar ratio of AM, DAAM and NIPAM units was 15.3:10.7:74.0, which was slightly different than the feeding ratio due to the steric effect [37]. GPC was performed to measure the molecular weight ($M_{n,GPC}$) and polydispersity index (PDI) of P_{70} (Figure S1b and Table S2). It was found that the $M_{n,GPC}$ was close to the theoretical value, and its PDI is 1.483, indicating a copolymer with a controllable molecular weight and molecular weight distribution [41].

PNIPAM is a typical temperature-responsive polymer that can be converted from soluble to insoluble as the temperature rises [39,42]. To confirm the temperature responsiveness of P_{70}, the apparent phenomenon of P_{70} solution was observed during the changing of the temperature. As shown in Figure S2, the solution is colorless and transparent at 25 °C, while the solution appeared milky white when the temperature increased to 40 °C, which indicated that the prepared copolymer had a temperature-responsive property [33]. To further determine the cloudy point of P_{70}, the transmittance of the solution at 600 nm along with the temperature was tested. As shown in Figure S2, the transmittance of the solution started to decrease when the temperature was increased to 35 °C, and the transmittance dropped to nearly 0% as the temperature increased to 37.5 °C. The cloudy point of P_{70} was 36.3 °C, which is the midpoint temperature at the curve of transmittance with increasing of the temperature. The cloudy point of P_{70} was close to that of the human body temperature [43].

3.2. Synthesis and Characterization of AgNPs Hybrids Containing Hydrazide Groups

AgNPs hybrids with hydrazide groups were synthesized by surface functionalization and hydrazinolysis. XRD was performed to confirm the formation of AgNPs hybrids. As shown in Figure 1a, five diffraction peaks appeared in the XRD pattern at 2θ positions 38.1°, 44.2°, 64.4°, 77.4° and 81.5°, which are assigned to [1 1 1], [2 0 0], [2 2 0], [3 1 1] and [2 1 1] planes of the face-centered cubic (fcc) of AgNPs [44,45] in comparison with the JPDS card (no.: 04-0783). It also was found that there were other miscellaneous diffraction peaks with the increase in hydrazide bonds (such as AgNPs3), which might be attributed to too much hydrazide component having a negative effect on the crystallization of AgNPs3.

TEM was employed to visualize the morphology of AgNPs directly. As shown in Figure 1b–d, AgNPs containing hydrazide bonds exhibited a spherical shape and became smaller and more uniform in size distribution at the increasing of methyl 3-mercaptopropionate. This may be explained by AgNPs being wrapped with methyl 3-mercaptopropionate through a chemical bond between S ions and Ag atoms, and the increasing of the 3-mercaptopropionate ratio made more compounds on the AgNPs surface, which led to AgNPs hybrids being dispersed more homogeneously in the solution [46]. Therefore, the AgNPs size could be adjusted by changing the molar ratio of the compound to $AgNO_3$.

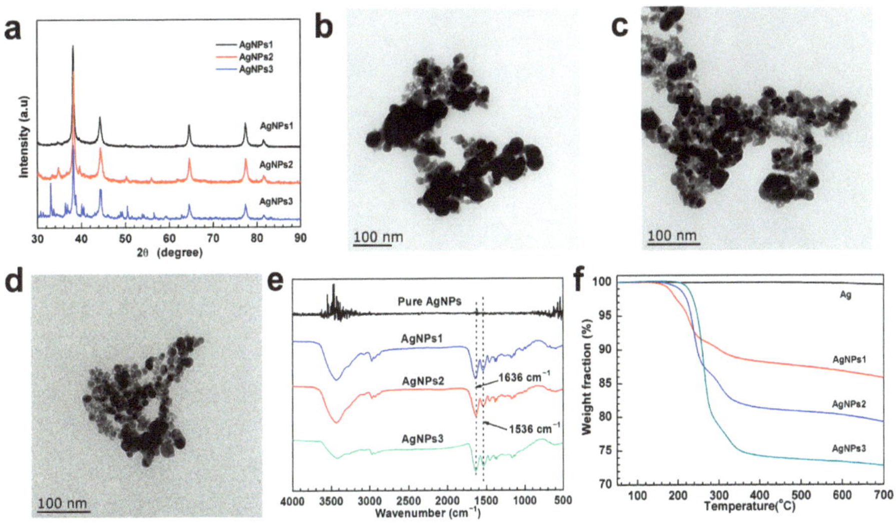

Figure 1. (**a**) XRD pattern of AgNPs with hydrazide groups; TEM images of AgNPs1 (**b**), AgNPs2 (**c**) and AgNPs3 (**d**); (**e**) FT-IR spectra of pure AgNPs and AgNPs hybrids; (**f**) TGA traces for pure AgNPs and AgNPs hybrids.

To check that hydrazide groups were introduced on AgNPs, FT-IR spectroscopy was employed to characterize the AgNPs hybrids. As shown in Figure 1e, AgNPs hybrids showed strong peaks at 1636 cm^{-1} and 1536 cm^{-1} in comparison with that of pure AgNPs, which were ascribed to the characteristic absorption band of C=O from ketone carbonyl groups and N-H of hydrazides [19,37,47], respectively. This suggests that hydrazide groups attached on the AgNPs surface. To determine the relative number of hydrazide groups on AgNPs, TGA measurements were performed. As given in Figure 1f, the weight loss of AgNPs under a nitrogen atmosphere happened in the temperature range from ca. 150 °C to ca. 380 °C, resulting from the decomposition of the organic component on AgNPs. One can find that the weight percentages of the organic component in AgNPs1, AgNPs2 and AgNPs3 are 13.7 wt%, 20.4 wt% and 26.9 wt%, respectively, providing additional evidence of hydrazide groups on AgNPs. The mole number of hydrazide groups per gram of AgNPs could be calculated by dividing the molecular weight of the organic component. The values for AgNPs1, AgNPs2 and AgNPs3 are 1.14 mmol, 1.70 mmol and 2.24 mmol, respectively, which suggests that the content of hydrazide groups on AgNPs can be regulated by feeding ratios.

3.3. Preparation and Characterization of Dynamic Hydrogels with AgNPs

The dynamic acylhydrazone can be formed between ketone and hydrazide [19,39], which served as the crosslinking for hydrogels. Thus, the copolymer P_{70} containing ketone groups was mixed with AgNPs hybrids, resulting in the formation of dynamic hydrogel HAg1, HAg2 and HAg3. As shown in Figure 2a, the mixture of P_{70} and AgNPs hybrids displayed as gels and could support their weight in a vial when the pH of the solutions was adjusted to ca. 6. To confirm the formation of the hydrogels, dynamic rheology measurements were performed. Figure 2b shows the dynamic modulus versus the frequency at a 1% strain at 25 °C. One can find that the storage modulus (G') of the three hydrogels was consistently higher than their loss modulus (G'') in the tested frequency range, indicative of the formation of hydrogels [13,19]. In addition, G' of HAg1, HAg2 and HAg3 were ca. 126 Pa, ca. 227 Pa and ca. 395 Pa, i.e., G' increased gradually with the increment of hydrazide groups on the AgNPs surface. This should be attributed to the greater the number of hydrazide groups in one nanoparticle is, the greater the amount of polymer chains bonded together, forming a more robust network and leading to a higher modulus.

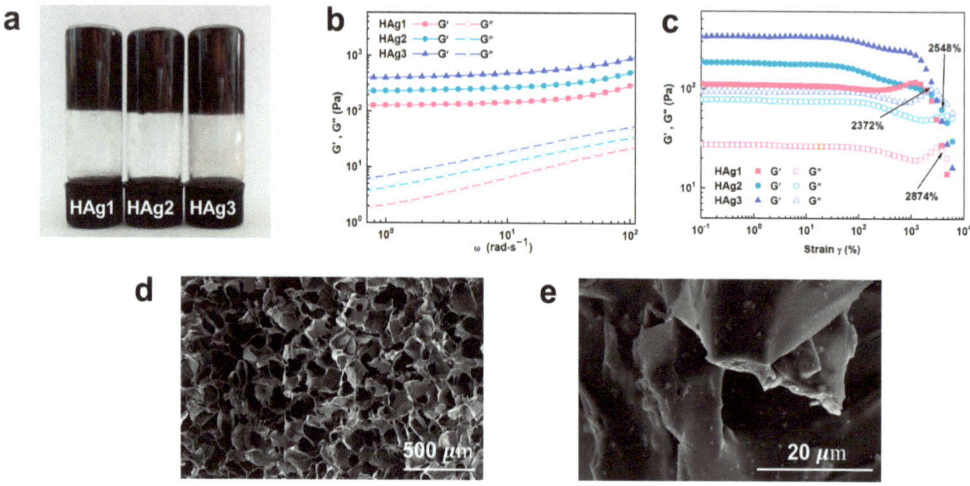

Figure 2. (**a**) Image of the obtained hydrogels; (**b**) G' and G'' of the hydrogels as functions of frequency (ω); (**c**) Strain amplitude sweep of the hydrogels with a fixed frequency of 1 Hz at 25 °C; (**d**) SEM image of HAg1 and (**e**) its magnified SEM image on skeleton.

In addition, strain amplitude sweeps for obtained hydrogels were measured at a fixed frequency of 1 Hz at 25 °C. As illustrated in Figure 2c, the G' of the hydrogels was greater than G'' at the initial strain region. When the strain reached a critical value, G'' overlapped with G', suggesting that the network of the hydrogel was destroyed. The critical strain values of HAg1, HAg2 and HAg3 were 2877 ± 13%, 2545 ± 9% and 2373 ± 14%, respectively, which indicates that the hydrogels could withstand large deformation. However, the critical strain value decreased on the contrary, though the G' increased from HAg1 to HAg3. This should be attributed to the chains in the network gradually becoming tight due to the increment of hydrazide groups in one nanoparticle, which would increase the stiffness of the hydrogel and concomitantly might reduce the movement of polymer chains. This led to the disruption of hydrogels at a lower strain.

SEM was carried out to observe the microstructure of the hydrogels. As shown in Figure 2d, HAg1 displayed a uniform three-dimensional porous structure, further indicating the formation of hydrogel between AgNPs hybrids and copolymer. Moreover, AgNPs hybrids were uniformly distributed in the skeleton of hydrogel (Figure 2e). HAg2 and HAg3 also showed similar porous structures (Figure S3). However, the pore size increased, and its skeleton became thicker from HAg1 to HAg3 (Figures 2d and S3), which further indicated that the increment of hydrazide groups in nanoparticles would react with more polymer chains. In order to further verify the formation of hydrogels with AgNPs, the prepared hydrogels were characterized by elemental analysis. The EDS spectra of the hydrogels is shown in Figure S4. One can find that the hydrogels contained C, N, O, S and Ag elements, which originated from AgNPs hybrids and the copolymer. However, it should be noted that the content of Ag gradually decreased from HAg1 to HAg3. This result should be attributed to the feeding ratio of AgNPs hybrids in hydrogel decreasing (Table S3) with the increasing of hydrazide groups on the AgNPs surface due to the ketone groups remaining constant.

To confirm the formation of hydrogels via acylhydrazone crosslinking, FT-IR was performed to analyze the precursors and hydrogel. The characteristic absorption peak of the C=O of ketones in the polymer appeared at 1720 cm^{-1} (Figure S5). However, when the AgNPs1 were mixed with the copolymer to form HAg1, the characteristic absorption peak of C=O at 1720 cm^{-1} disappeared, while the characteristic absorption of the C=N bond of acylhydrazone appeared at 1665 cm^{-1} [37], indicating the formation of the acylhydrazone crosslinking in the network. To demonstrate the formation of hydrogels based on AgNPs, the hydrogels were analyzed by XRD. As shown in Figure S6, five diffraction peaks of AgNPs appeared at 2θ of 38.1°, 44.2°, 64.4°, 77.4° and 81.5°, consistent with that of AgNPs hybrids. This result indicated that AgNPs in hydrogels showed a face-centered cubic structure, and AgNPs were not destroyed during the formation of hydrogels. In addition, AgNPs in hydrogels remained stable even after storing at room temperature for 1 month, as it was found that XRD patterns of the hydrogels were similar to that of initial hydrogel samples (Figure S7).

3.4. Self-Healing and Injectable Properties of the Hydrogels

According to previous reports [19,37], hydrogels constructed with acylhydrazone bonds have self-healing properties. Therefore, the self-healing properties of hydrogels were visually examined. As shown in Figure 3a, a circular hydrogel HAg1 was cut into half. Then, the two semicircle gels were spliced together for 15 min without any external intervention. It was found that they could heal completely and be gently stretched with forceps, indicating the self-healing of the hydrogel. Moreover, step-strain measurements were carried out to verify the self-healing property. As shown in Figure 3b, when the applied oscillatory shear strain was stepped from 1% to 3000% and maintained for 100 s, the G'' slightly surpassed G', while they immediately restored to their initial values after the strain returned to 1%. Similarly, when a larger strain (3500% and 4000%) was applied to the hydrogel, G'' became greater than G', indicating that the network was destroyed and appeared as a liquid-like state. However, G' and G'' were quickly recovered when the strain returned to 1%, which implies the automatic repair of the hydrogel network [11,30]. The self-healable ability of hydrogels should be attributed to the reversibility of acylhydrazone bonds (Scheme 1), which could reform at the interfaces of broken hydrogels [15,16,19]. Furthermore, tensile stress–strain tests were conducted to assess their self-healing efficiency. The tensile stress–strain curves before and after self-healing hydrogels are shown in Figure S8. The self-healing efficiencies of the three hydrogels were 92.99%, 92.29% and 94.83% (Figure 3c), respectively, which proved that the hydrogels have high self-healing efficiency and their mechanical properties could be recovered after being damaged. In addition, the hydrogels also exhibited high self-healable abilities after storing at room temperature for two weeks. As shown in Figure S9, the self-healing efficiencies were still over 85%. However, it should be pointed out that the fracture stresses of the hydrogels were slightly higher than those of freshly prepared samples (Figure S9a), while the self-healing efficiencies decreased slightly (Figure S9b). This should be attributed to the evaporation of water, though they were kept under seal, reducing the movement of polymer chains.

As mentioned above, the hydrogel network could be disrupted at a large shear strain, implying the shear-thinning behavior of hydrogel. Thus, the variation of the gel's viscosity with the increasing of the shear rate was measured. As shown in Figure 3d, the viscosity of HAg1 decreased with the increasing of the shear rate, which resulted from the disruption of the acylhydrazone crosslinking under high shear rates [39]. This shear-thinning behavior bestowed an injectable property on the hydrogel. It was observed that the hydrogel could be extruded through a syringe (Figure 3d, inset).

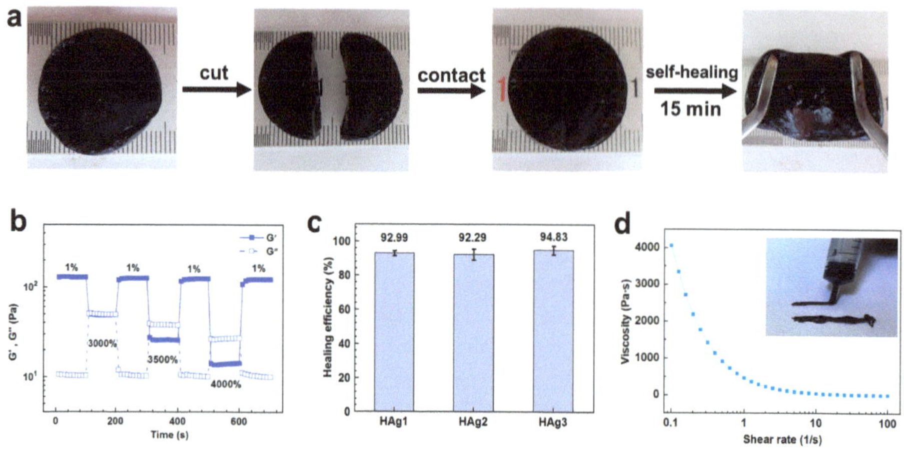

Figure 3. (**a**) Self-healing process of HAg1; (**b**) Step-strain measurements of HAg1 at 25 °C; (**c**) The self-healing efficiency of the hydrogels; (**d**) Viscosity and shear-thinning behavior of HAg1. The inset image was HAg1 injected with a conventional syringe.

3.5. Thermo-Responsive and Sol–Gel Transition of Hydrogels

As the prepared copolymer exhibited a thermo-responsive property, the temperature responsiveness of its hydrogels was examined. When the hydrogels were placed at 37 °C for 5 min, it was observed that the volume of the hydrogels shrunk due to the collapsion of polymer chains [39] (Figure 4a). When the hydrogels were returned to room temperature, the hydrogels could slowly return to their original state, indicative of the thermo-responsive property of the hydrogels. Dynamic rheological measurements were performed to reveal the temperature responsiveness of the hydrogel. As depicted in Figure 4b, G' was invariably greater than G'', and the modulus increased dramatically at around 34 °C to 50 °C, suggesting that the hydrogel network was strengthened. This was mainly ascribed to the collapsion of polymer chains.

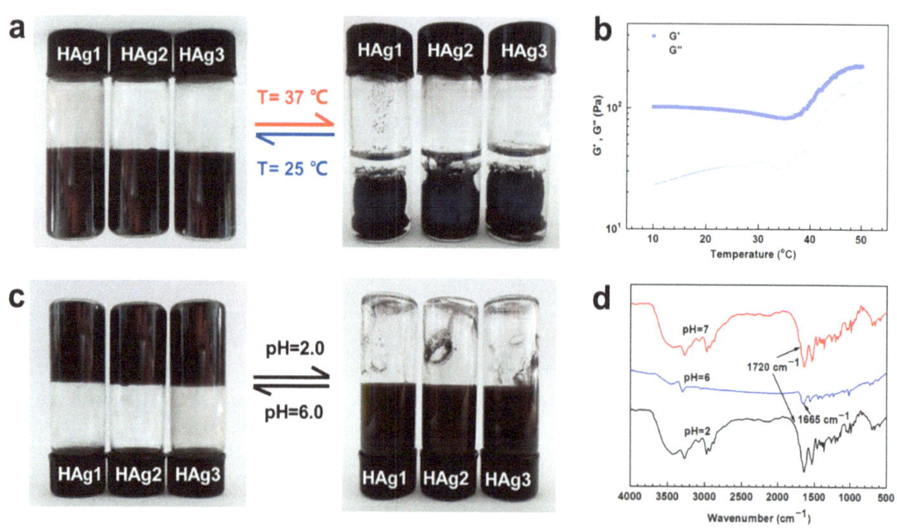

Figure 4. (**a**) Thermo-responsive performances of the hydrogels; (**b**) The modulus variation of HAg1 upon the increasing of temperature; (**c**) Photographs of the sol–gel transition of hydrogels; (**d**) FT-IR spectra of the mixture of the copolymer and AgNPs1 at different pH.

In addition, the hydrogels exhibit a reversible sol–gel transition. As shown in Figure 4c, when HCl (4 M) was added to the hydrogels to adjust the pH of the mixture to 2.0, the hydrogels transformed into free-flowing solutions. The mixture could recover hydrogels, as the pH was tuned to 6, suggesting the reversible sol–gel transition of hydrogels by altering the pH. This transition can be attributed to the network with reversible dissociation and regeneration of acylhydrazone crosslinking between hydrazides of AgNPs hybrids and the ketone groups of copolymer. As illustrated in Scheme 1, acylhydrazone was generated at a high pH, while it could dissociate at a low pH, leading to the hydrogels with a reversible sol–gel transition. FT-IR spectroscopy was used to confirm the mechanism. As shown in Figure 4d, the mixture of AgNPs1 and P_{70} did not become hydrogel at pH 7, and the absorption peak of ketone C=O appeared at 1720 cm^{-1}. When the pH was adjusted to 6, the characteristic absorption peak of the ketone C=O at 1720 cm^{-1} disappeared, while the characteristic absorption peak of the acylhydrazone C=N was found at 1663 cm^{-1}, indicating that acylhydrazone bond was formed. When pH was adjusted to 2, C=N at 1663 cm^{-1} disappeared, and the peak of ketone C=O was observed at 1720 cm^{-1} again, which means that the hydrazone bond dissociated.

3.6. Antibacterial Assay of Hydrogels

It is well known that AgNPs have wide-spectrum antimicrobial properties [33]. Thus the antimicrobial properties of the obtained hydrogels were examined. Here, *S. aureus* was chosen as a representative bacterium, as *S. aureus* has high lethality and sturdy resistance to antibiotics and is one of the major causes of hospital-acquired infections [48]. The hydrogel (H0) of copolymer crosslinked with hexanedihydrazide rather than AgNPs hybrids served as control. As shown in Figure 5a, the inhibition zone of H0 is zero, suggesting that hydrogel without AgNPs did not have an inhibitory effect on *S. aureus*. In contrast, the inhibition zones of HAg1, HAg2 and HAg3 were 24.05 mm, 25.48 mm and 23.97 mm, respectively, which indicated that the hydrogels with AgNPs were able to kill *S. aureus* [49]. Furthermore, the morphology changes of *S. aureus* on hydrogels was observed by SEM. It can be seen that a large number of *S. aureus* grew on the H0 surface and had smooth surfaces (Figure 5b). In contrast, there were very few bacteria on the surface of the three hydrogels with AgNPs hybrids, and they show wrinkled or cracked cell surfaces (Figures 5c and S10). Those results indicate that the hydrogels had good antibacterial properties. This should be attributed to Ag could be release slowly from the hydrogels and killed bacteria.

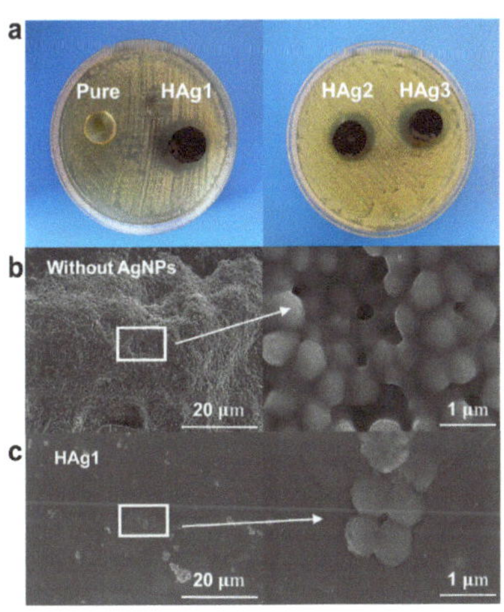

Figure 5. (**a**) Photographs of the inhibition zones of the hydrogels for *S. aureus*; SEM images morphology of *S. aureus* on H0 (**b**) and HAg1 (**c**).

3.7. Conductivity and Sensing Performance of Hydrogels

In order to determine whether the prepared hydrogels with AgNPs have conductivity, a series circuit was designed. As shown in Figure 6a, when HAg1 was placed in the circuit and the LED bulb was lit, it indicated the conductivity of HAg1. The conductivity of the hydrogels was measured (Figure 6b). The conductivities of HAg1, HAg2 and HAg3 were 6.85 S/m, 3.54 S/m and 1.97 S/m, respectively. The conductivity of hydrogels should be mainly attributed to electronically conductive AgNPs, as they have high conductivity [29]. Meanwhile, the ionic conduction also existed in hydrogels due to the presence of ions such as H^+ and Cl^- [29], which resulted from the addition of small amounts of HCl solution. Among them, HAg1 had the highest conductivity, as HAg1 contained the most AgNPs (Table S3), which is higher than that of repost hydrogels with AgNPs [33,50]. Based on its high conductivity, HAg1 was tested to use as a strain sensor. The real-time change in resistance with tensile strain was measured. As displayed in Figure 6c, the resistance rates ($\Delta R/R_0$) increased with the increasing of tensile strain. The curve of ΔR/R0 versus the applied strain was linearly fitted with a high-regression coefficient (R^2). The gauge factor (GF), used to evaluate the sensitivity of the sensor [6], reached up to 2.14. Further, the hydrogel exhibited a fast response, and the response time and recovery time were 40 ms and 80 ms during the stretching and recovering processes (Figure 6d), respectively, which are less than that of human skin (~0.1 s) [51]. Apart from fast response, the sensor exhibited repeatable sensing during cyclic stretching. As shown in Figure 6e, $\Delta R/R_0$ changes periodically at 10% strain and remained stable after 30 cycles of the stretching–recovering process. Those results suggest that the HAg1 sensor had accurate and rapid detection of the strain signal, which may be used to monitor the human body's movement. Thus, the sensor adhered to the volunteer's wrist to detect its bending activities, as shown in Figure 6f. The $\Delta R/R_0$ values remained almost steady during the repeated bending process, indicating stable and durable sensitivity. Interestingly, the strain sensor could detect the characteristic signals of speech when it was attached to the throat, When the volunteer said "OK" and "goodbye" repeatedly, the distinguishable and reproducible signals were recorded (Figure 6g), indicating that the HAg1 sensor can be potentially utilized for speech recognition.

Apart from strain sensing, the hydrogel exhibited a temperature detection capability due to the temperature responsiveness of the hydrogel. Figure 6h shows the temperature/resistance relation of HAg1 between 27 °C and 41 °C. The $\Delta R/R_0$ values had a positive correlation with the temperature, i.e., the resistance increased with the increment of temperature. This should result from AgNPs being gradually wrapped by the collapsed polymeric chains after temperature responsiveness, which impeded electron transport and led to the increasing of resistance. Based on the $\Delta R/R_0$ versus temperature, there are two temperature GFs, i.e., the GF values are 7.54%/°C and 22.05%/°C during the temperature ranges of 27~33 °C and 33~41 °C, respectively, which is higher than that of the previously reported temperature-sensing materials [52,53]. Moreover, the $\Delta R/R_0$ values at the cooling process mostly overlapped with that of heating, and the temperature GFs during cooling were also close to that of heating, suggesting that temperature sensing was reversible. In comparison with the previously reported hydrogels crosslinked with dynamic covalent bonds or containing AgNPs [54–57] (Table S4), HAg1 simultaneously exhibited antimicrobial activity, high conductivity and thermomechanical sensing.

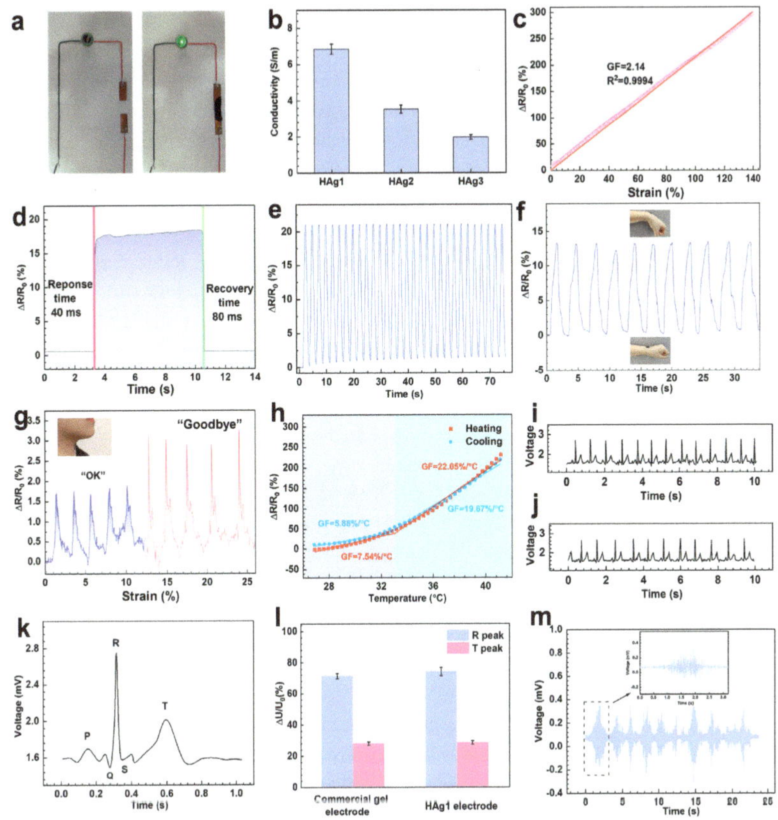

Figure 6. (**a**) Photographs of the series circuit with the LED connected to HAg1; (**b**) Conductivity of the hydrogels; (**c**) Relative resistance ($\Delta R/R_0$) of HAg1 sensor with increment of tensile strain. The red line is the fitting line; (**d**) Response time and recovery time of HAg1 sensor; (**e**) The response stability of HAg1 sensor for 30 cycles at a strain of 10%; (**f**) $\Delta R/R_0$ curves of the sensor with periodical bending of the wrist; (**g**) The variation of $\Delta R/R_0$ during speech; (**h**) $\Delta R/R_0$ change as a function of temperature; (**i**,**j**) Commercial Ag/AgCl gel electrodes and HAg1 electrodes for ECG signals detection; (**k**) Single ECG signal recorded by HAg1; (**l**) T/R peak ratio and baseline deviation for HAg1 electrodes and commercial Ag/AgCl gel electrodes; (**m**) EMG signals of HAg1 electrodes recorded on the extensor muscle.

Furthermore, the hydrogel was used as the electrode for ECG signal detection. Here, a commercial Ag/AgCl gel electrode was used as control. HAg1 electrode pairs were attached to the human skin, and ECG signals were recorded continuously (Figure 6j), which was similar to that of a commercial electrode (Figure 6i). Compared with that of a commercial electrode (Figure S11), the P wave, QRS complex and T wave of the ECG signal were more clearly detected by HAg1 hydrogel (Figure 6k). Moreover, the ratio of T and R peak values (T/R ratio) was closer to 1/3 (Figure 6l), indicating good quality for sensing [26]. The results showed that the HAg1 hydrogel electrodes were better able to detect the ECG signals in comparison with commercial gel electrodes. Furthermore, HAg1 could also serve as electrodes to record EMG signals when the volunteer gripped an object (5 kg) (Figure 6m).

4. Conclusions

We prepared well-defined thermo-responsive copolymer P(NIPAM-co-DAAM-co-AM) by RAFT polymerization and synthesized AgNPs with hydrazide groups by surface functionalization and hydrazinolysis. The hybrid hydrogels were generated by mixing the copolymer and AgNPs via the formation of acylhydrazone bonds between ketone groups and hydrazides. Based on dynamical acylhydrazone crosslinking, the obtained hydrogels exhibited sol–gel transition, self-healable and injectable properties. And the hydrogels displayed thermo-responsive behavior. Moreover, the hydrogels had antimicrobial properties and high conductivity (6.85 S/m). The fabricated hydrogel-based sensors exhibited strain and temperature sensing, and the sensors would detect human motion and speech. They also could be used as electrodes for ECG and EMG signals detection. With these attractive characteristics, the dynamic hydrogels may find potential applications in antibacterial wearable electronics, health monitoring and speech recognition.

Supplementary Materials: The following supporting information can be downloaded at: https://www.mdpi.com/article/10.3390/polym17010054/s1, Table S1: Feeding ratio of AgNPs with hydrazide bonds. Table S2: Molecular weight, polydispersity of the P70 measured by GPC. Table S3: Components of hydrogels; Table S4: Performances and functions of our hydrogels with AgNPs compared with the previously reported hydrogels with dynamic covalent bonds or containing AgNPs. Figure S1: (a) 1H NMR spectrum of P70 in D2O; (b) GPC traces of P70. Figure S2: Temperature dependence of optical transmittance for aqueous solution of P70. The insets images are P70 at 30 °C and 40 °C, respectively. Figure S3: SEM images of HAg2 (a) and HAg3 (b). Figure S4: EDS diagrams of silver nanoparticle hydrogels: (a) HAg1, (b) HAg2, (c) HAg3. Figure S5: FT-IR spectra of the polymer, AgNPs1 and HAg1. Figure S6: XRD patterns of the hydrogels. Figure S7: XRD patterns of the hydrogels after storing at room temperature for 1 month. The hydrogels were kept at room temperature under seal (wrapped with plastic wrap). Figure S8: Stress–strain curves of the original and the self-healed hydrogels. Figure S9: (a) Stress–strain curves of the hydrogels and the self-healed hydrogels after storing at room temperature for two weeks, and (b) the corresponding self-healing efficiencies. The hydrogels were kept at room temperature under seal (wrapped with plastic wrap). Figure S10: SEM images of the S. aureus morphology on hydrogels for 6 h: (a) HAg2, (b) HAg3. Figure S11: Single ECG signal recorded by commercial Ag/AgCl gel electrodes.

Author Contributions: Investigation and writing—original draft, T.Q.; investigation, X.L.; data curation, N.Z.; formal analysis, J.H. and Y.N.; writing—original draft, W.X.; supervision, writing—review and editing and funding acquisition, Z.G.; resources and writing—review and editing, B.C. All authors have read and agreed to the published version of the manuscript.

Funding: This project is supported by the Natural Science Foundation of Fujian Province, China (2022J01900, 2023J01916) and the National Natural Science Foundation of China (21802041).

Institutional Review Board Statement: Not applicable.

Data Availability Statement: The original contributions presented in this study are included in the article/Supplementary Material. Further inquiries can be directed to the corresponding authors.

Conflicts of Interest: The authors declare no conflicts of interest.

References

1. Hamidi, M.; Azadi, A.; Rafiei, P. Hydrogel nanoparticles in drug delivery. *Adv. Drug. Deliver. Rev.* **2008**, *60*, 1638–1649. [CrossRef] [PubMed]
2. Koehler, J.; Brandl, F.P.; Goepferich, A.M. Hydrogel wound dressings for bioactive treatment of acute and chronic wounds. *Eur. Polym. J.* **2018**, *100*, 1–11. [CrossRef]
3. Ying, B.; Liu, X. Skin-like hydrogel devices for wearable sensing, soft robotics and beyond. *iScience* **2021**, *24*, 103174. [CrossRef] [PubMed]

4. Chen, Y.; Zhang, Y.; Li, H.; Shen, J.; Zhang, F.; He, J.; Lin, J.; Wang, B.; Niu, S.; Han, Z. Bioinspired hydrogel actuator for soft robotics: Opportunity and challenges. *Nano Today* **2023**, *49*, 101764. [CrossRef]
5. Liu, W.; Xie, R.; Zhu, J.; Wu, J.; Hui, J.; Zheng, X.; Huo, F.; Fan, D. A temperature responsive adhesive hydrogel for fabrication of flexible electronic sensors. *NPG Flex. Electron.* **2022**, *6*, 68. [CrossRef]
6. Zhang, Y.; Tan, Y.; Lao, J.; Gao, H.; Yu, J. Hydrogels for flexible electronics. *ACS Nano* **2023**, *17*, 9681–9693. [CrossRef]
7. Gao, Y.; Zhang, Z.; Ren, X.; Jia, F.; Gao, G. A hydrogel sensor driven by sodium carboxymethyl starch with synergistic enhancement of toughness and conductivity. *J. Mater. Chem. B* **2022**, *10*, 5743–5752. [CrossRef]
8. Zhu, W.; Zhang, J.; Wei, Z.; Zhang, B.; Weng, X. Advances and progress in self-healing hydrogel and its application in regenerative medicine. *Materials* **2023**, *16*, 1215. [CrossRef]
9. Tang, Z.; Yang, J.; Li, S.; Wu, Z.; Kanti Mondal, A. Anti-swellable, stretchable, self-healable, shape-memory and supramolecular conductive TA-based hydrogels for amphibious motion sensors. *Eur. Polym. J.* **2024**, *211*, 113034–113043. [CrossRef]
10. Zazoum, B.; Batoo, K.M.; Khan, M.A.A. Recent advances in flexible sensors and their applications. *Sensors* **2022**, *22*, 4653. [CrossRef]
11. Liu, L.; Chen, Y.; Zhao, C.; Guo, M.; Wu, Y.; Li, Y.; Xiang, D.; Li, H.; Wang, L.; Zhang, X. Highly sensitive, super high stretchable hydrogel strain sensor with underwater repeated adhesion and rapid healing. *Polymer* **2023**, *285*, 126317. [CrossRef]
12. Wei, Z.; Yang, J.H.; Liu, Z.Q.; Xu, F.; Zhou, J.X.; Zrínyi, M.; Osada, Y.; Chen, Y.M. Novel biocompatible polysaccharide based self-healing hydrogel. *Adv. Funct. Mater.* **2015**, *25*, 1352–1359. [CrossRef]
13. Chen, H.; Wang, P.; Zheng, C.; Zhang, L. Thermal-responsive and 3D printable hydrogel based on an acylhydrazine-terminated dynamic covalent bond and Pluronic F127. *ACS Appl. Polym. Mater.* **2022**, *4*, 6312–6321. [CrossRef]
14. Li, G.; Shi, Z.; Zong, H.; Zhang, K.; Yan, S.; Yin, J. Injectable, self-healing poly(amino acid)-hydrogel based on phenylboronate ester bond for osteochondral tissue engineering. *Biomed. Mater.* **2023**, *18*, 055001. [CrossRef]
15. Chen, M.; Ren, X.; Dong, L.; Li, X.; Cheng, H. Preparation of dynamic covalently crosslinking keratin hydrogels based on thiol/disulfide bonds exchange strategy. *Int. J. Biol. Macromol.* **2021**, *182*, 1259–1267. [CrossRef]
16. Qiao, L.; Liu, C.; Liu, C.; Zong, L.; Gu, H.; Wang, C.; Jian, X. Self-healing, pH-sensitive and shape memory hydrogels based on acylhydrazone and hydrogen bonds. *Eur. Polym. J.* **2022**, *162*, 110838. [CrossRef]
17. Liu, Y.; Hsu, S.H. Synthesis and biomedical applications of self-healing hydrogels. *Front. Chem.* **2018**, *6*, 449. [CrossRef]
18. Li, Q.; Liu, C.; Wen, J.; Wu, Y.; Shan, Y.; Liao, J. The design, mechanism and biomedical application of self-healing hydrogels. *Chin. Chem. Lett.* **2017**, *28*, 1857–1874. [CrossRef]
19. Guo, Z.; Ma, W.; Gu, H.; Feng, Y.; He, Z.; Chen, Q.; Mao, X.; Zhang, J.; Zheng, L. pH-switchable and self-healable hydrogels based on ketone type acylhydrazone dynamic covalent bonds. *Soft Matter* **2017**, *13*, 7371–7380. [CrossRef]
20. Deng, G.; Li, F.; Yu, H.; Liu, F.; Liu, C.; Sun, W.; Jiang, H.; Chen, Y. Dynamic hydrogels with an environmental adaptive self-healing ability and dual responsive sol–gel transitions. *ACS Macro Lett.* **2012**, *1*, 275–279. [CrossRef]
21. Hou, F.; Xi, B.; Wang, X.; Yang, Y.; Zhao, H.; Li, W.; Qin, J.; He, Y. Self-healing hydrogel with cross-linking induced thermo-response regulated light emission property. *Colloids Surf. B-Biointerfaces* **2019**, *183*, 110441. [CrossRef] [PubMed]
22. Liu, Y.; Liu, Y.; Wang, Q.; Han, Y.; Chen, H.; Tan, Y. Doubly dynamic hydrogel formed by combining boronate ester and acylhydrazone bonds. *Polymers* **2020**, *12*, 487. [CrossRef] [PubMed]
23. Zhou, Y.; Fei, X.; Tian, J.; Xu, L.; Li, Y. A ionic liquid enhanced conductive hydrogel for strain sensing applications. *J. Colloid Interface Sci.* **2022**, *606*, 192–203. [CrossRef] [PubMed]
24. Dai, Y.; Wai, S.; Li, P.; Shan, N.; Cao, Z.; Li, Y.; Wang, Y.; Liu, Y.; Liu, W.; Tang, K.; et al. Soft hydrogel semiconductors with augmented biointeractive functions. *Science* **2024**, *386*, 431–439. [CrossRef]
25. Anjali, J.; Jose, V.K.; Lee, J.M. Carbon-based hydrogels: Synthesis and their recent energy applications. *J. Mater. Chem. A.* **2019**, *7*, 15491–15518. [CrossRef]
26. Zhou, L.; Li, Y.; Xiao, J.; Chen, S.W.; Tu, Q.; Yuan, M.S.; Wang, J. Liquid metal-doped conductive hydrogel for construction of multifunctional sensors. *Anal. Chem.* **2023**, *95*, 3811–3820. [CrossRef]
27. Zhou, C.; Wu, T.; Xie, X.; Song, G.; Ma, X.; Mu, Q.; Huang, Z.; Liu, X.; Sun, C.; Xu, W. Advances and challenges in conductive hydrogels: From properties to applications. *Eur. Polym. J.* **2022**, *177*, 111454. [CrossRef]
28. Mo, F.; Zhou, P.; Lin, S.; Zhong, J.; Wang, Y. A review of conductive hydrogel-based wearable temperature sensors. *Adv. Healthc. Mater.* **2024**, *13*, 2401503. [CrossRef]
29. Chen, Z.; Chen, Y.; Hedenqvist, M.S.; Chen, C.; Cai, C.; Li, H.; Liu, H.; Fu, J. Multifunctional conductive hydrogels and their applications as smart wearable devices. *J. Mater. Chem. B.* **2021**, *9*, 2561–2583. [CrossRef]
30. El-Sayed, N.S.; Moussa, M.A.; Kamel, S.; Turky, G. Development of electrical conducting nanocomposite based on carboxymethyl cellulose hydrogel/silver nanoparticles@polypyrrole. *Synth. Met.* **2019**, *250*, 104–114. [CrossRef]
31. Han, X.; Lv, Z.; Ran, F.; Dai, L.; Li, C.; Si, C. Green and stable piezoresistive pressure sensor based on lignin-silver hybrid nanoparticles/polyvinyl alcohol hydrogel. *Int. J. Chem. Biomol. Eng.* **2021**, *176*, 78–86. [CrossRef] [PubMed]

32. Fan, Q.; Nie, Y.; Sun, Q.; Wang, W.; Bai, L.; Chen, H.; Yang, L.; Yang, H.; Wei, D. Nanocomposite hybrid biomass hydrogels as flexible strain sensors with self-healing ability in harsh environments. *ACS Appl. Polym. Mater.* **2022**, *4*, 1626–1635. [CrossRef]
33. Wang, X.; Wang, Z.; Wang, X.; Shi, L.; Ran, R. Preparation of silver nanoparticles by solid-state redox route from hydroxyethyl cellulose for antibacterial strain sensor hydrogel. *Carbohydr. Polym.* **2021**, *257*, 117665. [CrossRef] [PubMed]
34. Chen, K.; Wang, F.; Hu, Y.; Liu, M.; Liu, P.; Yu, Y.; Feng, Q.; Xiao, X. Highly stretchable, sensitive, and durable Ag/tannic acid@graphene oxide-composite hydrogel for wearable strain sensors. *ACS Appl. Polym. Mater.* **2022**, *4*, 2036–2046. [CrossRef]
35. Han, Y.; Cao, Y.; Lei, H. Dynamic Covalent Hydrogels: Strong yet Dynamic. *Gels* **2022**, *8*, 577. [CrossRef] [PubMed]
36. Lai, J.T.; Filla, D.; Shea, R. Functional polymers from novel carboxyl-terminated trithiocarbonates as highly efficient RAFT agents. *Macromolecules* **2002**, *35*, 6754–6756. [CrossRef]
37. He, Y.; Luo, Y.; Liu, T.; Zhang, Y.; Guo, Z.; Zhang, J. Preparation and properties of temperature-responsive hydrogels based on acylhydrazone reversible covalent bonds. *J. Funct. Polym.* **2022**, *35*, 93–100.
38. Boyer, C.; Bulmus, V.; Davis, T.P.; Ladmiral, V.; Liu, J.; Perrier, S. Bioapplications of RAFT polymerization. *Chem. Rev.* **2009**, *109*, 5402–5436. [CrossRef]
39. Lin, C.; Chen, L.; He, Y.; Xiang, W.; Nie, Y.; Cai, B.; Guo, Z. Injectable, self-healing and degradable dynamic hydrogels with tunable mechanical properties and stability by thermal-induced micellization. *RSC Adv.* **2024**, *14*, 16207–16217. [CrossRef]
40. Wang, S.; Xiong, P.; Liu, X.; Meng, F.; Ma, Q.; Song, C.; Zhang, Q.; Che, Y. A Hydroxypropyl Methyl Cellulose-Based Graft Copolymer with Excellent Thermothickening and Anti-salt Ability for Enhanced Oil Recovery. *Energy Fuels* **2022**, *36*, 2488–2502. [CrossRef]
41. Ma, W.; Yang, X.; Liu, H.-B.; Guo, Z.R.; Zhang, J.L.; Kang, G.Y.; Yu, C.Y.; Wei, H. Fabrication of thermo and pH-dual sensitive hydrogels with optimized physiochemical properties via host-guest interactions and acylhydrazone dynamic bonding. *React. Funct. Polym.* **2023**, *184*, 105513. [CrossRef]
42. Cao, M.; Wang, Y.; Hu, X.; Gong, H.; Li, R.; Cox, H.; Zhang, J.; Waigh, T.A.; Xu, H.; Lu, J.R. Reversible thermoresponsive peptide–PNIPAM hydrogels for controlled drug delivery. *Biomacromolecules* **2019**, *20*, 3601–3610. [CrossRef] [PubMed]
43. Liu, F.; Urban, M.W. Recent advances and challenges in designing stimuli-responsive polymers. *Prog. Polym. Sci.* **2010**, *35*, 3–23. [CrossRef]
44. Mehta, B.; Chhajlani, M.; Shrivastava, B. Green synthesis of silver nanoparticles and their characterization by XRD. *J. Phys. Conf. Ser.* **2017**, *836*, 012050. [CrossRef]
45. Guzmán, M.G.; Dille, J.; Godet, S. Synthesis of silver nanoparticles by chemical reduction method and their antibacterial activity. *Int. J. Chem. Biomol. Eng.* **2009**, *2*, 104–111.
46. Khan, Z.; Al-Thabaiti, S.A.; Obaid, A.Y.; Al-Youbi, A.O. Preparation and characterization of silver nanoparticles by chemical reduction method. *Colloids Surf. B-Biointerfaces* **2011**, *82*, 513–517. [CrossRef]
47. Guo, Z.; Gu, H.; He, Y.; Zhang, Y.; Xu, W.; Zhang, J.; Liu, Y.; Xiong, L.; Chen, A.; Feng, Y. Dual dynamic bonds enable biocompatible and tough hydrogels with fast self-recoverable, self-healable and injectable properties. *Chem. Eng. J.* **2020**, *388*, 124282. [CrossRef]
48. Yang, Y.; Ma, L.; Cheng, C.; Deng, Y.; Huang, J.; Fan, X.; Nie, C.; Zhao, W.; Zhao, C. Nonchemotherapic and robust dual-responsive nanoagents with on-demand bacterial trapping, ablation, and release for efficient wound disinfection. *Adv. Funct. Mater.* **2018**, *28*, 1705708. [CrossRef]
49. Nadtoka, O.; Virych, P.; Bezugla, T.; Doroschuk, V.; Lelyushok, S.; Pavlenko, V.; Yeshchenko, O.; Kutsevol, N. Antibacterial hybrid hydrogels loaded with nano silver. *Appl. Nanosci.* **2021**, *12*, 629–636. [CrossRef]
50. Cai, J.; Zhang, X.; Liu, W.; Huang, J.; Qiu, X. Synthesis of highly conductive hydrogel with high strength and super toughness. *Polymer* **2020**, *202*, 122643. [CrossRef]
51. Johansson, R.S.; Flanagan, J.R. Coding and use of tactile signals from the fingertips in object manipulation Tasks. *Nat. Rev. Neurosci.* **2009**, *10*, 345–359. [CrossRef] [PubMed]
52. Kim, J.; Lee, M.; Shim, H.J.; Ghaffari, R.; Cho, H.R.; Son, D.; Jung, Y.H.; Soh, M.; Choi, C.; Jung, S. Stretchable silicon nanoribbon electronics for skin prosthesis. *Nature Commun.* **2014**, *5*, 5747. [CrossRef] [PubMed]
53. Fan, J.A.; Yeo, W.H.; Su, Y.; Hattori, Y.; Lee, W.; Jung, S.Y.; Zhang, Y.; Liu, Z.; Cheng, H.; Falgout, L.; et al. Fractal design concepts for stretchable electronics. *Nature Commun.* **2014**, *5*, 3266. [CrossRef]
54. Ren, Z.; Ke, T.; Ling, Q.; Zhao, L.; Gu, H. Rapid self-healing and self-adhesive chitosan-based hydrogels by host-guest interaction and dynamic covalent bond as flexible sensor. *Carbohydr. Polym.* **2021**, *273*, 118533. [CrossRef]
55. Liu, X.; Ren, Z.; Liu, F.; Zhao, L.; Ling, Q.; Gu, H. Multifunctional self-healing dual network hydrogels constructed via host–guest interaction and dynamic covalent bond as wearable strain sensors for monitoring human and organ motions. *ACS Appl. Mater. Interfaces* **2021**, *13*, 14612–14622. [CrossRef]

56. Shekh, M.I.; Zhu, G.; Xiong, W.; Wu, W.; Stadler, F.J.; Patel, D.; Zhu, C. Dynamically bonded, tough, and conductive MXene@oxidized sodium alginate: Chitosan based multi-networked elastomeric hydrogels for physical motion detection. *Int. J. Biol. Macromol.* **2023**, *224*, 604–620. [CrossRef]
57. Liu, H.; Wang, X.; Cao, Y.; Yang, Y.; Yang, Y.; Gao, Y.; Ma, Z.; Wang, J.; Wang, W.; Wu, D. Freezing-tolerant, highly sensitive strain and pressure sensors assembled from ionic conductive hydrogels with dynamic cross-links. *ACS Appl. Mater. Interfaces* **2020**, *12*, 25334–25344. [CrossRef]

polymers

MDPI

Article

Analysis of Residual Stress at the Interface of Epoxy-Resin/Silicon-Wafer Composites During Thermal Aging

Jianyu Wu, Fangzhou Chen, Jiahao Liu, Rui Chen, Peijiang Liu, Hao Zhao * and Zhenbo Zhao *

Science and Technology on Reliability Physics and Application of Electronic Component Laboratory, China Electronic Product Reliability and Environmental Testing Research Institute, Guangzhou 511370, China; wujianyu@ceprei.com (J.W.); chenfangzhou@ceprei.com (F.C.); liujiahao@ceprei.com (J.L.); chenr@ceprei.com (R.C.); liupeijiang@ceprei.com (P.L.)

* Correspondence: zhaohao1@ceprei.com (H.Z.); zhaozb@ceprei.com (Z.Z.)

Abstract: During the thermal aging process of epoxy resin, microcracks, interfacial delamination, and warpage are the key factors leading to semiconductor device damage. Here, epoxy-resin specimens (EP-Ss) and epoxy-resin/silicon-wafer composites (EP-SWs) were prepared to analyze the distribution of residual stress (RS) in epoxy resin and its thermal aging process changes. The uniaxial tensile approach and Raman spectroscopy (RAS) showed that the peak shift of aliphatic C-O in EP-Ss was negatively correlated with the external stress, and that the stress correlation coefficient was -2.76×10^{-2} cm^{-1}/MPa. Then, RAS was used to evaluate the RS distribution of EP-SWs, obtaining a high-resolution stress-distribution image of 50×50 pixels and revealing a strong stress concentration at the interface between the epoxy resin and the silicon wafer. Additionally, Fourier transform infrared spectroscopy (FTIR), Differential scanning calorimetry (DSC), Field-emission scanning electron microscopy (FE-SEM), and RAS were used to analyze the chemical composition, molecular structure, interfacial microstructure, and RS of the epoxy resin during the thermal aging process. With the increase in the thermal aging time, the epoxy resin underwent secondary curing, the RS at the interface changed from tensile stress to compressive stress, and cracks were formed. The results illuminate the effect of the thermal aging process on the interface-failure mechanism of composite materials, aiding in the reliability evaluation and safety design of semiconductor devices.

Keywords: epoxy resin; Raman spectroscopy; residual stress; thermal aging; crack

Academic Editor: Severo Raul Fernandez-Vidal

Received: 8 December 2024
Revised: 23 December 2024
Accepted: 26 December 2024
Published: 28 December 2024

Citation: Wu, J.; Chen, F.; Liu, J.; Chen, R.; Liu, P.; Zhao, H.; Zhao, Z. Analysis of Residual Stress at the Interface of Epoxy-Resin/ Silicon-Wafer Composites During Thermal Aging. *Polymers* **2025**, *17*, 50. https://doi.org/10.3390/polym17010050

1. Introduction

Thermosetting materials like epoxy resin are three-dimensional, highly cross-linked networks formed by cross-linking reactions between epoxy-resin monomers or oligomers and curing agents [1–4]. Epoxy resin has excellent thermal, mechanical, and electrical properties and is widely used in various applications, including adhesives, coatings, electronic packaging, and insulation [5–7]. In recent years, the manufacturing of packaging devices, from the simplest plastic encapsulants to advanced packaging in heterogeneous integration, involves a curing process, which inevitably introduces residual stress (RS) [8,9]. Additionally, in a multi-material structure bonded by different materials, owing to the nonuniform chemical and physical shrinkage caused by the thermal aging effect during high-temperature service, large amounts of RS may also be generated. A sufficiently large amount of RS may cause microcracks, interfacial delamination, warpage, or damage to the epoxy resin, which is gradually becoming one of the main challenges affecting the reliability of packaging devices [10–14]. Therefore, evaluating the RS of the material structure is crucial for understanding the impact of thermal aging on the reliability of the composite structure.

In one sense, the initial stress of the epoxy resin-based multi-material structure comes from the thermal RS generated during curing. During the curing process, the epoxy resin shrinks when cooled to room temperature. At this time, the metal or nonmetal substrate has a lower coefficient of thermal expansion and shrinks much less than the epoxy resin, resulting in the generation of RS at the interface between the epoxy resin and the substrate [15–17]. In another sense, in a thermal-oxidative environment, the epoxy resin in the multi-material structure undergoes an oxidation reaction, resulting in changes in the density of the polymer (due to the grafting of oxygen atoms along the polymer chain) and mass changes (a large amount of volatile products are formed by a chain breakage near the chain end, ultimately leading to mass loss), finally causing the epoxy resin to undergo volume shrinkage. Owing to the inherent inertness of the substrate, the substrate exhibits negligible deformation at temperatures below 300 °C. Different degrees of oxidative deformation can be observed between the shrinking epoxy-resin matrix and the substrate, thus changing the RS at the interface [18–20]. With the increase in the duration of the thermal aging process, the RS at the interface further accumulates, causing warpage and interfacial peeling of the material structure products [21]. Therefore, understanding the evolution law of the RS during the thermal aging process is an important research topic for exploring the correlation between thermal aging and structural damage of the composite material.

Currently, many techniques are available for characterizing residual stress, including X-ray diffraction (XRD), Raman spectroscopy (RAS), ultrasonic acoustics, and strain gauges [16]. Compared to ultrasonic acoustics and strain gauges, which can only obtain residual stresses in large areas, XRD and Raman spectroscopy can accurately measure residual stresses in micro areas. However, X-ray diffraction is limited to residual stress measurement of materials with crystallinity and is not suitable for amorphous materials such as epoxy resins. Therefore, RAS is currently an important multi-scale and nondestructive approach for characterizing the RS of amorphous polymer-based materials with high resolution. Based on the change in vibrational energy in polymer molecules under mechanical stress, RAS can confirm the displacement value of sensitive bands and then calculate the RS, so it has been widely used to understand the mechanical properties of various polymers [22–25]. Abiko et al. employed RAS to evaluate the RS distribution of epoxy-resin/aluminum composites [23]. Through Raman imaging, it was found that compared with room-temperature curing, a more intense stress concentration was found at the interface between the epoxy resin and the aluminum. Wu et al. imaged the RS of an epoxy-based thermosetting material by RAS and found that the RS of the thermosetting material containing a highly dynamic thiocarbamate bond decreased by 44% after annealing at 30 °C [24]. These studies have shown that the RS caused by differences in the curing process and physical properties of the resin of the polymer-based material can be detected and analyzed by Raman imaging. Therefore, Raman imaging is beneficial for better understanding the evolution of the micromechanics and complex failure processes of the composite structure, including interfacial debonding, matrix cracking, warpage, and local stress distribution. However, its related applications in the thermal aging of polymer-based materials have been rarely reported.

Here, epoxy-resin specimens (EP-Ss) and epoxy-resin silicon-wafer composites (EP-SWs) were prepared. The uniaxial tensile approach and RAS were used to study the relationship between the stress and peak shift of the EP-Ss. Then, the EP-SWs were placed in an oven for thermal aging test at 105 °C. The microscopic RS distribution of the EP-SWs during the thermal aging process at different times was measured. The distribution uniformity and evolution law of the RS at the interface of the composite structure were analyzed. Additionally, Fourier transform infrared spectroscopy (FTIR), Differential scanning calorimetry (DSC), and Field-emission scanning electron microscopy (FE-SEM) were used to analyze the chemical composition, molecular structure, and interfacial microstructure of the epoxy resin after the thermal aging process.

2. Materials and Methods

2.1. Materials

The bisphenol-A diglycidyl, 4,4′-methylenedianiline, and N,N-dimethylformamide used in this work were all purchased from Shanghai Aladdin Biochemical Technology Co., Ltd. (Shanghai, China). The silicon wafers were purchased from Shijiazhuang Jing Yan Electronic Technology Co., LTD (Shijiazhuang, China). All the above samples were used as received, without purification treatment.

2.2. Sample Preparation

2.2.1. Preparation of Epoxy-Resin Specimens (EP-Ss)

The EP-Ss for uniaxial tensile approach were prepared as follows. The bisphenol-A diglycidyl ether monomer and 4,4′-methylenedianiline curing agent were dissolved in a flask containing 2 mL of N,N-dimethylformamide solution at a weight ratio of 4.5:1 to obtain a mixed solution. The mixed solution was then cast in a rectangular mold (L80 × W5 × T2 mm) and cured in a vacuum at 60 °C for 24 h to obtained EP-Ss.

2.2.2. Preprocessing of Silicon Wafers

Silicon wafers measuring 10 mm × 10 mm × 0.6 mm were transferred to a beaker containing 30 mL of chloroform and ultrasonically cleaned for 30 min. Then, the silicon wafer was transferred to a plasma cleaner for atmospheric-pressure treatment for 5 min.

2.2.3. Preparation of EP-SWs

First, the bisphenol-A diglycidyl ether monomer and 4,4′-methylenedianiline curing agent were dissolved in a flask containing 5 mL of N,N-dimethylformamide solution at a weight ratio of 4.5:1. The mixed solution was mechanically stirred in a vacuum environment at 60 °C for 30 min to remove bubbles, and the DMF solution was added to the mixed solution, thereby preparing a prepolymer solution. Then, the prepolymer solution was poured into a silicone mold with a silicon wafer attached to the bottom and cured in a vacuum at 80 °C for 8 h and at 100 °C for 2 h. After curing, the heating was stopped, and the temperature in the oven was slowly cooled to room temperature. After demolding, EP-SWs were obtained by cutting and polishing. All samples were cuboids of 10 mm in length, 10 mm in width, and 5 mm in height.

2.3. Thermal Aging Process

EP-SWs were isothermally oxidized and aged in an air-circulating oven, which could continuously deliver external air to the oven to maintain a fresh oxygen environment around the EP-SWs. To exclude the moisture effect, all the specimens were baked in an oven at 60 °C for 2 h. Subsequently, under high-temperature atmosphere conditions, the cured EP-SWs were placed in the oven. The thermal aging was conducted based on the standard IEC 60505 using an oven with a stable temperature of 105 °C [26]. The specimens were taken out regularly, and the time span was 2, 6, 14, and 25 d, respectively.

2.4. Characterization and Measurement

2.4.1. RS Measurement

Chemical Structure Analysis

To analyze the changes in chemical bonds of the epoxy resin during curing and thermal aging, we carried out a Fourier transform infrared spectroscopy measurement in the attenuated total reflection (ATR) mode (Nicolet IS50 + Continuum, Thermo Fisher Scientific Co., Ltd., Waltham, MA, USA) with 16 accumulative scans, a resolution of 2 cm^{-1}, and a scanning range of 500–4000 cm^{-1}. Additionally, the analysis process of the infrared spectrum was carried out using Spectrum software (PerkinElmer, Glen Waverley, VIC, Australia).

Glass Transition Measurement

A differential scanning calorimetry measurement was carried out using a DSC Q1000 (METTLER TOLEDO, Switzerland). The sample with a mass range of 3–5 mg was sealed in an aluminum pan and heated from 0 to 250 °C at a ramp rate of 10 °C/min under a nitrogen flow of 50 mL/min. The results were interpreted using TA analysis software. DSC analysis was carried out to check the total curing degree of the sample and measure the glass transition temperature value of the aged sample. The Tg was measured during the second heating ramp at the inflection point of the thermogram (i.e., after removing the thermal history of the sample).

Micromorphology Characterization

Before imaging the microregion, the sample measurement area was sprayed with platinum via an ion sputtering coating to improve the resolution. An Apreo 2 field-emission scanning electron microscope produced by Thermo Fisher Scientific was used to observe the microscopic morphology of the EP-SW.

3. Results and Discussion

3.1. Relationship Between Peak Shift and Stress

As shown in Figure 1, EP-Ss and EP-SWs were prepared by a nucleophilic addition reaction, using bisphenol-A diglycidyl ether as the matrix and 4,4′-methylenedianiline as the curing agent. The Raman spectrum of the EP-Ss in the range of 1000–3200 cm^{-1} is shown in Figure 2a. Since the narrower the width of the Raman peak, the greater the improvement in the measurement accuracy of the peak shift, combined with the stress measurement approaches reported in the existing literature, the peaks at 3078 cm^{-1}, 1618 cm^{-1}, and 1123 cm^{-1} with narrower half-width peaks were selected for stress measurement and analysis. These three peaks correspond to the aromatic C-H stretching vibration, aromatic C=C stretching vibration, and aliphatic C-O stretching vibration, respectively. After continuously collecting the Raman spectra at the same point, we finally obtained the accurate peak positions by curve fitting using the Voigt function (Gaussian + Lorentzian function), as shown in Figure 2b–d.

To confirm the influence of the laser-irradiation power intensity on the Raman peak position, we measured the EP-Ss with different laser powers and fitted the peak positions, as shown in Figure 3. It could be observed that, owing to the local temperature rise in the measurement area caused by laser irradiation, the peak position decreased with the increase in the laser power. However, as long as the laser power was constant, this thermal shift was expected to be constant. Additionally, with the increase in the laser power, the intensity of the Raman characteristic peak also increased, and the deviation of the peak position obtained by Voigt fitting was smaller. Therefore, to ensure the reproducibility and accuracy of the measurement data, we used the maximum laser power for measurement in this work.

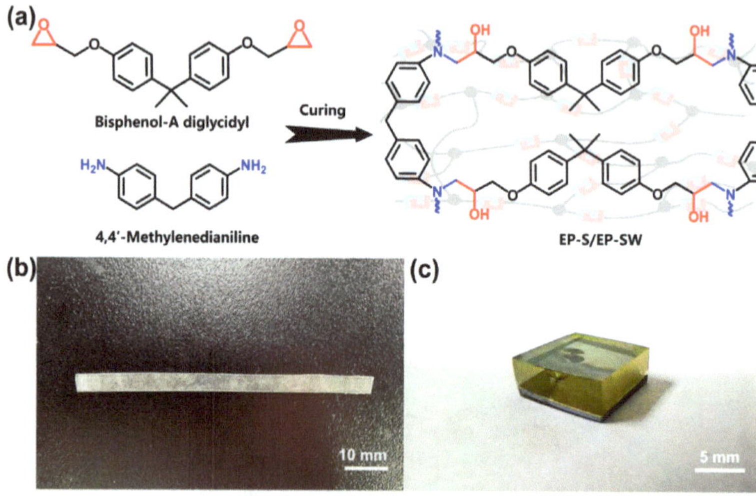

Figure 1. (**a**) Schematic illustration of preparation of the epoxy resin. (**b**) Image of EP-S. (**c**) Image of EP-SW.

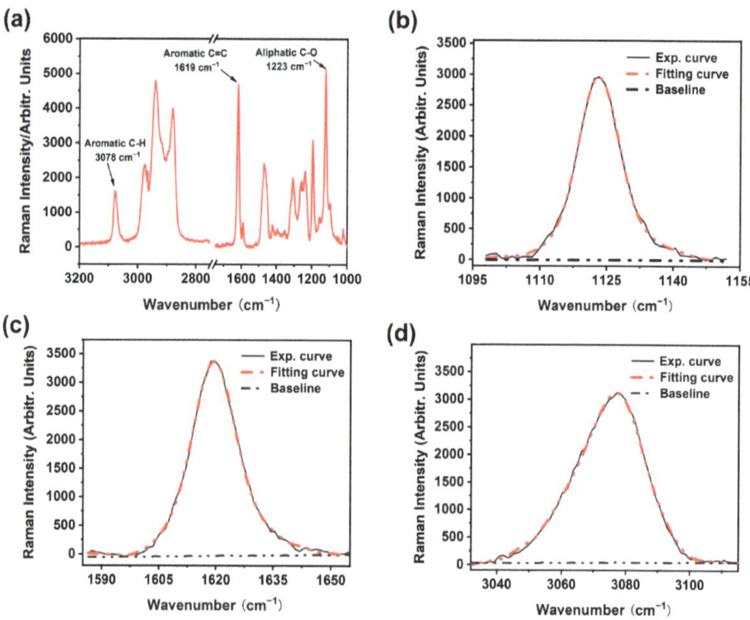

Figure 2. (**a**) Raman spectra of epoxy resin; (**b**) 1123 cm^{-1} (aliphatic C-O stretching vibration) peak and the results fit of the data; (**c**) 1619 cm^{-1} peak (aromatic C=C stretching vibration) and the results fit of the data; (**d**) 3078 cm^{-1} peak (aromatic C-H stretching vibration) and the results fit of the data.

Based on the ability of RAS to detect the local change in molecular vibration energy, by measuring the corresponding relationship between strain and Raman wave number and combining the linear relationship between the stress and strain of elastic mechanical materials, researchers can obtain the Raman frequency shift–stress factor, and the relationship between the material stress and Raman frequency shift can be established. It has been experimentally demonstrated that when using RAS to measure the stress of amorphous materials, according to Hooke's law, the relationship between the peak shift and stress is shown in Equation (1) [22]:

$$\Delta v = v_{\sigma} - v_0 = S^{\sigma} \times \Delta\sigma, \quad (1)$$

where Δv represents the peak shift; $\Delta\sigma$ represents the stress; v_{σ} and v_0 are the peak positions with and without applied stress, respectively; and S^{σ} is the Raman mechanical coefficient. In the uniaxial tensile approach, a small uniaxial tensile platform is used to apply tensile load to the epoxy resin, and the laser is focused on the sample surface to obtain the Raman spectrum. Figure 4 shows the peak shift diagram of EP-Ss. The peak shift reflected the difference in the relative peak positions before and after the application of stress. The absolute value of the peak shift decreased linearly with the increase in the load. When an external compressive stress load was applied, owing to the shortening of the molecular bond length, the vibration frequency increased, and the spectral band shifted in the high-frequency direction (the wave number becomes larger); conversely, when the epoxy resin was subjected to a tensile stress load, the spectral band shifted in the low-frequency direction (the wave number became smaller). A regression line was obtained by the least square approach, and all the data points were located within the 95% prediction interval.

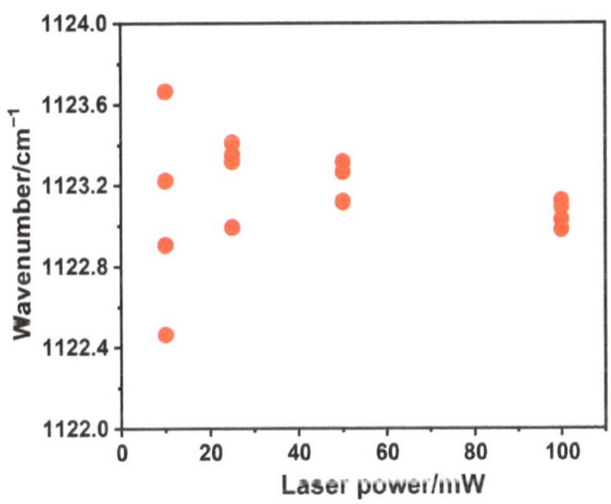

Figure 3. Peak position as a function of illumination laser power.

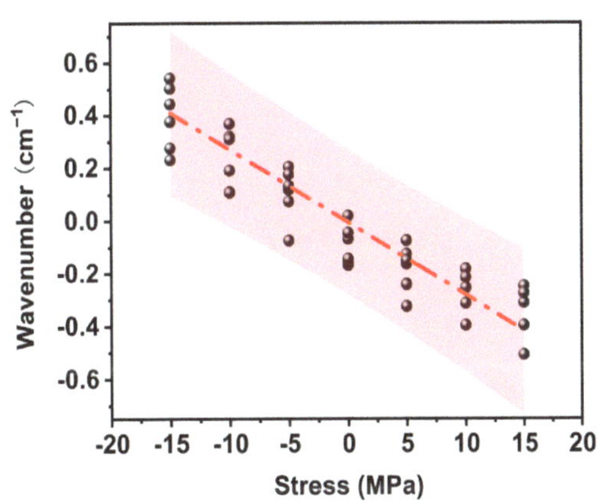

Figure 4. The straight line of the relationship between the frequency shift of the 1223 cm^{-1} peak of the epoxy resin and the stress caused by the load in the uniaxial tensile approach represents the least-square fitting regression line of the data points.

From the slope of the regression line, the Raman mechanical coefficient of the 1123 cm^{-1} peak was -2.76×10^{-2} cm^{-1}/MPa (Figure 5a). Additionally, the stress-peak shift detection was also carried out for the stretching vibration of aromatic C=C at 1619 cm^{-1} and aromatic C-H at 3078 cm^{-1}. The aliphatic C=C peak did not show stress sensitivity associated with stress (Figure 5b). The Raman mechanical coefficient of the 3078 cm^{-1} peak was -1.08×10^{-2} cm^{-1}/MPa. In the literature, the Raman mechanical coefficient has been reported at -1.00×10^{-2} cm^{-1} /MPa of the stretching vibrations of aromatic C-H of epoxy resin. The difference in constants is due to the different modes of vibration produced by the protons in the aromatic ring or the aromatic ring skeleton. By observing the stretching vibrations of aromatic C-H and aliphatic C-O, it can be seen that the Raman stress correlation coefficients were all negative, which is in line with the corresponding relationship between the measured strain and Raman wave number (Figure 5c). Comparing the stress coefficients of aromatic C-H and aliphatic C-O, we found that the stress sensitivity of aromatic C-H was less than half, indicating that the aliphatic C-O peak can be used for stress measurement of the epoxy resin and can provide more accurate data measurement results. Combining the above results, aliphatic C-O was selected as the Raman peak for stress measurement in this work.

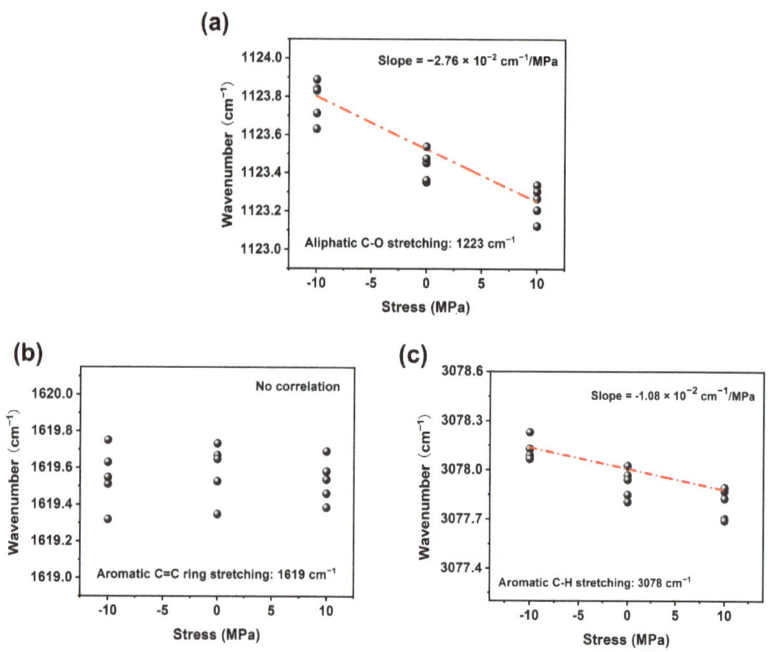

Figure 5. (**a**) Raman frequency shift–stress relationship diagram of the 1223 cm^{-1} aliphatic C-O stretching vibration peak. (**b**) Raman frequency shift–stress relationship diagram of the 1619 cm^{-1} aliphatic C=C stretching vibration peak. (**c**) Raman frequency shift–stress relationship diagram of the 3078 cm^{-1} aliphatic C-H stretching vibration peak.

3.2. Imaging of RS Distribution

RAS was used to measure the RS of the sample after curing. The schematic diagram of the Raman imaging test process of RS and measurement area are shown in Figure 6a,b. The X-axis was defined as the distance parallel to the bonding interface. The Y-axis was defined as the vertical distance from the EP-SW interface. The same Raman mechanical coefficient (-2.76×10^{-2} cm^{-1}/MPa) was used to measure the RS of the unaged epoxy resin. Then, the measurement area was characterized by RAS with a scanning step size of 1 μm. By measuring the 1123 cm^{-1} peak shift and calculating the RS of each point, the distribution image of RS in the region of 50 × 50 μm^2 could be obtained, as shown in Figure 6c. Compared with the literature, the distribution of residual stress in epoxy resin shows similar result [23]. Furthermore, it can be seen that at the interface (X = 0), a large tensile stress (11.64 Mpa) of the epoxy resin was detected because the epoxy resin matrix was restrained by the silicon wafer. As the distance between the measuring position and the interface increased, the residual stress decreased gradually. Interestingly, at a distance of 25 microns from the interface, the epoxy resin showed zero stress, which was caused by matrix shrinkage and silicon chip restraint. When the measurement position continues to increase, the matrix shrinkage occupies the dominant factor and presents compressive stress (5.40 Mpa).

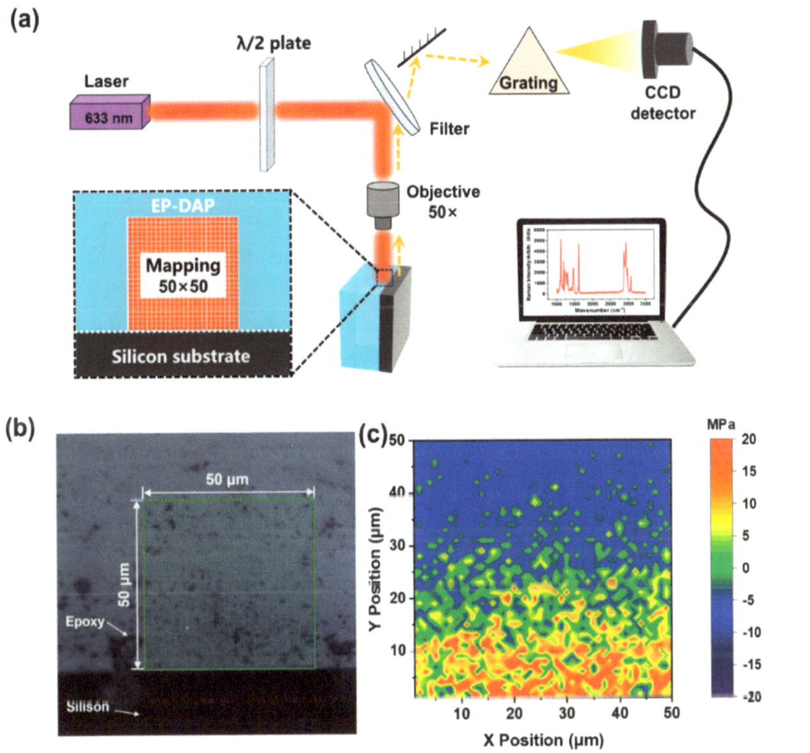

Figure 6. (**a**) Schematic diagram of the residual stress (RS) testing process. (**b**) Image of the Raman spectroscopy measurement area. (**c**) Image of RS distribution of epoxy resin at the interface.

3.3. Thermal Aging Analysis

Under dry conditions, the cured EP-SW was exposed at 105 °C for 25 days and detected regularly using RAS and ATR-FTIR, respectively. As shown in Figure 7a, the characteristic peaks of epoxy resin at different aging times are basically consistent, which does not prove that the epoxy resin undergoes thermal degradation reaction. This phenomenon may be due to the fact that thermal degradation of polymer produced polar groups, whereas RAS was only sensitive to non-polar groups in the molecular chain. To demonstrate that the polymer undergone thermal degradation during thermal aging, ATR-FTIR was used to characterize the epoxy resin. In Figure 7b, the 914 cm^{-1} band corresponded to the epoxy group, and the peak height of the band gradually decreased and disappeared with the extension of high-temperature loading time, which could be attributed to the secondary cross-linking of the epoxy resin during the thermal aging process. As shown in Figure 7c, the detection of the epoxy amine infrared fingerprint region showed that the main changes caused by thermal oxidation occurred in two peaks within the wave number range of 1570–1860 cm^{-1}. The peak at 1663 cm^{-1} usually came from the amide group created by aging on the molecular chain. The band at 1726 cm^{-1} usually came from the vibration of the carbonyl group produced by auto-oxidation [27]. The schematic diagram of the thermal degradation mechanism of epoxy resin is shown in Figure 7d. The growth of both bands was referred to the thermo-oxidative degradation of the epoxy chains.

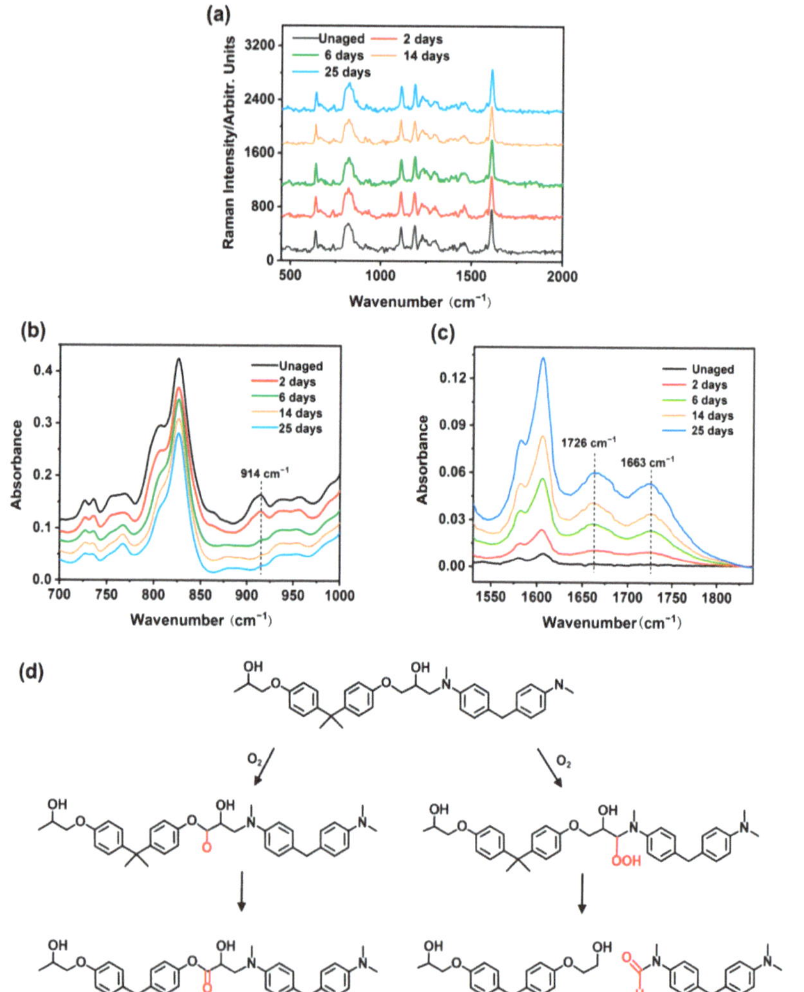

Figure 7. (**a**) Raman spectra of EP-SWs with different aging times in the wavenumber range of 450–2000 cm^{-1}. (**b**) FTIR spectra of EP-SWs with different aging times in the wavenumber range of 700–1000 cm^{-1}. (**c**) FTIR spectra of EP-SWs with different aging times in the wavenumber range of 1570–1860 cm^{-1}. (**d**) Schematic diagram of thermal degradation mechanism of epoxy resin.

The interface position between epoxy resin and silicon wafer with different thermal aging times was observed using scanning electron microscopy. Before thermal aging, the epoxy resin at the interface was fully bonded to the silicon wafer without any cracks, indicating the formation of a strong and stable composite structure. When the sample was baked at 105 °C for 14 days, although the interface was still relatively flat, a small crack appeared at the interface. This phenomenon indicated that RS within the matrix have an impact on the morphology of the interface. After 6 days of aging, a crack with a width of 3 μm was formed at the interface position, as shown in Figure 8. This might be due to the increase in the cross-linking density of the epoxy-resin matrix during secondary curing, resulting in shrinkage stress at the interface, making the epoxy resin unable to withstand high stress and peel off to form cracks. After 25 days, the crack significantly expanded, indicating that the RS of the material continuously accumulated and led to the further expansion of the crack.

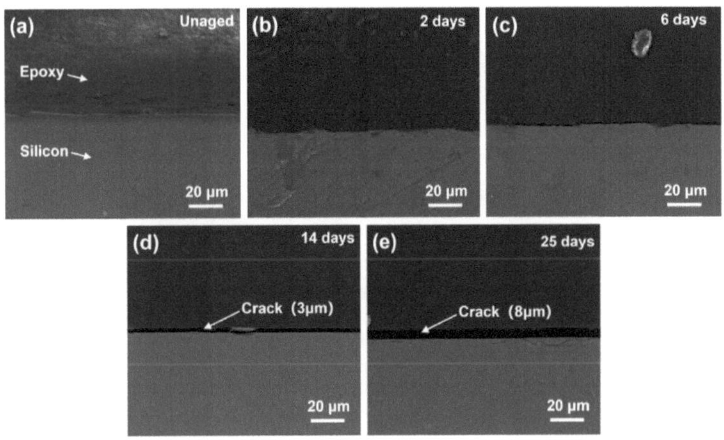

Figure 8. FE-SEM images of (**a**) unaged, (**b**) aged for 2 days, (**c**) aged for 6 days, (**d**) aged for 14 days, and (**e**) aged for 25 days.

DSC curves of different aging times are shown in Figure 9. Endothermic reactions were observed in EP-SWs both before and after thermal aging over a wide temperature range, and differences existed between the glass transition temperatures of aged and nonaged epoxies. The obvious glass transition temperatures (Tgs) were 99.57 °C, 113.50 °C, 120.70 °C, 122.86 °C, and 123.09 °C for the samples that were not aged, aged for 2 d, aged for 6 d, aged for 14 d, and aged for 25 d, respectively. It can be observed that the Tg of the whole system gradually increased with the increase in aging time, and that the Tg value tended to be stable after 14 d of aging, which is considered to be the result of the joint action of cross-linking and chain-breaking effect of epoxy resin [28].

To verify the evolution law of the interfacial RS, we selected cross sections of the unaged, aged for 6 d, and aged for 25 d EP-SWs for Raman imaging. Figure 10 shows the distribution of the RS of the epoxy resin near the position of the silicon wafer. According to the results of the Raman imaging, owing to the mismatch of the thermal expansion coefficients between the epoxy resin and the silicon wafer, the unaged EP-SWs formed tensile stress at the interface, and this stress state changed to compressive stress as the distance from the interface increased. With the increase in the thermal aging time, owing to the increase in the cross-linking density of the matrix caused by the secondary curing inside the epoxy resin, the shrinkage stress at the interface was further increased. Notably, after 25 d of aging, owing to the cracks formed at the epoxy-resin/silicon-wafer interface, the matrix was no longer constrained by the silicon wafer at all, and the entire area uniformly presented a shrinkage stress.

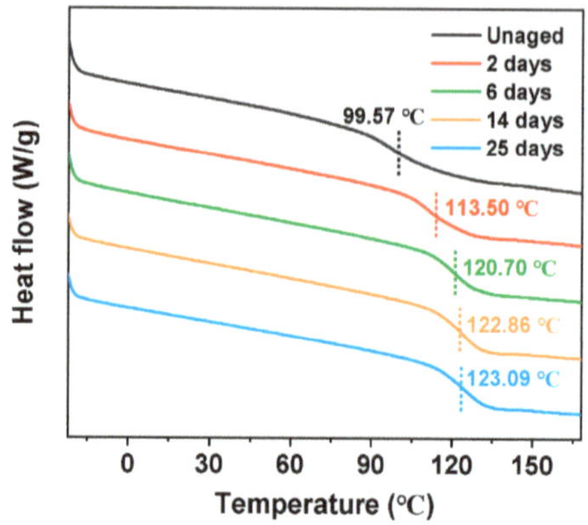

Figure 9. DSC curves of epoxy-resin samples at different aging times during the thermal aging process at 105 °C.

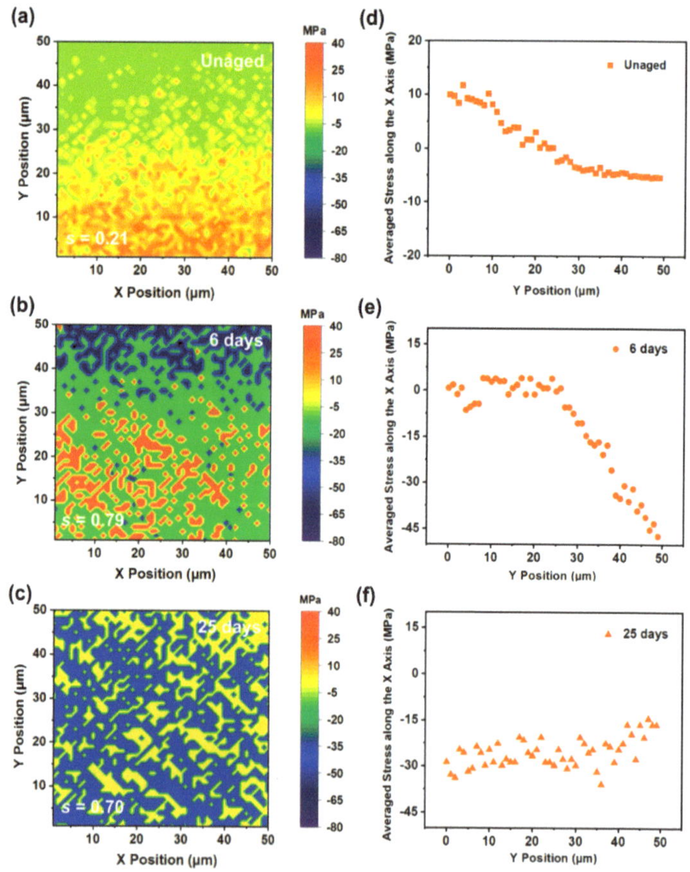

Figure 10. (**a**) Distribution of RS at the interface of unaged (EP-SW). (**b**) Distribution of RS at the interface of EP-SW aged for 6 days. (**c**) Distribution of RS at the interface of EP-SW aged for 25 days. (**d**) Average stress along the X-axis of the interface of unaged epoxy resin obtained by Raman imaging. (**e**) Average stress along the X-axis of the interface of epoxy resin aged for 2 days obtained by Raman imaging. (**f**) Average stress along the X-axis of the interface of epoxy resin aged for 25 days.

Additionally, the parameter s, which represents the standard deviation of the sample, was used to quantify the uniformity of the RS distribution, as shown in Equation (2):

$$s^2 = \frac{\sum_{i=1}^{n}(x_i - x)^2}{n-1} \quad (2)$$

where s^2 is the sample variance, n is the number of test points, x_i is the fitted value of the Raman shift at the i-th point, and x is the average value. Before aging, since the cross-linking curing inside the epoxy resin was not yet complete, the distribution of tensile stress at the interface was not obvious, and the value of s was 0.21. After aging at 105 °C for 6 d and 25 d, the epoxy resin showed significant stress differences owing to the peeling off of the EP-SW adhesive interface, resulting in matrix shrinkage and molecular chain fracture with values of 0.79 and 0.70, respectively.

To compare the stress change trends in the depth direction of the interfaces (Y-axis) of different aging times of EP-SW, we averaged the relative stresses on the surfaces of the epoxy resins in the X-axis direction. As shown in Figure 10d, the RS increased as the distance to the silicon interface decreased, and a tensile stress of 9.9 MPa was detected on the resin surface. With the increase in the time of the thermal aging process, the RS at the interface further accumulated, and the shrinkage stress formed inside the matrix gradually intensified. Finally, in the epoxy resins shown in Figure 10e,f, the average shrinkage stress on the resin surface of the entire area was approximately 25.5 MPa. The above results further confirm that the formation and expansion of cracks was caused by the increase in the shrinkage stress inside the matrix.

4. Conclusions

In this paper, we prepared epoxy-resin/silicon-wafer composites with a bilayer structure. The RS of the epoxy-resin/silicon-wafer composites during the curing and thermal aging process was measured with a Raman spectrometer. The distribution of the RS on the cross section of the epoxy-resin/silicon-wafer composites was measured and calculated, and two-dimensional RS images on the epoxy-resin interface were acquired. The stress images clearly show that, stress concentration occurred near the interface of the resin/silicon wafer composites after curing process, because the thermal contraction amplitude of the epoxy resin was greater than that of the silicon wafer when cooling to room temperature after molding. As the exposure time in the thermal aging environment at 105 °C continued to increase, the stress at the interface gradually transformed into shrinkage stress and gradually increased, eventually leading to the formation of cracks. Meanwhile, FTIR was used to analyze the molecular structure of the epoxy resin after thermal aging. The results show that with the increase in the thermal aging time, secondary curing occurred in the epoxy resin, the RS at the interface changed from tensile stress to compressive stress, and cracks were formed. Therefore, in this work, research on the characterization approach of micro-region RS was carried out by combining Raman spectroscopy, revealing the distribution state of the interfacial RS. Additionally, the development of the reliability-evaluation technology for RS and the clarification of the correlation between the evolution of RS and its crack initiation have important application research value and practical significance for evaluating the technology of local thermal RS at the interface of multi-material structures.

Author Contributions: Methodology, J.W. and P.L.; Validation, F.C. and R.C.; Investigation, J.W. and J.L.; Writing—original draft, J.W. and F.C.; Writing—review and editing, Z.Z. and H.Z. All authors have read and agreed to the published version of the manuscript.

Funding: This research was funded by the Opening Project of Science and Technology on Reliability Physics and Application Technology of Electronic Component Laboratory (22D07).

Institutional Review Board Statement: Not applicable.

Data Availability Statement: The original contributions presented in the study are included in the article, and further inquiries can be directed to the corresponding authors.

Acknowledgments: The authors would like to express their sincere gratitude to the Opening Project of Science and Technology on Reliability Physics and Application Technology of Electronic Component Laboratory.

Conflicts of Interest: The authors declare no conflicts of interest.

References

1. Liu, J.; Zhang, L.; Shun, W.; Dai, J.; Peng, Y.; Liu, X. Recent development on bio-based thermosetting resins. *J. Polym. Sci.* **2021**, *59*, 1474–1490. [CrossRef]
2. Zhang, Y.; Liu, X.; Wan, M.; Zhu, Y.; Zhang, K. Recent development of functional bio-based epoxy resins. *Molecules* **2024**, *29*, 4428. [CrossRef]
3. Amirbeygi, H.; Khosravi, H.; Tohidlou, E. Reinforcing effects of aminosilane-functionalized graphene on the tribological and mechanical behaviors of epoxy nanocomposites. *J. Appl. Polym. Sci.* **2019**, *136*, 47410. [CrossRef]
4. Mostovoi, A.S.; Kurbatova, E.A. Controlling the properties of epoxy composites filled with brick dust. *Russ. J. Appl. Chem.* **2017**, *90*, 246–256. [CrossRef]
5. Zhang, Y.; Peng, C.; Wang, G.; Huang, X.; Yu, Y. Hydroxyl-free epoxy thermoplastics from active esters with low dielectric constant and water sorption. *ACS Appl. Electron. Mater.* **2024**, *6*, 4578–4586. [CrossRef]
6. Capricho, J.C.; Fox, B.; Hameed, N. Multifunctionality in epoxy resins. *Polym. Rev.* **2020**, *60*, 1–41. [CrossRef]
7. Ellis, B. Introduction to the chemistry, synthesis, manufacture and characterization of epoxy resins. In *Chemistry and Technology of Epoxy Resins*; Ellis, B., Ed.; Springer: Dordrecht, The Netherlands, 1993; pp. 1–36.
8. Sui, H.; An, D.; Yang, K.; Zhang, Z.; Zhang, X.; Zhou, F.; Li, J.; Wu, K. Anomalous residual stress behavior and corresponding dielectric properties of epoxy resin induced by physical aging. *IEEE Trans. Dielectr. Electr. Insul.* **2023**, *30*, 1787–1794. [CrossRef]
9. Bauhofer, A.A.; Krödel, S.; Rys, J.; Bilal, O.R.; Constantinescu, A.; Daraio, C. Harnessing photochemical shrinkage in direct laser writing for shape morphing of polymer sheets. *Adv. Mater.* **2017**, *29*, 1703024. [CrossRef]
10. Tian, W.; Chen, X.; Zhang, G.; Chen, Y.; Luo, J. Delamination of plasticized devices in dynamic service environments. *Micromachines* **2024**, *15*, 376. [CrossRef]
11. Pan, C.-T.; Wang, S.-Y.; Yen, C.-K.; Ho, C.-K.; Yen, J.-F.; Chen, S.-W.; Fu, F.-R.; Lin, Y.-T.; Lin, C.-H.; Kumar, A.; et al. Study on delamination between polymer materials and metals in IC packaging process. *Polymers* **2019**, *11*, 940. [CrossRef] [PubMed]
12. Liu, L.; Li, E.; Luo, X. Study on delamination variation trend of two plastic packaging devices in combination reliability test. In Proceedings of the 2017 18th International Conference on Electronic Packaging Technology (ICEPT), Harbin, China, 16–19 August 2017; pp. 231–234.
13. Wang, Y.; Ding, Y.; Yin, Y. Reliability of wide band gap power electronic semiconductor and packaging: A review. *Energies* **2022**, *15*, 6670. [CrossRef]
14. Watanabe, Y.; Yamaguchi, H.; Enomoto, T.; Ogawa, K.; Kobayashi, T.; Omiya, M. Evaluation of thermal fatigue crack propagation in underfill resin materials for electronic packages. *Fatigue Fract. Eng. Mater. Struct.* **2022**, *45*, 1349–1360. [CrossRef]
15. Li, M.; Min, Z.; Shi, Z.; Wang, Q.; Huang, W.; Shao, K. Research on the thermal oxygen aging mechanism of the composite curing system epoxy asphalt with different curing agent proportions on the molecular scale. *J. Appl. Polym. Sci.* **2023**, *140*, e54308. [CrossRef]
16. Izumi, A.; Kakara, T.; Otsuki, M.W.; Shudo, Y.; Koganezawa, T.; Shibayama, M. In situ residual stress analysis in a phenolic resin and copper composite material during curing. *Polymer* **2019**, *182*, 121857. [CrossRef]
17. Sham, M.-L.; Kim, J.-K. Evolution of residual stresses in modified epoxy resins for electronic packaging applications. *Compos. Part A* **2004**, *35*, 537–546. [CrossRef]
18. Kravchenko, O.G.; Li, C.; Strachan, A.; Kravchenko, S.G.; Pipes, R.B. Prediction of the chemical and thermal shrinkage in a thermoset polymer. *Compos. Part A* **2014**, *66*, 35–43. [CrossRef]
19. Middleton, J.; Hoffman, J.; Burks, B.; Predecki, P.; Kumosa, M. Aging of a polymer core composite conductor: Mechanical properties and residual stresses. *Compos. Part A* **2015**, *69*, 159–167. [CrossRef]
20. Cysne Barbosa, A.P.; Fulco, A.P.P.; Guerra, E.S.S.; Arakaki, F.K.; Tosatto, M.; Costa, M.C.B.; Melo, J.D.D. Accelerated aging effects on carbon fiber/epoxy composites. *Compos. Part B* **2017**, *110*, 298–306. [CrossRef]
21. Sim, K.-B.; Lee, T.-H.; Han, G.-Y.; Kim, H.-J. Thermal expansion and mechanical properties of urethane-modified epoxy bonded CFRP/steel joints at low and high temperatures for automotive. *Compos. Struct.* **2023**, *322*, 117426. [CrossRef]

22. Nakamura, H.; Sato, C.; Ikegami, K. Internal stress induced during curing process of adhesively bonded parts of resin and metal. *Zairyo* **1997**, *46*, 820–825.
23. Abiko, K.; Kato, Y.; Hohjo, H.; Kishida, Y.; Sudo, E. Raman imaging of residual stress distribution in epoxy resin and metal interface. *J. Raman Spectrosc.* **2020**, *51*, 193–200. [CrossRef]
24. Wu, K.; An, D.; Zhang, Z.; Zhao, G.; Cui, C.; Zhou, F.; Li, J. Relief of residual stress in bulk thermosets in the glassy state by local bond exchange. *Macromol. Rapid Commun.* **2024**, *45*, 2300735. [CrossRef]
25. Gouadec, G.; Colomban, P. Raman Spectroscopy of nanomaterials: How spectra relate to disorder, particle size and mechanical properties. *Prog. Cryst. Growth Charact. Mater.* **2007**, *53*, 1–56. [CrossRef]
26. Deng, Y.; Wang, Q.; Ma, J.; Oh, J.T.; Chen, Z. The combined impact of voids and thermal aging on the mechanical reliability of epoxy resin evaluated by statistical analysis. *Polym. Degrad. Stab.* **2023**, *215*, 110455. [CrossRef]
27. Morsch, S.; Liu, Y.; Lyon, S.B.; Gibbon, S.R.; Gabriele, B.; Malanin, M.; Eichhorn, K.J. Examining the early stages of thermal oxidative degradation in epoxyamine resins. *Polym. Degrad. Stab.* **2020**, *186*, 108030. [CrossRef]
28. Ng, J.X.Y.; Cao, Q.; Golovanevskiy, V.; Farhat, H.; Iannuzzi, M.; Pojtanabuntoeng, T. A structured approach to the determination of residual stresses in an epoxy-amine coating and the influence of thermal ageing on the thermomechanical properties, residual stresses and the corresponding failure modes of coating. *Prog. Org. Coat.* **2024**, *186*, 108030. [CrossRef]

polymers

Article

Flexible PAN/P25 Multi-Porous Nanotubular Electrospun Membrane Constructed by a Facile Ethylene Glycol Solvothermal Induction with Excellent Photocatalytic Degradation and Sterilization Performance

Yiwen Miao [1], Chenghao Zhang [2], Ya Sun [2], Chunlei Wang [2,*], Juntao Yan [2], Sunhua Deng [3] and Ruan Chi [4]

1 College of Chemical and Biomolecular Engineering, The Hong Kong University of Science and Technology, Hong Kong 999077, China; ymiaoaa@connect.ust.hk
2 College of Chemistry and Environmental Engineering, Wuhan Polytechnic University, Wuhan 430023, China; sunya@whu.edu.cn (C.Z.); sunya230@whpu.edu.cn (Y.S.); 12111@whpu.edu.cn (J.Y.)
3 College of Construction Engineering, Jilin University, Changchun 130021, China; dengsh@jlu.edu.cn
4 Hubei Three Gorges Laboratory, Yichang 443000, China; 12112@whpu.edu.cn
* Correspondence: wangcl@whpu.edu.cn; Tel.: +86-27-83943956; Fax: +86-27-83937409

Abstract: A series of flexible polyacrylonitrile/TiO_2 (PAN/P25) multi-porous nanotubular membranes were successfully constructed by facile electrospinning combined with an ethylene glycol solvothermal induce strategy. The effects of P25 dosage and solvothermal time on the morphology of samples were systematically investigated, which were characterized in terms of surface morphology, microstructure, specific surface area, thermal analysis, wettability, photoelectrochemical and fluorescence spectra. Rhodamine B (RhB) and *Escherichia coli* (*E. coli*) were employed as simulated pollutants to evaluate photocatalytic degradation and antibacterial properties of the PAN/P25-3 multi-porous nanotubular membrane. The PAN/P25-3 membrane exhibited the highest photocatalytic degradation efficiency, with 96.1% degradation of RhB within 120 min under a xenon lamp light source and a photocatalytic inactivation rate of 95.8% for *E. coli* under 365 nm monochromatic light irradiation. The photocatalytic degradation mechanism of the PAN/P25-3 multi-porous nanotubular membrane for RhB was deduced from the results of 3D-EEM fluorescence and scavenger experiments of reactive species. Additionally, the cyclic photodegradation experiments demonstrated that the PAN/P25-3 membrane maintained excellent stability and photocatalytic performance after multiple degradation cycles, confirming its potential for sustainable wastewater treatment applications.

Keywords: electrospinning; multi-porous nanotubular membrane; photocatalytic degradation and sterilization; solvothermal induction

Citation: Miao, Y.; Zhang, C.; Sun, Y.; Wang, C.; Yan, J.; Deng, S.; Chi, R. Flexible PAN/P25 Multi-Porous Nanotubular Electrospun Membrane Constructed by a Facile Ethylene Glycol Solvothermal Induction with Excellent Photocatalytic Degradation and Sterilization Performance. *Polymers* **2024**, *16*, 3484. https://doi.org/10.3390/polym16243484

Academic Editor: Andrea Sorrentino

Received: 13 November 2024
Revised: 3 December 2024
Accepted: 6 December 2024
Published: 13 December 2024

1. Introduction

Organic and bacterial contaminants have brought about serious water environment pollution [1]. The emission of organic dyes from the textile industry has caused health problems for living organisms [2]. Furthermore, bacterial contamination such as *Escherichia coli* (*E. coli*) in water has also resulted in potential risks to human drinking water [3]. In this case, many water treatment technologies have been developed to purify water, such as ozone disinfection, chlorination, UV irradiation, electrolysis, physical adsorption, and photocatalytic degradation [4–7]. It has been proven that solar-driven photocatalytic oxidation technology is widely explored for water purification [8]. Semiconductor-based photocatalysts have attracted extensive attention for their ability to eliminate organic pollutants, inactivate harmful bacteria, and produce clean energy [9]. This technology is considered a promising water treatment method due to the absence of sterilization by-products [10,11]. As is known, commercially available TiO_2 (P25) nano-photocatalyst is widely used due to its high stability, low cost and eco-friendliness [12,13]. However, the P25 nano-photocatalyst is prone to agglomerate due to its high surface energy, which reduces the photocatalytic activity [14]. In addition, it is difficult to separate and recover the P25 nano-photocatalyst from the treated solution, which limits the cyclic utilization [15,16]. In order to overcome this problem, people make it easier to recycle by loading magnetic materials in photocatalysts or using electrospinning.

Due to the large aspect ratio, large specific surface area, controllable chemical composition, good morphology and high porosity [17–20], electrospinning technology is considered to be a simple, universal, economical and effective method to fabricate one-dimensional nanofiber and composite nanofiber membranes containing inorganic particles [21,22]. The electrospun polyacrylonitrile (PAN) nanofiber possesses a one-dimensional nanostructure with good chemical stability and excellent flexibility [23], which can be utilized as a substrate to immobilize inorganic nanoparticles to construct functional membrane and avoid separating the photocatalyst from water. Pan et al. have combined the different inorganic materials with electrospun PAN, making the composite materials have different properties, such as enhanced Raman scattering [24] and adsorption [25–27]. Chen et al. fabricated PAN/P25 nanofibers by the electrospinning method, which possessed the enhanced piezoelectric photocatalysis performance toward Rhodamine B via polar functional group engineering [28]. P25 nanoparticles can be added to the spinning solution because of their good stability. However, many P25 nanoparticles are encapsulated into the fiber during the electrospinning process, which will reduce the exposure of active sites; thus, the photocatalytic activity of P25 is weakened. Therefore, how to make P25 nanoparticles more exposed and realize facile recycling becomes a challenge. Ramasundaram et al. have fixed P25 nanoparticles on steel mesh by electric spraying combined with a high-temperature hot-pressing method (350 °C, 100 MPa) [29]. Romas et al. have immobilized P25 nanoparticles within zinc acetate/PVA nanofibers by an electrostatically modified electrospinning process, which is calcinated at 600 °C to obtain polycrystalline ZnO and ZnO/P25 [30]. These methods often require more stringent conditions or complex equipment, so it is urgent that an economical and simple method be found.

Herein, the flexible PAN/P25 multi-porous nanotubular membranes are firstly fabricated by a facile electrospinning method combined with an ethylene glycol solvothermal induction strategy, which possesses excellent photocatalytic degradation of dyes and photocatalytic sterilization performance. We discussed the morphological changes caused by the solvothermal treatment time. Additionally, the chemical, physical and photoelectric properties of the samples were characterized. The multi-porous nanotubular structure improves the light utilization efficiency and the transmission efficiency of e^- and h^+. At the same time, the flexible PAN/P25 multi-porous nanotubular membrane has stable cycle performance. This multi-porous nanotubular functional membrane will be effectively used in the field of water pollution treatment in the future.

2. Experimental

2.1. Materials

Poly (vinyl pyrrolidone) (PVP, Mw = 1,300,000) and polyacrylonitrile (PAN, Mw = 150,000) were purchased from Aladdin Regent Company, America. N,N-dimethylformamide (DMF), and Ethylene glycol (EG) were obtained from Shanghai Chemical Regent Company, China. Rhodamine B (RhB) was obtained from Tianjin Guangfu Chemical Reagents Company, China. P25 (TiO_2) was purchased from Degussa, and *E. coli* (ATCC 8739) was provided by the research group.

2.2. Preparation of PAN/PVP/P25 Fiber Membrane

As shown in Figure 1, X g P25 (X = 0.1, 0.15, 0.2, 0.25) was ultrasonically dispersed in 6 mL DMF, then 0.6 g PAN and 0.6 g PVP were added sequentially and the uniform spinning solution was obtained after magnetic stirring for about 6 h. After that, the spinning solution was transferred to a syringe with a steel needle, which was connected to a direct current high voltage of 12 kV; the receiver was placed 15 cm away from the tip of the steel needle.

The temperature and humidity were 25 °C and 45%, respectively. Finally, the as-spun fiber membrane was placed in an oven at 60 °C for 2 h to remove the residual solvent. The products were named PAN/PVP/P25-0.1, PAN/PVP/P25-0.15, PAN/PVP/P25-0.2 and PAN/PVP/P25-0.25 fiber membrane, respectively. As a control, the PAN/PVP fiber membrane was fabricated at the same conditions without introducing the P25 particles.

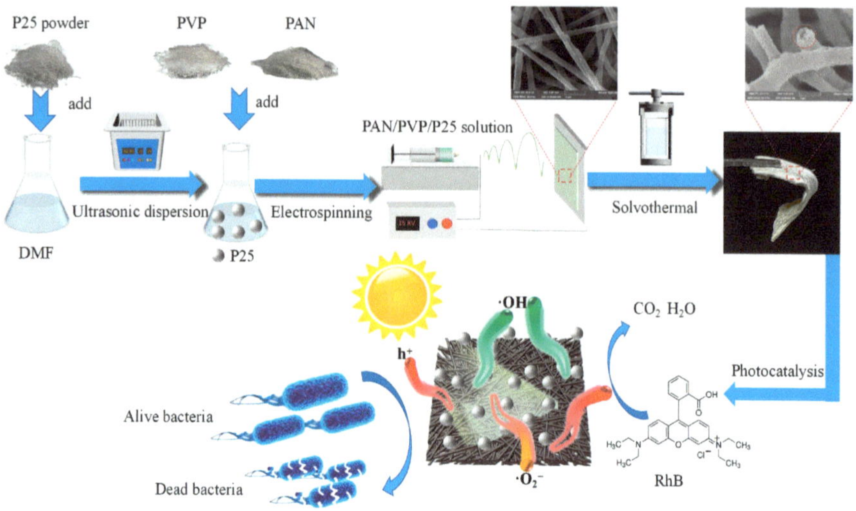

Figure 1. Schematic representation of the preparation and application of PAN/P25 multi-porous nanotubular membrane.

2.3. *Preparation of Flexible PAN/P25 Multi-Porous Nanotubular Membrane*

A flexible PAN/P25 multi-porous nanotubular membrane was achieved by the ethylene glycol (EG) solvothermal induction strategy. The as-spun PAN/PVP/P25-0.2 fiber membrane was cut into square pieces with a side length of 3 cm and transferred to the stainless steel high-pressure reactor containing EG. The reaction was performed at 180 °C for Y h (Y = 2, 3, 4). After cooling to room temperature, the obtained sample was repeatedly washed with deionized water and confronted with freeze-drying for 12 h. Finally, the PAN/P25 multi-porous nanotubular membranes were obtained, which were named PAN/P25-2, PAN/P25-3 and PAN/P25-4 multi-porous nanotubular membranes, respectively. The detailed formulation is shown in Table 1.

Table 1. Recipes of PAN/P25 multi-porous nanotubular membranes.

Sample	PAN (g)	PVP (g)	P25 (g)	Solvothermal Time (h)
PAN/PVP/P25-0.1	0.6	0.6	0.1	—
PAN/PVP/P25-0.15	0.6	0.6	0.15	—
PAN/PVP/P25-0.2	0.6	0.6	0.2	—
PAN/PVP/P25-0.25	0.6	0.6	0.25	—
PAN/PVP	0.6	0.6	0	—
PAN/P25-2	0.6	—	0.2	2
PAN/P25-3	0.6	—	0.2	3
PAN/P25-4	0.6	—	0.2	4

2.4. *Photocatalytic Experiment*

The photocatalytic activities of the obtained photocatalysts were evaluated by the degradation of RhB in an aqueous solution. The entire degradation reaction process was carried out in an opaque 100 mL jacket beaker under a light source with 2 °C condensed water throughout the jacket of the jacket beaker. Firstly, 50 mg of PAN/PVP/P25 or 29 mg PAN/P25-3 (to ensure the equivalent of P25) catalytic membrane was added in 50 mL of RhB (10 mg/L) solution, and the adsorption–desorption equilibrium state was reached after dark reaction for 30 min without turning on the light source. Then, we turned on the xenon lamp light source (A 300 W PLS-SXE300 xenon lamp, Beijing Perfect Light Technology Co., LTD, Beijing, China) for the degradation experiment. The suspension was taken every 30 min, and the absorbance A_t at 554 nm of the supernatant was recorded. The final residual quantity C_t/C_0 was approximated by A_t/A_0, where C_t and A_t represent the concentration and absorbance at time t, and C_0 and A_0 represent the initial concentration and absorbance. The degradation rate is calculated by $(1 - C_t/C_0) \times 100\%$.

2.5. Photocatalytic Disinfection of *E. coli*

Firstly, *E. coli* was inoculated in a Luria-Broth (LB) nutrient solution, cultured in a shaker at 37 °C for 12 h, centrifuged and collected, washed two times with sterile saline, and then re-suspended in fresh sterile saline. The final *E. coli* concentration was adjusted to 3×10^6 CFU/mL. In a typical experiment, 10 mg of PAN/P25-3 nanotubular membrane was immersed into 10 mL of E. coil suspension. The photocatalytic sterilization performance of the PAN/P25-3 multi-porous nanotubular membrane was evaluated by a xenon lamp with a DT365 filter. Within a certain time interval, 500 µL suspension was taken and diluted with sterile saline. In order to determine the cell density of viable *E. coli*, 100 µL diluted solution was daubed to nutrient agar and incubated at 37 °C for 12 h. We then determined the number of viable bacteria. Each experiment was repeated three times.

2.6. Characterization

Fourier transform infrared (FT-IR) spectra were recorded on a Nicolet Instruments Research Series 5PC FTIR. The X-ray diffraction (XRD) patterns were obtained using a Shimadzu XRD-7000 X-ray diffractometer with monochromatized Cu Kα radiation. The XRD measurement was conducted with a 2θ range of from 10° to 80°, a scan speed of 5°/min, and a step size of 0.02°. The surface morphologies and composition of samples were examined using SEM (TESCAN MAIA 3 LMH, Bratislava, Czech Republic). TEM images were recorded by an FEI Tecnai F20 microscope operated at 200 kV. The Brunauer-Emmett-Teller (BET) specific surface area of the material was obtained by measuring the nitrogen adsorption–desorption isotherms using a nitrogen adsorption apparatus (ASAP2020, Norcross, GA, USA). Thermogravimetric (TG) was performed by heating the nanofiber membranes from 25 °C to 800 °C at a heating rate of 10 °C/min in the air atmosphere (Netzsch STA 449 F3, Selb, Bavaria, Germany). Static WCA tests for products were carried out via a contact angle meter (Krüss DSA25, Hamburg, Germany). The UV–vis diffused reflectance spectra (DRS) measurements of the samples were carried out using a TU-1901 system with $BaSO_4$ as the reflectance standard. Photoluminescence (PL) spectra of the samples were detected on a fluorescence spectrophotometer (Hitachi F-7000, Tokyo, Japan) with an excitation wavelength of 200 nm. Electrochemical signals were recorded by a CHI660B electrochemical analyzer (Chenhua, Shanghai, China). Three-dimensional excitation–emission matrix (3DEEM) fluorescence spectra were recorded using a Fluoromax-4 spectrofluorometer (Zolix SmartFluo-Pro, Beijing, China), and both the excitation wavelength (Ex) and emission wavelength (Em) were in the range of 200–700 nm. The absorption spectra of organic dyes and antibiotics solutions were obtained by a UV–vis spectrophotometer (SHIMDZUUV-3600, Kyoto, Japan).

3. Results and Discussion

3.1. Composition and Crystal Structure

In order to evaluate the physical and chemical changes of the composites, FT-IR and XRD are used for spectral analysis. Figure 2A shows the PAN/PVP fiber membrane, PAN/PVP/P25 fiber membrane and PAN/P25-3 multi-porous nanotubular membrane. Observing curve (a), the wide absorption peak near 3400 cm^{-1} is attributed to the stretching vibration of -OH, which is due to the adsorbed water on the sample surface. The absorption bands of the spectrum observed at 3000–2900 cm^{-1} and 1454 cm^{-1} are due to stretching and deformation vibrations of C-H bonds, respectively. At 2245 cm^{-1}, the stretching vibration of the C≡N bond in PAN is presented [31–34]. The obvious peaks near 1450 cm^{-1} and 1250 cm^{-1} are indexed into C=C and C-N stretching vibrations. In addition, there is an obvious peak at 1680 cm^{-1}, which is attributed to the C=O stretching vibration in the PVP molecule. Compared with curve (a), curves (b) and (c) have a new broad peak in 850–550 cm^{-1}, which belongs to the Ti-O-Ti stretching bond and proves the existence of P25. Figure 2A (curve b) shows a decrease in the peak intensity at 1680 cm^{-1} for PAN/PVP/P25-0.2, even before the solvothermal process, which may suggest partial removal or changes in the PVP structure prior to the reaction. Figure 2B shows the XRD patterns. Curve (a) has no obvious peak. In curve (b), the anatase phase of P25 corresponds to 25.4°, 37.7°, 47.8°, 54.8°, 62.5° and 75° and the rutile phase of P25 corresponds to 27.3°, 36°, 41°, 53.8° and 74.9° [35,36], which confirms the existence of P25 in the composite film, but the peak intensity is weak. This is because most P25 nanoparticles are wrapped in the PAN fiber. Moreover, the XRD peak at approximately $2\theta \approx 17°$ observed in both the PAN/PVP and PAN/P25-3 patterns likely originates from the (100) reflection of the semi-crystalline structure of PAN. It is worth noting that the peak strength in curve (c) increases significantly. This is because more P25 nanoparticles are largely exposed to the surface of the PAN/P25 multi-porous nanotube after the removal of the PVP component, which is mainly ascribed to the role of the EG solvothermal induction process.

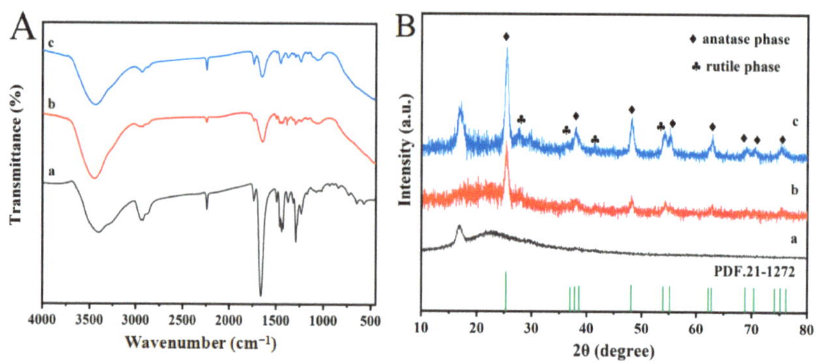

Figure 2. (**A**) FT-IR and (**B**) XRD patterns of (a) PAN/PVP; (b) PAN/PVP/P25-0.2; (c) PAN/P25-3.

3.2. Morphology of PAN/P25 Multi-Porous Nanotubular Membrane

The surface morphology evolution of PAN/PVP fibrous membranes is observed by the SEM. As shown in Figure 3(A1–A4), the PAN/PVP fiber membrane possesses a smooth surface with a diameter of 0.32 μm. For the PAN/PVP/P25-0.2 fiber membrane in Figure 3(B1–B3), some P25 nanoparticles are uniformly loaded on the surface of the fiber membrane and the surface has become rough. Moreover, the fiber diameter increases from 0.32 μm to 0.72 μm because the viscosity of the spinning solution becomes larger, which makes it difficult for the electrostatic force to stretch the fiber more thoroughly. As depicted in Figure 3(C1–C4), it is interesting that the PAN/P25-3 multi-porous nanotubular structure is achieved after the ethylene glycol solvothermal induction treatment, the nanotubular structure is marked with a red circle, and there are many pores and P25 nanoparticles exposed to the surface of the PAN/P25 multi-porous nanotube (Red circle mark). The formation of multi-porous nanotubes is mainly dependent on the phase separation phenomenon between PAN and PVP [37]. Due to the different viscosity of the two components of PAN and PVP, the PAN component is mainly located outside the fibers, and the PVP component is mainly located inside the PAN/PVP fibers during the electrospinning process. Moreover, most of the PVP component is removed from the fibers due to the solubility of PVP in EG solvent during the EG solvothermal induction reaction. In addition, because

PAN is rich in a large amount of -CN, which has the ability to trap metal ions, P25 is still preserved on the fiber. Moreover, the multi-porous nanotube diameter increases from 0.72 μm to 0.90 μm; the reason for this is that the removal of PVP leads to fiber swelling during the EG solvothermal treatment. The unique multi-porous nanotubular structure of PAN/P25-3 is favorable for the exposure of P25 nanoparticles and the light-harvesting and utilization by means of multiple reflections and scattering, yielding more electrons and holes. Furthermore, the multi-porous nanotubular structure can shorten the diffusion path of pollutants by passing through the hole in the tube wall and provide more active sites for contact with pollutants. As a consequence, the above two aspects are beneficial to the improvement of photocatalytic performance.

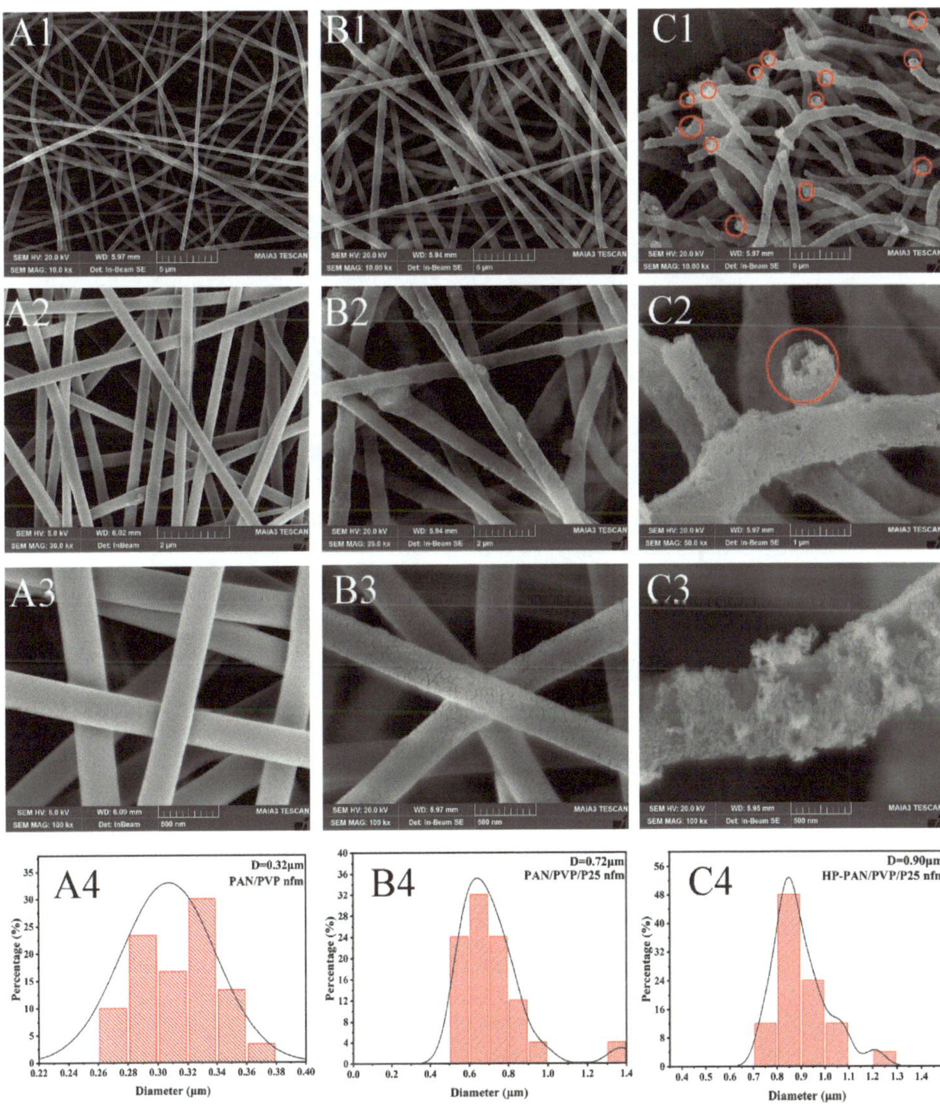

Figure 3. SEM images and diameter distribution of (**A1–A4**) PAN/PVP; (**B1–B4**) PAN/PVP/P25-0.2; (**C1–C4**) PAN/P25-3.

TEM images in Figure 4A,B further confirm the one-dimensional and multi-porous hollow structure of PAN/P25-3. Moreover, it is vividly shown that the exposed P25 nanoparticles are located on the multi-porous nanotube surface in the enlarged TEM in Figure 4C,D. Both the increase in holes and the exposure of the P25 nanoparticles are beneficial to enhance the specific surface area and provide more active sites and flow channels, which can facilitate the organic dyes or bacteria in sewage contaction with the P25 photocatalyst. In addition, the light inside the holes can be refracted multiple times to improve the light utilization rate, thereby improving the photocatalytic efficiency.

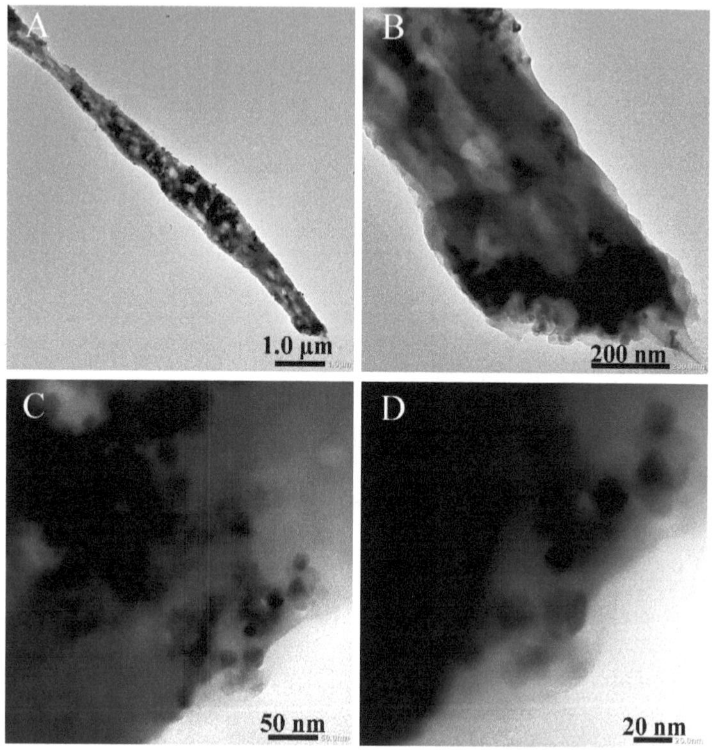

Figure 4. TEM images of PAN/P25-3 multi-porous nanotubular membrane.

3.3. Effect of Solvothermal Time on the Morphology of PAN/P25

The surface morphology changes of the membranes at different times were observed by SEM. The PAN/PVP/P25-0.2 fibrous membrane without solvothermal treatment is shown in Figure 5(A1,A2); a small number of P25 nanoparticles are loaded on the surface of the fiber, most of them are coated inside the fiber, which makes it have a low photocatalytic efficiency. After solvothermal treatment for 2 h (Figure 5(B1,B2)), the holes on the surface of the fiber are attributed to the dissolution of PVP, but it does not cause the nanotubular structure, which is still unable to achieve the highest utilization of light. With the extension of the solvent heat treatment time, the pores on the surface of the fiber gradually increase, and more P25 is exposed while the fibers form a hollow tubular structure (Figure 5(C1,C2)). However, too long a time (4 h) leads to the agglomeration of nanoparticles P25 (Figure 5(D2)), which may affect the photocatalytic property.

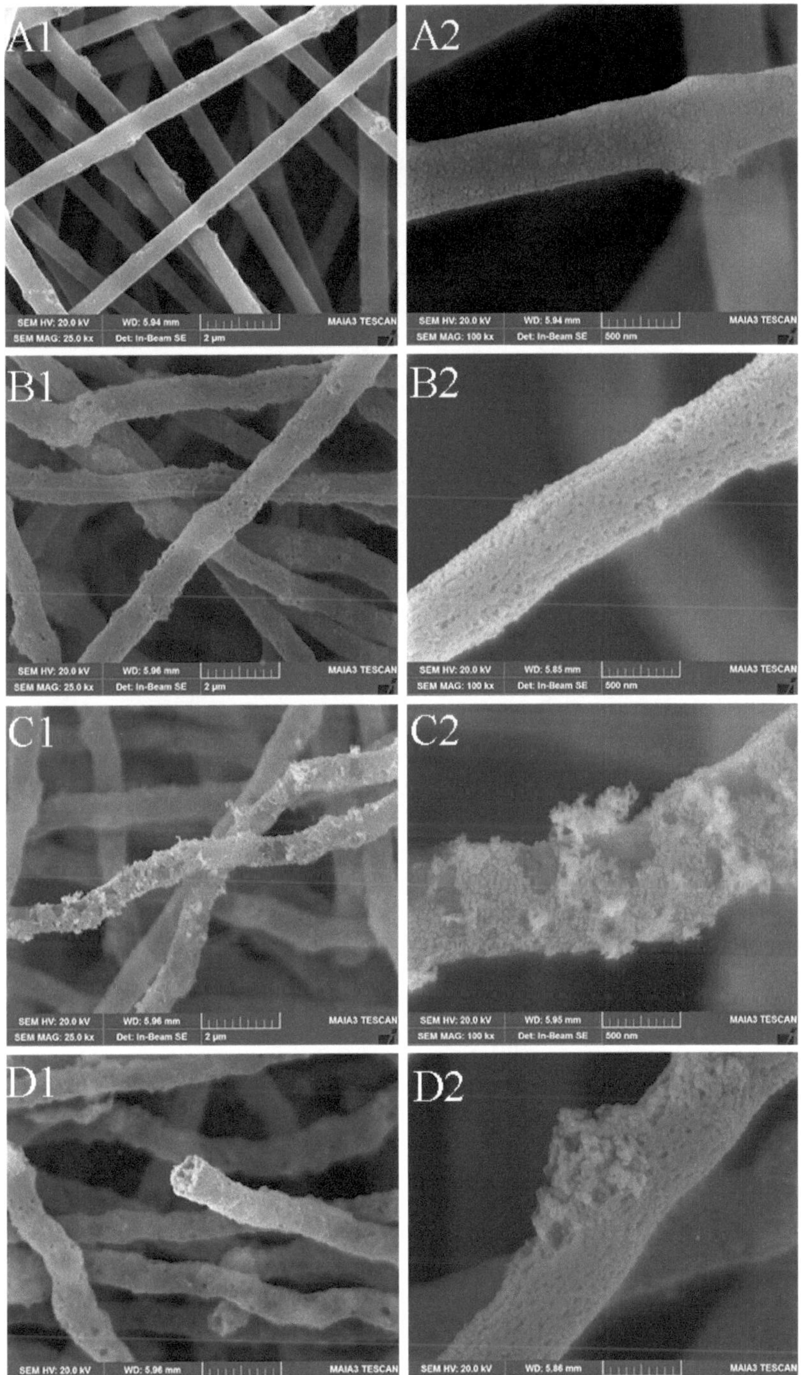

Figure 5. SEM images of (**A1**,**A2**) PAN/PVP/P25-0.2 and fiber membranes with different solvothermal reaction time (**B1**,**B2**) PAN/P25-2; (**C1**,**C2**) PAN/P25-3; (**D1**,**D2**) PAN/P25-4.

3.4. Nitrogen Sorption Isotherms

The physical properties of PAN/PVP/P25-0.2 and PAN/P25-3 are further characterized by the BET test. As shown in Figure 6, the N_2 adsorption–desorption isotherms and pore diameter of PAN/PVP/P25-0.2 fiber membrane and PAN/P25-3 multi-porous nanotubular membrane were evaluated to determine their specific surface area and the pore size. The PAN/PVP/P25-0.2 fiber membrane exhibits a type III isotherm, which indicates that the surface of the material is a non-porous or macroporous material. The measured pores data are attributed to the interlacing pores between the fibers. The PAN/P25-3 multi-porous nanotubular membrane exhibits a type III isotherm and H_3 type hysteresis loop ($0.8 < P/P_0 < 1.0$), which is one of the main characteristics of mesoporous materials [38], and its pore size distribution is relatively narrow (2–20 nm). Table 2 shows the specific data; the surface areas of the PAN/PVP/P25-0.2 fiber membrane and the PAN/P25-3 multi-porous nanotubular membrane are 12.29 $m^2 \cdot g^{-1}$ and 25.74 $m^2 \cdot g^{-1}$, respectively. The average pore size increased distinctly, ranging from 15.33 nm to 24.76 nm, and the pore volume also increased significantly, from 0.047 $cm^3 \cdot g^{-1}$ to 0.160 $cm^3 \cdot g^{-1}$. This increase is attributed to the removal of PVP, which creates a large number of holes. These holes are beneficial for increasing the active sites and enhancing light absorption.

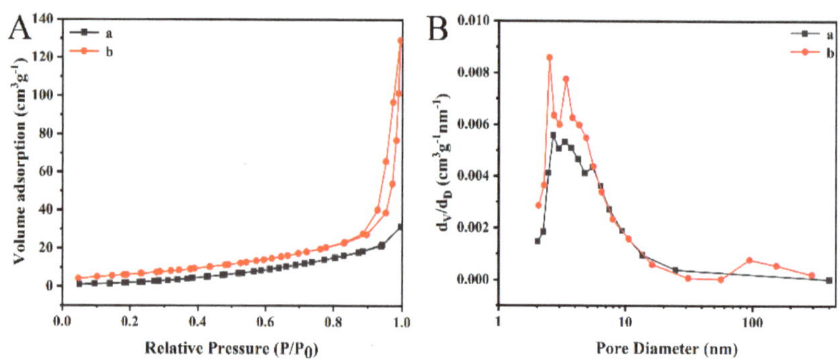

Figure 6. (**A**) N_2 adsorption–desorption isotherms and (**B**) pore size distribution curves of (a) PAN/PVP/P25-0.2; (b) PAN/P25-3.

Table 2. Physical properties of the photocatalytic membrane.

Sample	S_{BET} ($m^2 \cdot g^{-1}$)	Average Pore Size (nm)	Pore Volume ($cm^3 \cdot g^{-1}$)
PAN/PVP/P25-0.2	12.29 ± 0.01	15.33 ± 0.01	0.047 ± 0.001
PAN/P25-3	25.74 ± 0.01	24.76 ± 0.01	0.160 ± 0.001

3.5. Thermal Analysis

The TG analysis of the PAN/PVP fiber membrane, PAN/PVP/P25 fiber membrane and PAN/P25-3 multi-porous nanotubular membrane are shown in Figure 7. When the temperature reached 600 °C, PAN and PVP were all volatilized. As shown in observation curve (b), due to the hydrophilicity of PVP, the membrane surface adsorbs water in the air, and 3.41% of the mass loss around 100 °C is attributed to the water on the surface of the sample. The mass loss of 0.7% between 100 °C and 260 °C is attributed to residual DMF. The pyrolysis of PVP begins at 260 °C, and some PVP will cross-link with PAN [39]. At the same time, PAN is pyrolyzed at 280 °C. Therefore, 38.41% of the weight loss between 260 °C and 320 °C is ascribed to the partial decomposition of PVP and the loss of a small amount of ammonia and hydrogen cyanide during PAN cyclization [40]. Continuing to increase the temperature, the secondary weight loss of PAN starts due to carbonization and decomposition, but it is slower than the previous weight loss, which is due to the better heat resistance of PAN [41]. When the temperature rises to 600 °C, only P25 nanoparticles are left, and the residual weight of 14.20% (curve b) is close to the theoretical value (14.29%). Observing curve (c), due to the removal of PVP, the surface contains less adsorbed water, so there is almost no mass loss before 100 °C. The subsequent pyrolysis curve is similar to the trend of curve (a), and the final P25 content (24.59%) is close to the theoretical value (25%). It is suggested that the solvothermal process can remove the PVP component and hardly cause the shedding of P25 nanoparticles from the PAN/P25-3 multi-porous nanotubular membrane.

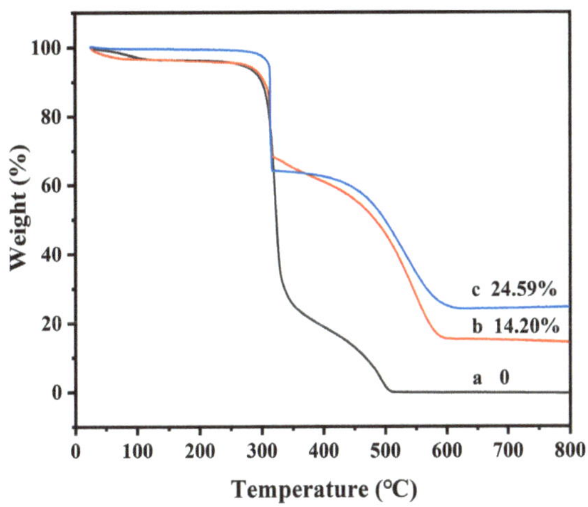

Figure 7. TG analysis of (a) PAN/PVP; (b) PAN/PVP/P25-0.2; (c) PAN/P25-3.

3.6. Wettability Test

As shown in Figure 8, the wettability of the PAN/PVP fiber membrane, PAN/PVP/P25-0.2 fiber membrane and PAN/P25-3 multi-porous nanotubular membrane were evaluated by measuring the contact angle between the surface of the nanofiber membrane and water. Due to the hydrophilicity of PVP, the PAN/PVP fiber membrane shows good hydrophilicity, and the contact angle is only 21.3°. Although the surface of the fiber became rough after loading P25, it still shows good hydrophilicity in Figure 8B. The contact angle of the PAN/P25-3 multi-porous nanotubular membrane slightly increases to 35.4° and still shows hydrophilic properties; this is probably because some hydrophilic groups of -OH that originated from the EG solvent are introduced into the PAN/P25-3 surface during the solvothermal process. This hydrophilicity is beneficial to the treatment of pollutants in water.

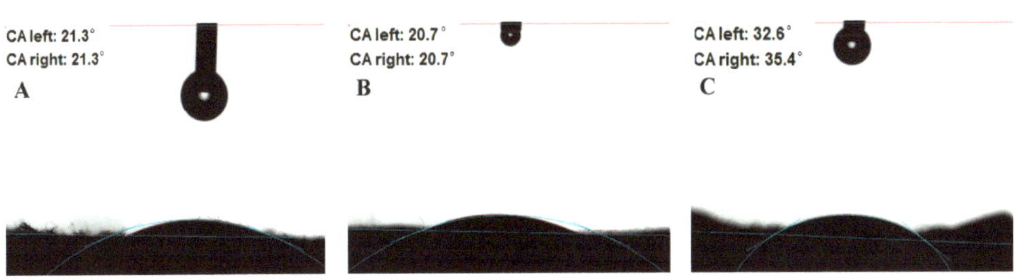

Figure 8. Water contact angle images: (**A**) PAN/PVP; (**B**) PAN/PVP/P25-0.2; (**C**) PAN/P25-3.

3.7. Optical Absorption and Photoelectrochemistry Analysis

DRS was used to study the optical absorption behavior of the PAN/PVP/P25-0.2 fiber membrane and PAN/P25-3 multi-porous nanotubular membrane; the results are shown in

Figure 9A. The absorption edges of the PAN/PVP/P25-0.2 fiber membrane and PAN/P25-3 multi-porous nanotubular membrane were at 387 nm and 391 nm, respectively, which is attributed to the intrinsic band gap absorption of P25 [42]. The higher light absorption value is because the porous structure is favorable for light to enter the inner wall, and the tubular structure is favorable for light to achieve multiple reflections in the inner wall, making the light utilization rate higher. At the same time, EG heat treatment may introduce functional groups such as -OH or -CHO to the surface of the material; this may also be another reason for the increased light absorption [20,22]. The obtained diffuse reflectance spectrum is converted into a Tauc diagram (Figure 9B) according to the following formula: $\alpha h\nu = A\,(h\nu - E_g)^{n/2}$, Where α, h, v, A and E_g represent the absorption coefficient, Planck's constant, optical frequency, direct leap constant and band gap energy, respectively. The value of n depended on the type of optical transition of semiconductors ($n_{direct} = 1$ and $n_{indirect} = 4$). The straight part of the figure is extended to the horizontal axis (y = 0) to obtain the band-gap energy (3.20 eV and 3.17 eV). The band-gap energy of the two samples did not change significantly, indicating that the P25 remained stable after solvent heat treatment.

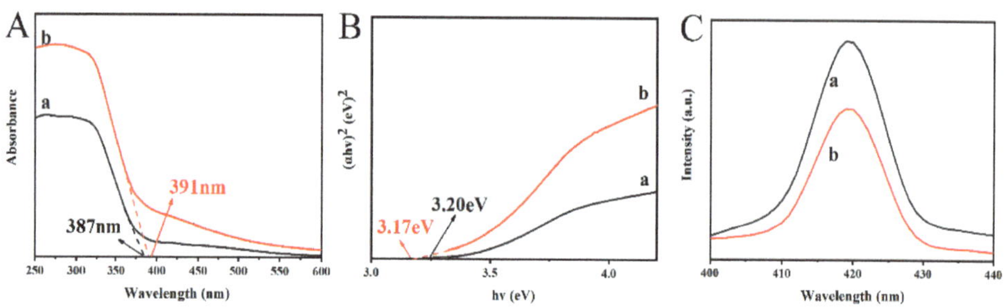

Figure 9. (**A**) DRS spectra; (**B**) plots of $(\alpha h\nu)^2$ versus energy ($h\nu$) and (**C**) PL spectra of (a) PAN/PVP/P25-0.2; (b) PAN/P25-3.

It is well known that the recombination of photogenerated electrons (e^-) and holes (h^+) is one of the important parameters determining photocatalytic activity [43]. The PL spectra (Figure 9C) of the samples were detected at 200 nm excitation wavelength to compare and analyze the recombination rate of e^- and h^+. It can be seen that the peak intensity of the PAN/P25-3 multi-porous nanotubular membrane at 420 nm was significantly lower than that of the PAN/PVP/P25-0.2 fiber membrane, which indicates that the photoinduced carrier separation efficiency was higher. This is because the recombination of e^- and h^+ occurs not only within a single P25 molecule but also between two or more adjacent molecules. The porous structure may improve the dispersion of P25, thereby reducing the recombination of e^- and h^+ between adjacent P25, thus reducing the recombination rate, and the multi-porous structure can shorten the diffusion path of e^- and h^+, thus reducing the recombination rate.

3.8. Photoelectrochemical Properties Test

Figure 10 shows the photoelectrochemical properties of the membrane. The arc radius of the PAN/P25-3 multi-porous nanotubular membrane is smaller, indicating that it has less resistance to free charge migration and a faster charge transfer rate [44]. Faster charge transfer is beneficial to promote the separation efficiency of photogenerated carriers, thereby improving the photocatalytic efficiency. Moreover, compared with the PAN/PVP/P25-0.2 fiber membrane, the photocurrent intensity of the PAN/P25-3 multi-porous nanotubular membrane also increased, further confirming the effective separation of photo-generated e^- and h^+ pairs.

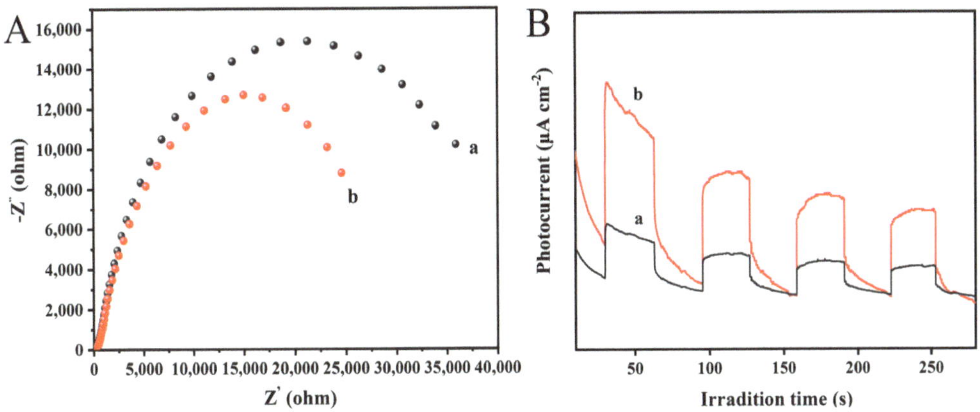

Figure 10. (**A**) EIS Nyquist plots and (**B**) photocurrents of (a) PAN/PVP/P25-0.2; (b) PAN/P25-3.

3.9. Photodegradation Activity and Stability

The performance of the obtained samples was studied by photocatalytic degradation of organic dyes, and RhB was selected as the object model. Figure 11 shows the photocatalytic degradation curve (A, C) and kinetic diagram (B, D) of RhB aqueous solution by different photocatalysts under simulated sunlight. All of the samples have degradation effects toward RhB, and the degradation characteristics conform to pseudo-first-order kinetics. The as-spun PAN/PVP/P25-0.2 fiber membrane was subjected to EG solvothermal treatment for different durations (2 h, 3 h and 4 h), resulting in the formation of multi-porous nanotubular membranes: PAN/P25-2, PAN/P25-3, and PAN/P25-4. These membranes exhibited improved photodegradation performance compared to the PAN/PVP/P25-0.2 membrane (Figure 11C), as the PAN/P25 multi-porous nanotubular membranes have more exposed P25 particles. In contrast, most of the P25 nanoparticles are incorporated within the fibers of the PAN/PVP/P25-0.2 membrane. With increasing treatment time, the photocatalytic degradation efficiency followed the following order: PAN/P25-3 > PAN/P25-4 > PAN/P25-2. Among them, the PAN/P25-3 multi-porous nanotubular membrane demonstrated the highest photocatalytic degradation efficiency of 96.1% for RhB within 120 min. According to the literature [45], the greater the rate constant k, the stronger the photocatalytic performance of the photocatalyst. The degradation rate (0.0240 min^{-1}) of the PAN/P25-3 multi-porous nanotubular membrane is four times that of the PAN/PVP/P25-0.2 fiber membrane (0.006 min^{-1}). This is because the multi-porous nanotubular membrane PAN/P25-3 has a larger specific surface area, higher light absorption, lower e^{-}-h^{+} recombination rate, smaller impedance and stronger photocurrent.

Stability is an important index affecting the practical application of the material [11]. PAN/P25-3 is subjected to five photocatalytic cycles. The used catalyst was removed, and the residual dye on the surface was washed with anhydrous ethanol and deionized water and then reused after freeze-drying. Figure 11E depicts the recycled photocatalytic performance of the PAN/P25-3 photocatalyst, which keeps its original properties of about 94% even after five runs. In addition, the XRD results (Figure 11F) show that the basic composition and structure have almost no change before and after use, and there was no significant change in mass before and after the reaction, which demonstrates that the PAN/P25-3 multi-porous nanotubular membrane has excellent stability.

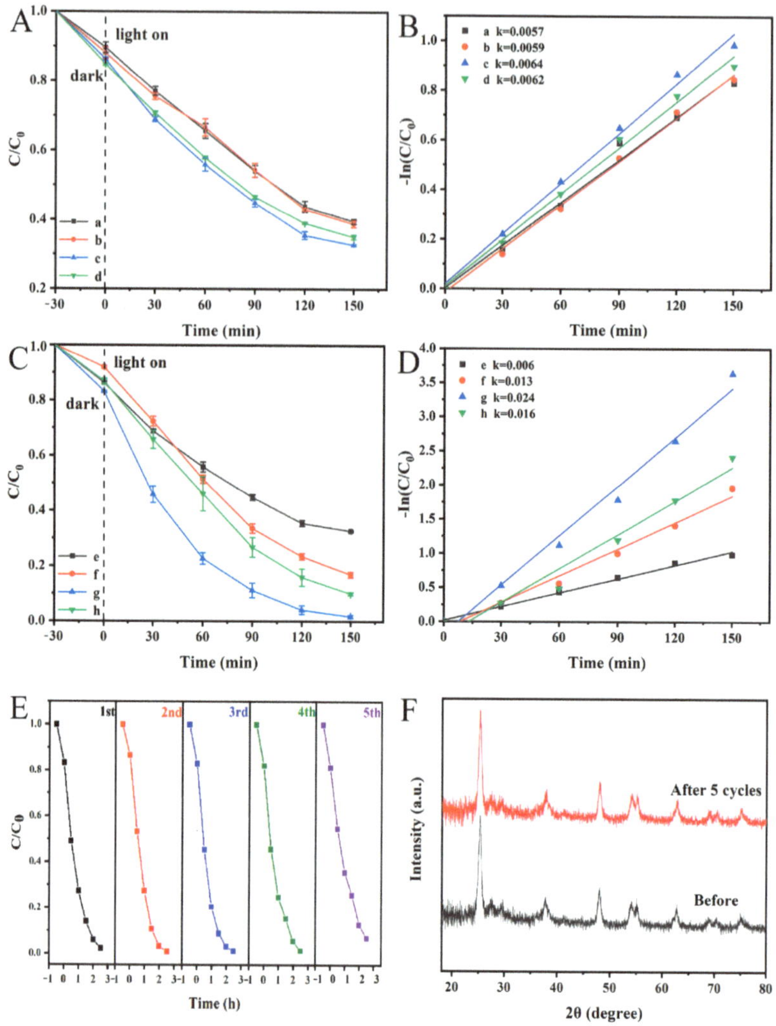

Figure 11. (**A**,**C**) Photocatalytic degradation and (**B**,**D**) kinetics curves for RhB aqueous solution over different photocatalysts under simulated solar light irradiation: (a) PAN/PVP/P25-0.1; (b) PAN/PVP/P25-0.15; (c,e) PAN/PVP/P25-0.2; (d) PAN/PVP/P25-0.25; (f) PAN/P25-2; (g) PAN/P25-3; (h) PAN/P25-4; (**E**) recycled photodegradation efficiency of PAN/P25-3; (**F**) XRD patterns of PAN/P25-3 before and after the recycling photocatalytic tests.

3.10. Photocatalytic Process and Mechanism

The photocatalytic PAN/P25-3 degradation process of the RhB solution was studied by the 3D-EEM fluorescence method. Initially, molecular fluorescence is attributed to the rigid planar structure of RhB, which can accelerate the conjugation effect of π electrons and improve fluorescence efficiency [46,47]. As shown in Figure 12A, the two fluorescence characteristic peaks of RhB after the adsorption–desorption equilibria are the (a) peak at Ex/Em of 550–575/620–660 nm and (b) peak at 225–275/575–650 nm [48]. The 3D-EEM fluorescence spectrum after adsorption equilibrium is shown in Figure 12B. The shape, position and intensity of the fluorescence characteristic peak are almost unchanged, indicating that RhB molecules are not decomposed during this process, just adsorbed on the surface of the photocatalyst. With the photocatalytic degradation time of 60 min (Figure 12C), the position of the fluorescence characteristic peak exhibits gradually red-shifted, indicating that the RhB molecules are decomposed. This phenomenon may be related to the generation of N-deethylated intermediates, in which macromolecules are converted into relatively small fragments and specific functional groups such as amines, hydroxyl groups and carbonyl groups are removed [49]. In addition, when the visible light irradiation time was extended to 150 min (Figure 12D), the intensity of the fluorescence characteristic peak in the residual solution gradually decreased, indicating that the formed n-deethylation intermediate may be further decomposed into small molecules. Therefore, the main process of RhB degradation is the dissociation of the conjugated chromophore structure of the RhB molecule and the mineralization of the corresponding intermediates.

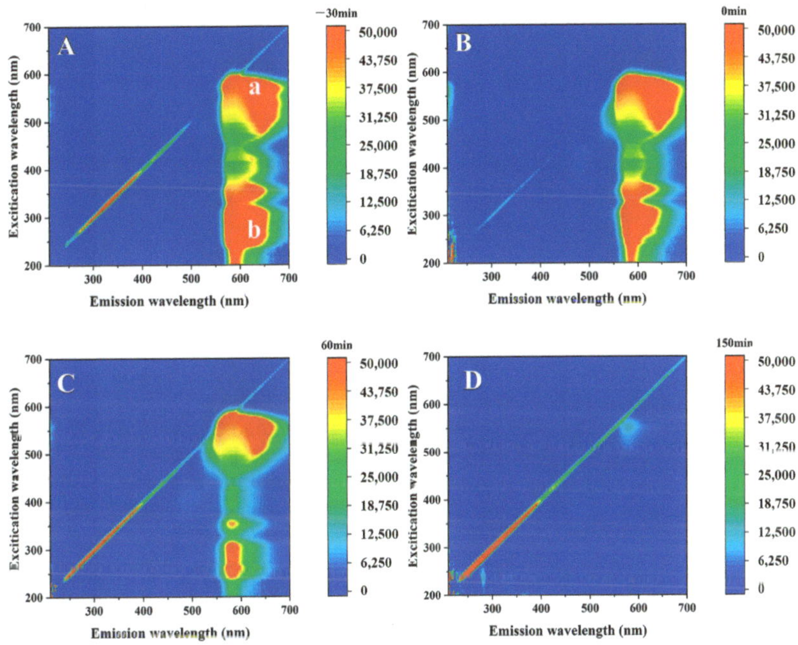

Figure 12. The 3D-EEM fluorescence spectra of RhB degradation by PAN/P25-3: (**A**) −30 min; (**B**) 0 min; (**C**) 60 min; (**D**) 150 min.

In order to further explain the photocatalytic mechanism of the PAN/P25-3 multi-porous nanotubular membrane, triethanolamine (TEOA), p-benzoquinone (p-BQ) and isopropanol (IPA) were used as the capture agents of photogenerated holes (h^+), superoxide radicals ($\cdot O_2^-$) and hydroxyl radicals ($\cdot OH$), respectively. As shown in Figure 13, the catalytic efficiency of PAN/P25-3 decreased to different degrees after the addition of three free radical trapping agents. These results show that the three kinds of free radicals have different effects on photocatalytic efficiency. Among them, the degradation rate of RhB is significantly reduced after TEOA is added, indicating that h^+ has been captured successfully, leading to a decrease in photocatalytic activity. However, the degradation rate of RhB decreases slightly in the presence of IPA and p-BQ, indicating that $\cdot OH$ and $\cdot O_2^-$ are not the main active species. In this case, it can be determined that h^+ is the main active species for the degradation of RhB by PAN/P25-3 multi-porous nanotubular membrane. $\cdot O_2^-$ and $\cdot OH$ are the secondary active substances.

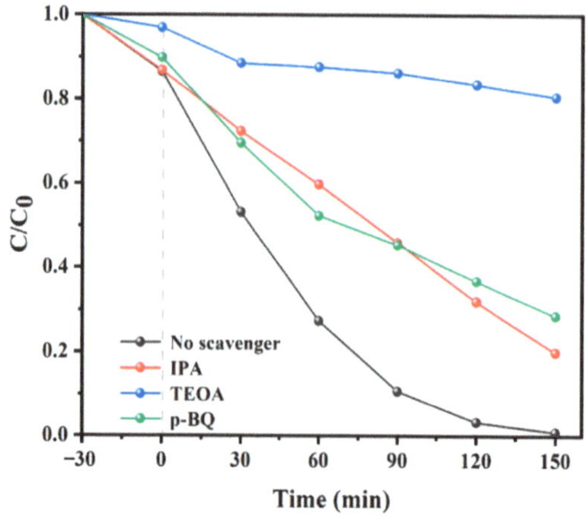

Figure 13. The active species trapping experiments during the photocatalytic degradation of RhB over PAN/P25-3 multi-porous nanotubular membrane.

3.11. Bacterial Inactivation Activity

The PAN/PVP fibrous membrane was initially obtained using the direct electrospinning method. When 0.2 g of P25 particles was dispersed in the PAN/PVP spinning solution, the PAN/PVP/P25-0.2 fibrous membrane was produced. The as-spun PAN/PVP/P25-0.2 fibrous membrane was then subjected to the EG solvothermal induction process for 3 h, resulting in the flexible PAN/P25-3 multi-porous nanotubular membrane. As shown in Figure 11C, PAN/P25-3 exhibited better photocatalytic degradation performance than the PAN/PVP/P25-0.2 membrane, which led to the selection of PAN/P25-3 for further investigation of its bacterial inactivation activity.

In addition, we also examined the photocatalytic sterilization performance of the PAN/P25-3 multi-porous nanotubular membrane under 365 nm light irradiation, as shown in Figure 14. Each experiment was repeated three times to ensure reliability and reproducibility. Under dark conditions, the survival number of *E. coli* cells does not decrease significantly after 3 h, which indicates that the material itself is non-toxic to *E. coli* and has just a weak adsorption. When only light irradiation at a wavelength of 365 nm is applied, the bacterial concentration decreases by 0.1 log within 3 h; this is because the band of light can make the bacterial oxygen free radicals, which cause oxidative stress and lead to bacterial death [50]. However, when the PAN/P25-3 multi-porous nanotubular membrane is used under light conditions, the inactivation rate of *E. coli* is significantly improved, and the bactericidal rate of *E. coli* is over 95%, which is obviously higher than the sum of the inactivation rate under light irradiation alone and only catalyst. It is indicated that the synergistic effect of the PAN/P25-3 multi-porous nanotubular membrane with light irradiation may play an important role. Under light conditions, PAN/P25-3 can produce different active free radicals, such as h^+, e^-, $\cdot O_2^-$, etc. Where h^+ and e^- can initiate redox reactions, and both can be converted into hydroxyl radicals, hydroxyl radicals can quickly attack and seize hydrogen atoms on viral and bacterial proteins or envelopes, inducing them to lose their normal physiological functions and leading to death. $\cdot O_2^-$ attacks on bacteria can cause a decrease in the activity of some antioxidant enzymes (catalase, superoxide dismutase) of bacteria and damage cell membranes, resulting in gradual oxidative damage and leakage of intracellular substances, especially proteins and DNA, the similar mechanism of photocatalytic bacterial inactivation is similar to that in reference [51].

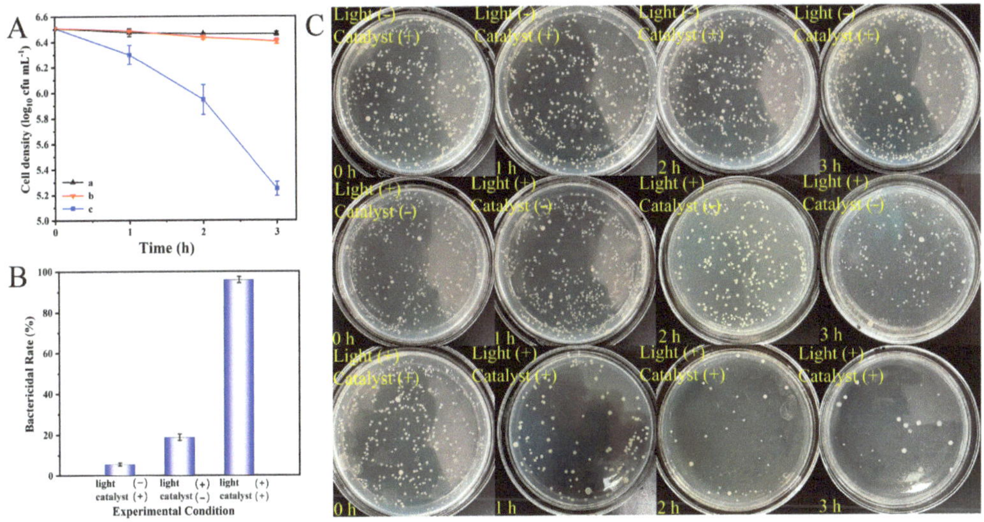

Figure 14. Photocatalytic bactericidal activity of PAN/P25-3 multi-porous nanotubular membrane with a dilution factor of 5000: (**A**) sterilization line under different conditions: (a) only PAN/P25-3; (b) only 365 nm light; (c) PAN/P25-3 with 365 nm light; (**B**) bactericidal rate of *E. coli* and (**C**) optical image of *E. coli* colonies.

4. Conclusions

In summary, a flexible PAN/P25 multi-porous nanotubular membrane was successfully prepared by the electrospinning and solvothermal methods. We explored the optimal spinning parameters and solvothermal reaction time. Interestingly, the solvothermal reaction not only transforms the fiber into a multi-porous nanotubular structure but also exposes the P25 wrapped in the fiber to the outside of the fiber. Compared with the PAN/PVP/P25-0.2 fiber membrane, the photocatalytic RhB degradation activity of the PAN/P25-3 multi-porous nanotubular membrane increased by four times. This enhanced photocatalytic performance can be attributed to the expanded specific surface area, more active edge exposure, higher light absorption efficiency, high light utilization, smaller impedance and stronger photocurrent response. After five cycles, the material still maintains good stability, which will be well applied in the environmental field. At the same time, the material also has photocatalytic sterilization ability; under 365 nm monochromatic light irradiation for 3 h the inactivation rate of *E. coli* was as high as 95.8%.

Author Contributions: Conceptualization, C.W.; methodology, C.Z.; software, Y.S.; formal analysis, Y.S.; investigation, C.Z. and J.Y.; resources, C.W.; writing—original draft preparation, Y.M. and J.Y.; writing—review and editing, Y.M.; visualization, S.D.; supervision, C.W.; project administration, C.W.; funding acquisition, R.C. All authors have read and agreed to the published version of the manuscript.

Funding: This work was supported by the Wuhan Metropolitan Area Collaborative Innovation Technology Project (2024070904020436), the Open and Innovation Fund of Hubei Three Gorges Laboratory (SK232007).

Institutional Review Board Statement: Not applicable.

Data Availability Statement: The data presented in this study are available on request from the corresponding author.

Conflicts of Interest: The authors declare no conflict of interest.

References

1. Patel, M.; Kumar, R.; Kishor, K.; Mlsna, T.; Pittman, C.U., Jr.; Mohan, D. Pharmaceuticals of emerging concern in aquatic systems: Chemistry, occurrence, effects, and removal methods. *Chem. Rev.* **2019**, *119*, 3510–3673. [CrossRef]
2. Du, X.; Wang, C.; Liu, J.; Zhao, X.; Zhong, J.; Li, Y.; Li, J.; Wang, P. Extensive and selective adsorption of ZIF-67 towards organic dyes: Performance and mechanism. *J. Colloid Interface Sci.* **2017**, *506*, 437–441. [CrossRef] [PubMed]
3. Schito, A.; Caviglia, D.; Piatti, G.; Alfei, S. A highly efficient polystyrene-based cationic resin to reduce bacterial contaminations in water. *Polymers* **2022**, *14*, 4690. [CrossRef] [PubMed]
4. Wei, F.; Lu, Y.; Shi, Q.; Chen, Z.; Li, K.; Zhang, T.; Shi, Y.; Xu, Q.; Hu, H. A dose optimization method of disinfection units and synergistic effects of combined disinfection in pilot tests. *Water Res.* **2022**, *211*, 118037. [CrossRef] [PubMed]
5. Polinarski, M.; Neves, A.C.; Fiorini, A.; Rosado, F.R.; Silva, E.A.; Alves, H.J. Ultraviolet radiation as an antimicrobial treatment in Brazilian diesel oil: Effect of biodiesel, sulfur, and water contents. *Fuel* **2022**, *308*, 122076. [CrossRef]
6. Tonanon, P.; Webster, R. Recent electrode and electrolyte choices for use in small scale water treatment applications—A short review. *Curr. Opin. Electrochem.* **2023**, *38*, 101211. [CrossRef]
7. Liu, A.; Chen, Y.; Chen, Z.; Liang, X.; Yan, J.; Li, B. Recycling of UO_2^{2+} expanded by metal-organic framework bearingpyrazinoquinoxaline tetracarboxylic acid: pH fluorescence sensing and photocatalytic activity. *Chem. Eng. J.* **2024**, *497*, 154946. [CrossRef]
8. Yan, J.; Chai, B.; Liu, Y.; Fan, G.; Song, G. Construction of 3D/2D $ZnFe_2O_4$/g-C_3N_4 S-scheme heterojunction for efficient photo-Fenton degradation of tetracycline hydrochloride. *Appl. Surf. Sci.* **2023**, *607*, 155088. [CrossRef]
9. Ding, D.; Liu, X.; Chen, J.; Xuan, R.; Peng, W.; Zhu, L.; Li, X.; Yan, J. Novel Bi_2O_2Se/BiOCl with high visible light photocatalytic activity for environmental remediation: Effective layer to layer interfacial charge transfer based on heterojunction. *Surf. Interfaces* **2024**, *47*, 104229. [CrossRef]
10. Yan, J.; Wei, L. 2D S-Scheme Heterojunction Photocatalyst. *Acta Phys.-Chim. Sin.* **2024**, *40*, 2312024. [CrossRef]
11. Yan, J.; Liu, J.; Sun, Y.; Ding, D.; Wang, C.; Sun, L.; Li, X. Exfoliation-induced O-doped g-C_3N_4 nanosheets with improved photoreactivity towards RhB degradation and H_2 evolution. *Inorg. Chem. Front.* **2022**, *9*, 1423. [CrossRef]
12. Liu, S.; Zhou, X.; Wei, C.; Hu, Y. Spatial directional separation and synergetic treatment of Cr(VI) and Rhodamine B mixed pollutants on three-layered Pd@MIL-101/P25 photocatalyst. *Sci. Total Environ.* **2022**, *842*, 156836. [CrossRef]
13. Deka, T.; Ningthoukhongjam, P.; Yadav, M.; Nair, R. Optimization of various photocatalytic reaction parameters of Degussa P25 under UV irradiation. *J. Indian Chem. Soc.* **2022**, *99*, 100747. [CrossRef]
14. Wang, C.; Ren, Y.; Chen, A.; Li, X.; Ding, D.; Yan, J.; Sun, L.; Chi, R. Facile electrospinning-electrospray method to fabricate flexible rice straw-derived cellulose acetate/TiO_2 nanofibrous membrane with excellent photocatalytic performance. *Int. J. Biol. Macromol.* **2024**, *282*, 137501. [CrossRef] [PubMed]
15. Zhou, D.; Chen, Z.; Yang, Q.; Dong, X.; Zhang, J.; Qin, L. In-situ construction of all-solid-state Z-scheme g-C_3N_4/TiO_2 nanotube arrays photocatalyst with enhanced visible-light-induced properties. *Sol. Energy Mater. Sol. Cells* **2016**, *157*, 399–405. [CrossRef]
16. Huang, M.; Xu, C.; Wu, Z.; Huang, Y.; Lin, J.; Wu, J. Photocatalytic discolorization of methyl orange solution by Pt modified TiO_2 loaded on natural zeolite. *Dyes Pigm.* **2008**, *77*, 327–334. [CrossRef]
17. Wang, C.; Hu, L.; Chai, B.; Yan, J.; Li, J. Enhanced light photocatalytic activity of electrospun nanofibrous TiO_2/g-C_3N_4 heterojunction photocatalyst under simulated solar light. *App. Surf. Sci.* **2018**, *430*, 243–252. [CrossRef]
18. Hu, L.; Yan, J.; Wang, C.; Chai, B.; Li, J. Direct electrospinning method for the construction of Z-scheme TiO_2/g-C_3N_4/RGO ternary heterojunction photocatalysts with remarkably ameliorated photocatalytic performance. *Chin. J. Catal.* **2019**, *40*, 458–469. [CrossRef]
19. Yan, J.; Liu, J.; Sun, Y.; Song, G.; Ding, D.; Fan, G.; Chai, B.; Wang, C.; Sun, L. Investigation on the preparation of rice straw-derived cellulose acetate and its spinnability for electrospinning. *Polymers* **2021**, *13*, 3463. [CrossRef] [PubMed]
20. Wang, C.; Tan, X.; Yan, J.; Chai, B.; Li, J.; Chen, S. Electrospinning direct synthesis of magnetic $ZnFe_2O_4$/ZnO multi-porous nanotubes with enhanced photocatalytic activity. *App. Surf. Sci.* **2017**, *396*, 780–790. [CrossRef]
21. Mo, J.; Hu, D.; Liu, X.; Sun, Y.; Li, X.; Wang, C.; Chai, B.; Ding, D.; Yan, J.; Sun, L. Eco-friendly ZIF-67/rice straw-derived cellulose acetate electrospun nanofiber mats for efficient CO_2 capturing and selectivity removal of methyl orange dye. *J. Environ. Chem. Eng.* **2022**, *10*, 108989. [CrossRef]
22. Yan, J.; Gao, S.; Wang, C.; Chai, B.; Li, J.; Song, G.; Chen, S. A facile electrospinning and direct annealing method for the fabrication of multi-porous $ZnFe_2O_4$ nanotubes with enhanced photocatalytic activity. *Mater. Lett.* **2016**, *184*, 43–46. [CrossRef]
23. Wang, C.; Ren, Y.; Chai, B.; Zhang, M.; Sun, Y.; Yan, J.; Sun, L.; Chi, R. Fabrication of rice-straw lignin/polyacrylonitrile composite membrane for lithium-ion battery separator. *Mater. Lett.* **2024**, *368*, 136687. [CrossRef]
24. Jia, P.; Chang, J.; Wang, J.; Zhang, P.; Cao, B.; Geng, Y.; Wang, X.; Pan, K. Fabrication and formation mechanism of Ag nanoplate decorated nanofiber mats and their application in SERS. *Chem. Asian J.* **2016**, *11*, 86. [CrossRef]
25. Luo, C.; Wang, J.; Jia, P.; Liu, Y.; An, J.; Cao, B.; Pan, K. Hierarchically structured polyacrylonitrile nanofiber mat as highly efficient lead adsorbent for water treatment. *Chem. Eng. J.* **2015**, *262*, 775–784. [CrossRef]
26. Wang, J.; Zhang, P.; Liang, B.; Liu, Y.; Xu, T.; Wang, L.; Cao, B.; Pan, K. Graphene oxide as an effective barrier on a porous nanofibrous membrane for water treatment. *ACS Appl. Mater. Inter.* **2016**, *8*, 6211–6218. [CrossRef] [PubMed]
27. Liang, B.; Zhang, P.; Wang, J.; Qu, J.; Wang, L.; Wang, X.; Guan, C.; Pan, K. Membranes with selective laminar nanochannels of modified reduced graphene oxide for water purification. *Carbon* **2016**, *103*, 94–100. [CrossRef]

28. Chen, J.; Xuan, R.; Dou, Y.; Xu, C.; Lu, T.; Jiang, Z.; Ding, D.; Wang, C.; Yan, J.; Yang, B. Enhanced Piezoelectric Photocatalysis performance of polymers/P25 nanofibers on rhodamine B remediation via polar functional group engineering. *ACS Appl. Polym. Mater.* **2024**, *6*, 13671–13680. [CrossRef]
29. Ramasundaram, S.; Seid, M.G.; Kim, H.-E.; Son, A.; Lee, C.; Kim, E.-J.; Hong, S.W. Binder-free immobilization of TiO_2 photocatalyst on steel mesh via electrospraying and hot-pressing and its application for organic micropollutant removal and disinfection. *J. Hazard. Mater.* **2018**, *360*, 62–70. [CrossRef]
30. Ramos, P.; Flores, E.; Sanchez, L.; Candal, R.; Hojamberdiev, M.; Estrada, W.; Rodriguez, J. Enhanced photoelectrochemical performance and photocatalytic activity of ZnO/TiO_2 nanostructures fabricated by an electrostatically modified electrospinning. *Appl. Surf. Sci.* **2017**, *426*, 844–851. [CrossRef]
31. Moskowitz, J.; Wiggins, J. Thermo-oxidative stabilization of polyacrylonitrile and its copolymers: Effect of molecular weight, dispersity, and polymerization pathway. *Polym. Degrad. Stab.* **2016**, *125*, 76–86.
32. Mohamed, A.; Osman, T.; Toprak, M.; Muhammed, M.; Yilmaz, E.; Uheida, A. Visible light photocatalytic reduction of Cr(VI) by surface modified CNT/titanium dioxide composites nanofibers. *J. Mol. Catal. A Chem.* **2016**, *424*, 45–53. [CrossRef]
33. Shakur, H.R.; Rezaee, E.S.K.; Abdi, M.R.; Azimi, G. Highly selective and effective removal of uranium from contaminated drinking water using a novel PAN/AgX/ZnO nanocomposite. *Microporous Mesoporous Mater.* **2016**, *234*, 257–266. [CrossRef]
34. Chumpol, J.; Siri, S. Electrospun cellulose acetate membrane for size separating and antibacterial screening of crude polysaccharides. *IET Nanobiotechnol.* **2016**, *10*, 405–410. [PubMed]
35. Dong, F.; Guo, S.; Wang, H.; Li, X.; Wu, Z. Enhancement of the visible light photocatalytic activity of C-doped TiO_2 nanomaterials prepared by a green synthetic approach. *J. Phy. Chem. C* **2011**, *115*, 13285–13292.
36. Zhang, L.; Tse, M.; Tan, O.; Wang, Y.; Han, M. Facile fabrication and characterization of multi-type carbon-doped TiO_2 for visible light-activated photocatalytic mineralization of gaseous toluene. *J. Mater.* **2013**, *1*, 4497–4507. [CrossRef]
37. Liu, C.; Tan, Y.; Liu, Y.; Shen, K.; Peng, B.; Niu, X.; Ran, F. Microporous carbon nanofibers prepared by combining electrospinning and phase separation methods for supercapacitor. *J. Energy Chem.* **2016**, *25*, 587–593.
38. Zhang, J.; Song, Y.; Kopeć, M.; Lee, J.; Wang, Z.; Liu, S.; Yan, J.; Yuan, R.; Kowalewski, T.; Bockstaller, M.R.; et al. Facile aqueous route to nitrogen-doped mesoporous carbons. *J. Am. Chem. Soc.* **2017**, *139*, 12931–12934.
39. Sun, D.; Qin, G.; Lü, M.; Wei, W.; Wang, N.; Jiang, L. Preparation of mesoporous polyacrylonitrile and carbon fibers by electrospinning and supercritical drying. *Carbon* **2013**, *63*, 585–588.
40. Kang, Y.; Chen, J.; Feng, S.; Zhou, H.; Zhou, F.; Low, Z.; Zhong, Z.; Xing, W. Efficient removal of high-temperature particulate matters via a heat resistant and flame retardant thermally-oxidized $PAN/PVP/SnO_2$ nanofiber membrane. *J. Membr. Sci.* **2022**, *662*, 120985. [CrossRef]
41. Grassie, N.; McGuchan, R. Pyrolysis of polyacrylonitrile and related polymers-I. Thermal analysis of polyacrylonitrile. *Eur. Polym. J.* **1970**, *6*, 1277–1291. [CrossRef]
42. Yu, J.; Yu, H.; Cheng, B.; Zhou, M.; Zhao, X. Enhanced photocatalytic activity of TiO_2 powder (P25) by hydrothermal treatment. *J. Mol. Catal. A Chem.* **2006**, *253*, 112–118. [CrossRef]
43. Liu, C.; Wang, L.; Xia, Z.; Chen, R.; Wang, H.; Liu, Y. Carbon hollow fibers with tunable hierarchical structure as self-standing supercapacitor electrode. *Chem. Eng. J.* **2022**, *431*, 134099. [CrossRef]
44. Duan, Y.; Li, X.; Lv, K.; Zhao, L.; Liu, Y. Flower-like g-C_3N_4 assembly from holy nanosheets with nitrogen vacancies for efficient NO abatement. *Appl. Surf. Sci.* **2019**, *492*, 166–176. [CrossRef]
45. Hailili, R.; Wang, Z.; Xu, M.; Wang, Y.; Gong, X.; Xu, T.; Wang, C. Layered nanostructured ferroelectric perovskite $Bi_5FeTi_3O_{15}$ for visible light photodegradation of antibiotics. *J. Mater. Chem. A* **2017**, *5*, 21275–21290. [CrossRef]
46. Guo, H.; Niu, C.; Wen, X.; Zhang, L.; Liang, C.; Zhang, X.; Guan, D.; Tang, N.; Zeng, G. Construction of highly efficient and stable ternary $AgBr/Ag/PbBiO_2Br$ Z-scheme photocatalyst under visible light irradiation: Performance and mechanism insight. *J. Colloid Interface Sci.* **2018**, *513*, 852–865. [CrossRef] [PubMed]
47. Ma, Y.; Jin, X.; Zhou, M.; Zhang, Z.; Teng, X.; Chen, H. Chemiluminescence behavior based on oxidation reaction of rhodamine B with cerium (IV) in sulfuric acid medium. *Anal. Chim. Acta* **2003**, *489*, 173–181. [CrossRef]
48. Jiang, Z.; Zhang, B.; Liang, A. A new sensitive and selective fluorescence method for determination of chlorine dioxide in water using rhodamine B. *Talanta* **2005**, *66*, 783–788. [CrossRef]
49. Qin, Y.; Li, H.; Lu, J.; Dong, H.; Liu, X.; Liu, Z.; Yan, Y. Synthesis of QDs self-modified $Bi_2MoO_6/Bi_4Ti_3O_{12}$ photocatalysts via controlling charge unidirectional flow for effective degradation of organic pollutants. *J. Mol. Liq.* **2019**, *286*, 110919. [CrossRef]
50. Akhavan, O.; Ghaderi, E. Photocatalytic Reduction of Graphene Oxide Nanosheets on TiO_2 Thin Film for Photoinactivation of Bacteria in Solar Light Irradiation. *J. Phys. Chem. C* **2009**, *113*, 20214–20220. [CrossRef]
51. Sun, H.; Li, G.; Nie, X.; Shi, H.; Wong, P.; Zhao, H.; An, T. Systematic Approach to In-Depth Understanding of Photoelectrocatalytic Bacterial Inactivation Mechanisms by Tracking the Decomposed Building Blocks. *Environ. Sci. Technol.* **2014**, *48*, 9412–9419. [CrossRef] [PubMed]

polymers

MDPI

Article

Facile In Situ Building of Sulfonated SiO_2 Coating on Porous Skeletons of Lithium-Ion Battery Separators

Lei Ding [1], Dandan Li [1], Sihang Zhang [2,*], Yuanjie Zhang [3], Shuyue Zhao [1], Fanghui Du [1] and Feng Yang [4]

1 Shandong Key Laboratory of Chemical Energy Storage and New Battery Technology, School of Chemistry and Chemical Engineering, Liaocheng University, No. 1, Hunan Road, Liaocheng 252000, China; dinglei921022@163.com (L.D.); lidandan10224@163.com (D.L.); zhshuyue09@163.com (S.Z.); dufanghui@lcu.edu.cn (F.D.)
2 School of Food Science and Engineering, Hainan University, 58 Renmin Avenue, Haikou 570228, China
3 Department of Chemistry and Biology, Liaocheng University Dongchang College, No. 266, North Outer Ring Road, Liaocheng 252001, China; yuanjiezhang2016@163.com
4 State Key Laboratory of Polymer Materials Engineering, College of Polymer Science and Engineering, Sichuan University, No.24 South Section 1, Yihuan Road, Chengdu 610065, China; yangfengscu@126.com
* Correspondence: sih.zhang@foxmail.com

Abstract: Polyolefin separators with worse porous structures and compatibilities mismatch the internal environment and deteriorate lithium-ion battery (LIB) combination properties. In this study, a sulfonated SiO_2 (SSD) composited polypropylene separator (PP@SSD) is prepared to homogenize pore sizes and in situ-built SSD coatings on porous skeletons. Imported SSD uniformizes pore sizes owing to centralized interface distributions within casting films. Meanwhile, abundant cavitations enable the in situ SSD coating to facilely fix onto porous skeleton surfaces during separator fabrications, which feature simple techniques, low cost, environmental friendliness, and the capability for continuous fabrications. A sturdy SSD coating on the porous skeleton confines thermal shrinkages and offers a superior safety guarantee for LIBs. The abundant sulfonic acid groups of SSD endow PP@SSD with excellent electrolyte affinity, which lowers Li^+ transfer barriers and optimizes interfacial compatibility. Therefore, assembled LIBs give the optimal C-rate capacity and cycling stability, holding a capacity retention of 82.7% after the 400th cycle at 0.5 C.

Keywords: lithium-ion battery separator; sulfonated SiO_2; in situ coating; pore size dispersion; cycling stability

Citation: Ding, L.; Li, D.; Zhang, S.; Zhang, Y.; Zhao, S.; Du, F.; Yang, F. Facile In Situ Building of Sulfonated SiO_2 Coating on Porous Skeletons of Lithium-Ion Battery Separators. *Polymers* **2024**, *16*, 2659. https://doi.org/10.3390/polym16182659

Academic Editor: Sheng-Heng Chung

Received: 25 August 2024
Revised: 17 September 2024
Accepted: 18 September 2024
Published: 20 September 2024

1. Introduction

Owing to the features of low self-discharge, excellent safety performance, stable cycle life, and high energy density, lithium-ion batteries (LIBs) have been widely used in energy storage devices including portable electronic devices and electric vehicles for more than 30 years since their first commercialization [1,2]. Four components of anode, cathode, separator, and electrolyte amalgamate to combine into LIBs. Separators are thus seated between two electrodes to avoid the physical contact of electrodes and meanwhile conduct Li^+, which requires separators to possess both absolute insulation and suitable porous structures [3,4]. Especially, porous structures govern ion transfer paths and affect battery performance even though chemically inert separators are not involved in electrode redox reactions [5]. Even and centralized pore size can ensure homogenized Li^+ flux and effectively stabilize the separator/electrode interface process [6,7]. Furthermore, features like flexibility, sufficient mechanical strength, low manufacturing cost, and electrolyte affinity are also vital for separators. Hence, separators prepared with polyolefin have become mainstream in the separator market nowadays [8,9].

However, the low surface energy of polyolefins generates instinctive hydrophobic properties and inevitably deteriorates electrolyte affinity. Since ions only migrate within the electrolyte-filled pores, poor electrolyte affinity determines partially lacking filling conditions and increases ion migration barriers [10–12]. Also, routine polyolefins own low melting points of about 130 °C and 165 °C for polyethylene and polypropylene. This limits weak thermal stability and tends to contract excessively at elevated temperatures, which increases the risk of short circuits caused by electrode contacts [13,14]. However, regular LIB operations produce massive heat and raise inner temperatures. The poor dimensional tolerance of separators to heating thus cannot ensure security for LIBs [15,16].

Building inorganic nano-ceramic layers, such as TiO_2, Al_2O_3, SiO_2, and ZrO_2 with various functionalizations, on separator top surfaces can construct sturdy skin skeletons to not only intensify electrolyte affinity but enhance thermal stability in actual separator industrial fabrications [17–26]. However, the matched binders are needed to ensure tight adhesions between surface coating and separators. Excess binders may seep into interstitial paths of nano-ceramics and separator surface pores, which chokes partial ion migration routes and minifies electrochemical performance. The superficial nano-ceramic coatings necessarily increase extra thickness and reduce battery energy density [27,28]. Since the nanoparticle coatings only appear on separator top surfaces, the electrolyte wettability and ion migration process within separator inner porous skeletons cannot be optimized [29–32]. In particular, existing research and commercial separators are coated offline. The inorganic nanoparticle coating is elaborately spread on the separator top surface after separator preparations or porous structure formations, which elevates facility costs and technological difficulty. Therefore, exploring a new coating separator manufacturing process to achieve the simultaneous in situ construction of ceramic coatings on the porous skeletons during pore formations is critical for the low-cost manufacturing of high-performance separators.

Presently, actual industrial separator manufacturing processes including dry processes and wet processes are based on the tensile-induced pore formation of casting films [33–36]. The dry process can be divided into uniaxial tensile and biaxial tensile. The wet process adopts casting film containing ultra-high molecular weight polyethylene and mineral oil and exerts biaxial drawing and solvent extraction to obtain a porous skeleton. The dry process, with uniaxial drawing, inflicts an ultra-high die draw ratio during extrusion, then undergoes annealing and uniaxial drawing to create acerose pores. Particularly, the dry process with biaxial drawing (DPBD) applies directly tensile on casting sheets, in which the β-crystal polypropylene lamellae within sheets are forced to separate due to incompact arrangements and further evolve into the final porous structure under biaxial drawing. Compared with other separator preparation processes, DPBD omits organic solvent extractions and annealing treatment and thus features eco-friendly, low-cost, and continuous productions, which owns tremendous potential for separator production [37,38]. However, the unique pore-forming process determines the scattered pore size [39,40] since lamellae with heterogeneous distribution present various deformation patterns, resulting in plentiful remaining coarse fibrils and worsening cavitation effects [41–43]. Blending nano-ceramics into β-crystal polypropylene casting sheets can improve interface distributions and thus visibly thin coarse fibrils owing to the stripping between nano-ceramics and lamellae. Meanwhile, ample cavities stem from direct separations of ceramic/lamellae interfaces and thus facilely fix the ceramic on the porous skeleton surface, with features including simple techniques, low cost, environmental friendliness, and the capability for continuous fabrications.

With these considerations in mind, nanometer-sized SiO_2 was first sulfonated (SSD) to optimize hydrophilicity in this research. Then, SSD-composited separators (PP@SSD) were prepared based on the DPBD to simultaneously homogenize pore sizes and in situ build SSD coatings on porous skeleton surfaces, featuring simple techniques, low cost, environmental friendliness, and the capability for continuous fabrications. Furthermore, the PP@SSD separator presented optimized thermal stability owing to the robust SSD coating on the porous skeleton. Abundant sulfonic acid groups on the SSD coating further endowed electrolyte affinity to the separator surface and inner pore walls, lowered Li+ transfer barriers, and optimized interfacial compatibility. Consequently, the PP@SSD separator gave assembled LIBs the optimal C-rate capacity and cycling stability. This approach offered practical guidance for the low-cost mass manufacturing of high-performance separators.

2. Experiments

2.1. Separator Preparation

An amount of 4 g of nano-SiO_2 (Hydrophilic-200, 7–40 nm, Shanghai Aladdin Biochemical Technology Co., Ltd., Shanghai, China) was added into a mixed solution containing 100 mL methanol and 2 mol concentrated sulfuric acid. The mixed solution successively experienced ultrasonic concussion for 100 min, suction filtration, vacuum drying at 100 °C for 24 h, and then trituration to obtain sulfonated SiO_2 (SSD). A casting film containing 10 wt% SSD, 89.7 wt% polypropylene (S801, Korea Petrochemical Ind. Co., Ltd., Shanghai, China), and 0.3 wt% β-crystal nucleating agent (NAB83, GCH Technology Co. Ltd., Guangzhou, China) was prepared based on the double-screw extruder (die temperature: 210 °C, casting temperature: 125 °C). Then, biaxial drawing was applied on casting films based on the Bruckner KARO IV biaxial stretcher (Bruckner Group, Siegsdorf, Germany) to prepare a porous PP@SSD separator, which contained the longitudinal drawing along the MD (drawing ratio: 300%, drawing temperature: 100 °C) and sequential drawing vertical to the MD (namely TD, drawing ratio: 300%, drawing temperature: 120 °C). For comparison, untreated SiO_2 (SD) and β-crystal polypropylene were composited to, respectively, produce pure PP and PP@SD separator.

2.2. Tests and Characterizations

The crystal morphologies of the casting films and the porous skeleton morphologies of the separators were observed by FEI Inspect F scanning electron microscopy (SEM). Before testing, casting films were first etched by the mixed acid (H_3PO_4:H_2SO_4 = 1:1, volume ratio, the concentrations of H_2SO_4 and H_3PO_4 used in this research are 85% and 98.3%) containing 1.5 wt% $KMnO_4$ for 24 h to remove the amorphous portion. The melting behaviors of casting films and separators are recorded by Mettler Toledo DSC3+ Differential scanning calorimetry (DSC, 25–210 °C, 20 °C/min). The DSC crystallinity ($X_{C,DSC}$) could be obtained based on the ratio of the sample melting enthalpy to that of 100% crystalline polypropylene. The DSC β-crystal content ($K_{\beta,DSC}$) was calculated by the ratio of the β-crystal crystallinity to $X_{C,\,DSC}$. The thermal stability of casting films was assessed on the Q-500 thermogravimetric (TG) analysis (30–800 °C, 15 °C/min). A D8 ADVANCE X-ray diffractometer (XRD) was used to gain XRD spectra. The XRD crystallinity ($X_{C,XRD}$) was obtained by calculating the diffraction area fractions of the crystalline phase. The XRD β-crystal content ($K_{\beta,XRD}$) was the area ratio of the β-phase to crystalline regions.

Separator porosity was acquired by $(V_T - V_S - V_P)/V_T$, where V_T is the total volume of tested separators, and V_S and V_P are volumes occupied by SSD and polypropylene. Gurley values were measured on the 4110N Gurley tester. The contact angle was tested on the KRUSS K100 (KRÜSS Scientific Instruments Co., Ltd., Hamburg, Germany) contact angle meter. Electrolyte uptake was calculated by $(W_a - W_i)/W_i$, where W_i is the initial separator weight and W_a is the separator weight after immersion in ethylene carbonate (EC): diethyl carbonate (DEC) = 1:1 (mass ratio) for 6 h. Electrolyte retention was obtained by $(W_r - W_i)/(W_a - W_i)$, where W_r is the weight of the separator placed in a sealed electronic scale at various times. Tensile and puncture properties were measured by a universal testing machine fitted with a heating chamber. The thermal shrinkage conditions of separators can be calculated by A_c/A_i, where A_i is the initial area of the separators, and A_c is the contractive separator area after being subjected to various temperatures for 30 min (measured by open source software "Image J", version number: 1.52).

A Reference 3000 (Gamry Instruments, Warminster, PA, USA) electrochemical workstation was used to measure separator electrochemical performance. The electrochemical stability window was recorded by the linear sweep voltammetry (LSV) of the Li-stainless steel (SS) cell (2–6 V, 5 mV/s). The alternating-current (AC) impedance of the SS/SS cell (10^6–10^{-2} Hz, 10 mV) was recorded to obtain bulk impedance (R_b). Ionic conductivity (σ) was thus measured by $\sigma = T/(R_b \cdot S)$, where T and S are the thickness and area of separators. Ion migration activation energy (E_a) was measured by $\sigma = \sigma_0 \cdot \exp(-E_a/RT)$, where σ_0, R, and T are the pre-factor, thermodynamic gas constant, and temperatures. Interface impedance (R_i) was assessed by the AC impedance of the Li/Li cell. Li^+ transfer number (t_+) was managed by $t_+ = I_s(\Delta V - I_iR_i)/I_i\ (\Delta V - I_sR_s)$, where ΔV is the chronoamperometry step potential (10 mV), R_s and I_s are the steady impedance and current on chronoamperometry plots, and I_i is the initial current of chronoamperometry. A CT3001A LAND (Wuhan LAND Electronic Co. Ltd., Wuhan, China) system was used to evaluate the C-rate discharge capacity retention and long-term effective cycling life of $LiCoO_2$/Li cells assembled with three separators (constant temperature conditions at 26 °C, voltage range: 2.75–4.25 V). The C-rate capacities of $LiCoO_2$/Li cells were recorded under the charge current of 0.2 C and discharge conditions of 0.2, 0.5, 1, 2, 4, and 8 C, respectively. Cycling tests were carried out at a constant charge–discharge current of 0.5 C. After assembly, the $LiCoO_2$/Li cells were left to set for 12 h and following activation (cycled three times at 0.1 C). Customized $LiCoO_2$ cathodes with 95.7 wt% active material, SS foil, lithium foil, and electrolyte (1 M $LiPF_6$ dissolved in solvents composed of EC, DEC, and dimethyl carbonate (DMC)) were offered by Kejing Material Technology Co., Ltd., Shenzhen, China.

3. Results and Discussion

3.1. Properties of Casting Films

The crystal morphology SEM images of three casting films are listed in Figure 1a–c. Abundant β-lamellae with a loose arrangement can be detected for the pure PP casting film, without any compact α-phase features. The XRD spectra of the PP film in Figure 1e present $X_{C,XRD}$ of 59.6% and high $K_{\beta,XRD}$ of 96.8%, which indicates pure β-lamellae in the PP film. Melt-recrystallizations of metastable β-phase at elevated temperatures generate the primary β-phase endothermic peak at 153.3 °C and inapparent α-phase melting peak at 166.4 °C on DSC heating scans (Figure 1d). Since XRD spectra show almost 100% $K_{\beta,XRD}$ content, the lowest $K_{\beta,DSC}$ (80.3%) of the PP film indicates the inferior β-lamellae thermal stability. Normal SD and sulfonated SSD are uniformly distributed between β-lamellae in the PP@SD (Figure 1b) and PP@SSD (Figure 1c) casting films. Tagged diffraction signals of β-phase (300) and (301) lattice planes at 16.1° and 21.2° appear on XRD spectra, together with the high $K_{\beta,DSC}$ of 97% and invisible α-phase diffraction signals, manifesting the undeteriorated β-phase after SD and SSD additions. TG heating plots (Figure 1f) display similar initial decomposition temperatures of about 420 °C for three films but higher weight retentions for PP@SD (10.5%) and PP@SSD (9.7%). The similar melting points of about 153 °C and higher $K_{\beta,DSC}$ for PP@SD (88.6%) and PP@SSD (87.9%) on DSC heating plots signify a more stable β-phase. Furthermore, the PP film presents the wide full width at half maximum of the β-phase melting peak (FWHM, 8.4 °C). Meanwhile, the narrower FWHMs of PP@SD (5.8 °C) and PP@SSD (6.7 °C) reflect centralized β-lamellae distributions, which can homogenize the deformation manners of casting films and consequent porous structure. The above mitigatory deformations can be proved by the gently drawing curves of PP@SD and PP@SSD (Figure 1g, drawing rate: 100 mm/min, drawing temperature: 100 °C). Profiting from the reinforcement of SSD, PP@SSD shows a higher yield stress of 11.8 MPa. Also, numerous interfaces between β-lamellae and SSD improve interface distributions and thus reduce stress concentration during the drawing process, exhibiting a smaller softening stress drop of 1.1 MPa and higher neck retentions for PP@SSD (Figure 1h,i).

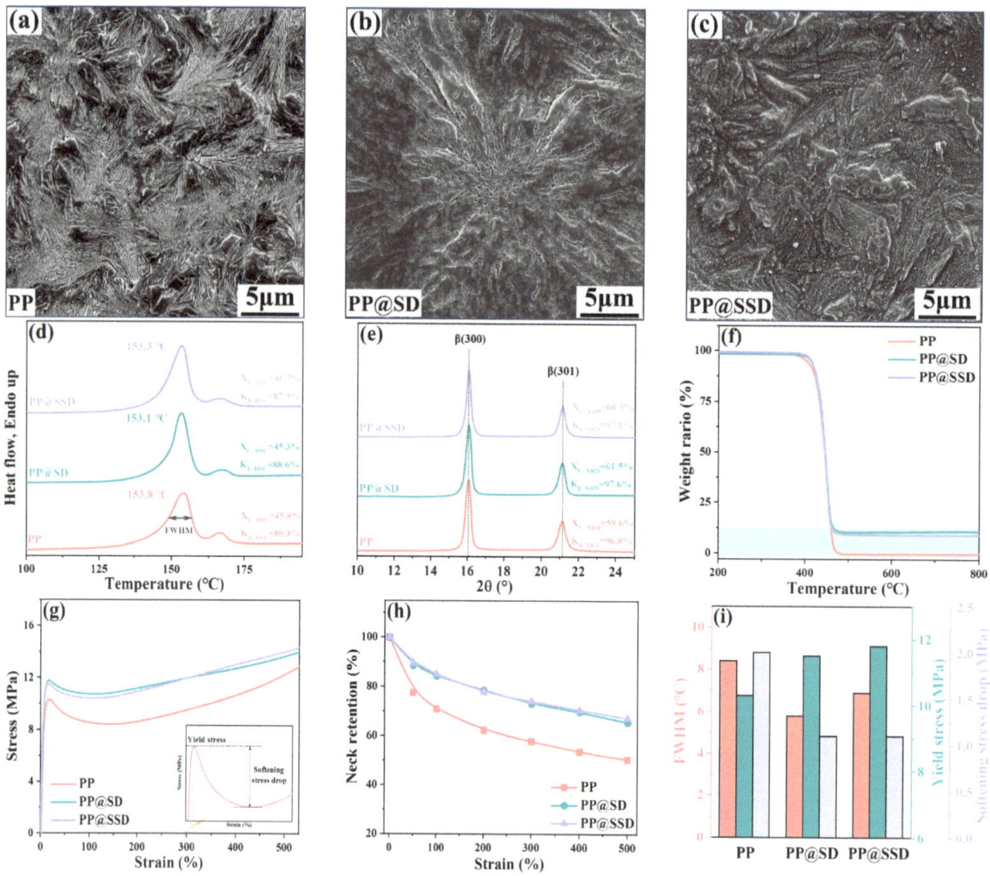

Figure 1. SEM images of three casting films: (**a**) PP, (**b**) PP@SD, and (**c**) PP@SSD. Key properties of casting films: (**d**) DSC heating scans, (**e**) XRD spectra, (**f**) TG plots, (**g**) drawing curves, (**h**) neck retention during drawing. (**i**) Crucial parameters of casting films (FWHM: full width at half maximum of β-phase melting peak, indicated in (**d**)).

3.2. Porous Structures

The separator morphologies (containing surface and cross-section) after the biaxial drawing are depicted in Figure 2a–c to verify the effects of the SSD on the porous structure. Pore size and fibril width distributions are listed in Figure 2d,e, accompanied by the porosity and Gurley value in Figure 2f. Longitudinal drawing compels β-lamellae along the TD to separate, while compacting β-lamellae along the MD to transform into dense α-phase fibrils. Subsequent transverse drawing can expand pore size but still retain numerous coarse fibrils on the surface. The above heterogeneous deformation modes thus generate obviously fractured fibrils and uneven pore size for the PP separator, which can be further proved by the broad pore size (mean size: 328 nm) and fibril width (mean width: 458 nm) distributions. Furthermore, the PP separator exhibits the typical multilayer stacked structure with a thickness of 18.8 μm due to the plane biaxial drawing. A normal porosity (42.5%) and a high Gurley value (312 s/100 mL) of the PP separator also signify the worse permeability. SD and SSD optimize interface distribution within casting films and relieve stress concentrations during drawing. Even if coarse fibrils emerge in the longitudinal drawing, the SD(SSD)/PP interfaces can be stripped again in the following transverse drawing, visibly narrowing fibril widths and concentrating pore size distribution.

Consequently, PP@SD and PP@SSD present uniform porous structures, together with similar mean pore sizes of 183 and 176 nm. The remarkable coarse fibrils cannot be detected on the surfaces of two separators. Average fibril widths also reduce to 236 and 243 nm, respectively. The higher porosity (PP@SD: 43.9%, PP@SSD: 43.8%) and lower Gurley value (PP@SD: 263 s/100 mL, PP@SSD: 261 s/100 mL) further certify the superior pore channel linearity, which can provide smoother paths for ion migrations.

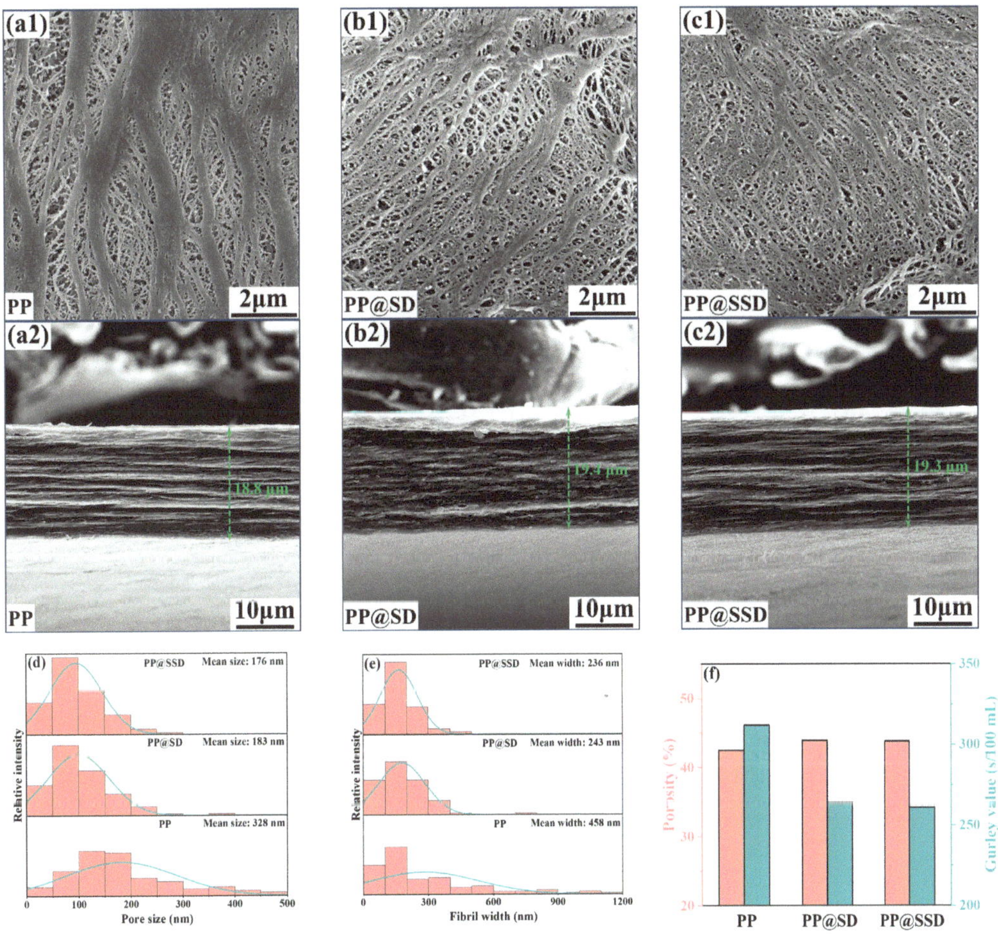

Figure 2. SEM images of three separators: (**a**) PP, (**b**) PP@SD, and (**c**) PP@SSD (1: surface morphology; 2: cross-section morphology). (**d**) Pore size distributions of separators. (**e**) Fibril width distributions of separators. (**f**) Porosity and Gurley value of separators.

3.3. Wettability, Thermal Stability, and Mechanical Properties

Better compatibility between separators and electrolytes improves electrolyte-filling processes into porous structures and reduces ion migration barriers [44–46]. The characteristic contact angle and electrolyte uptake are exhibited in Figure 3a. The PP separator shows a contact angle of 48.9° and electrolyte uptake of 88.3% because the low surface energy feature of polypropylene deteriorates electrolyte affinity. PP@SD and PP@SSD give superior electrolyte wettability owing to the SD and SSD coating on porous skeleton surfaces and thus reduce contact angles to 22.1° and 16.6°, maybe due to the transition from the Cassie–Baxter state to the Wenzel state. The stronger capillary intrusions further reinforce electrolyte uptake to 121.7% for PP@SD and reach a maximum of 149.6% for PP@SSD. Stronger capillary intrusions also endow PP@SSD with the fastest electrolyte absorption speed (Figure 3b), achieving electrolyte uptake of 133.1% after immersion for only 10 min. Furthermore, the highest electrolyte retention of PP@SSD (Figure 3c) demonstrates the best electrolyte retention capacity owing to the abundant sulfonic acid group.

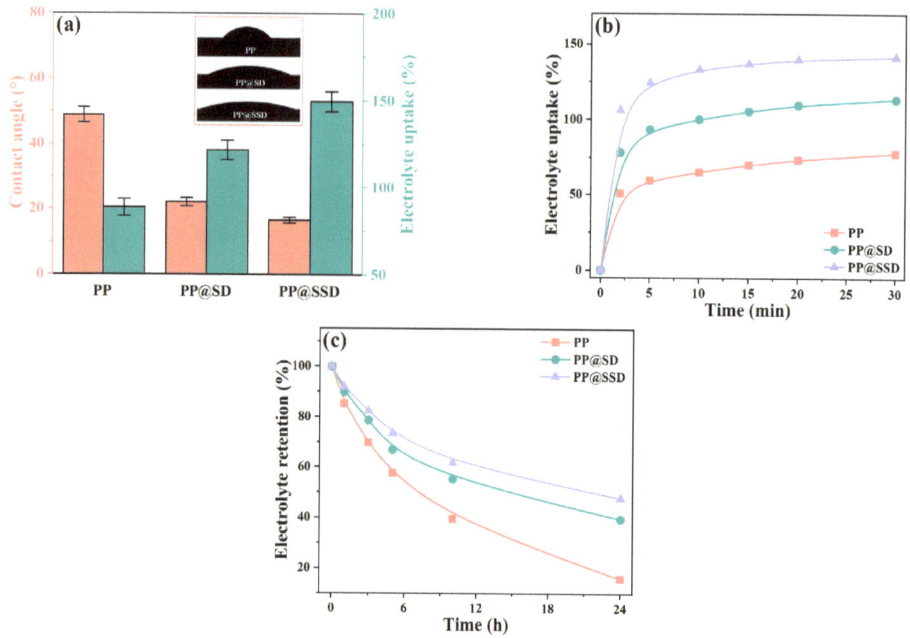

Figure 3. (**a**) Contact angle and electrolyte uptake of three separators. (**b**) Electrolyte uptake at various immersion times. (**c**) Electrolyte retention after immersion.

Separators encounter various conditions including tensile, coiling, compression, and puncture during LIB assembly and subsequent long-term operations [47,48]. Sufficient strengths are necessary for guaranteeing separator integrality. The tensile and puncture plots are listed in Figure 4a,b (drawing and puncture rate: 200 mm/min, temperature: 25 °C). The PP separator shows the drawing and puncture strength of 125.7 MPa and 276.5 g, which can cope with various encounters faced within LIBs. PP@SD and PP@SSD give a high drawing strength of 132.9 and 133.6 MPa owing to nanoparticle reinforcement effects. Meanwhile, the much higher puncture strength for PP@SD (360.8 g) and PP@SSD (377.2 g) can provide utilization potentiality for LIBs with high-security demands.

Stable thermal behaviors of separators act to prevent short circuits as LIBs are faced with thermal runaways. The DSC plots and thermal shrinkage states at various temperatures are displayed in Figure 4c,d. Since the metastable β-lamellae spontaneously translate into the α-phase under tensile and thermal stimuli, three separators show similar α-phase melting behaviors (PP: 169.1 °C, PP@SD: 169.4 °C, and PP@SSD: 169.2 °C), whereas the normal PP separator exhibits drastic thermal shrinkage of 68.7% at 165 °C. The contractions of PP@SD and PP@SSD are alleviated at elevated temperatures owing to the in situ SD and SSD coating on the porous skeleton surfaces, reaching 43.4% and 44.9% at 165 °C, respectively.

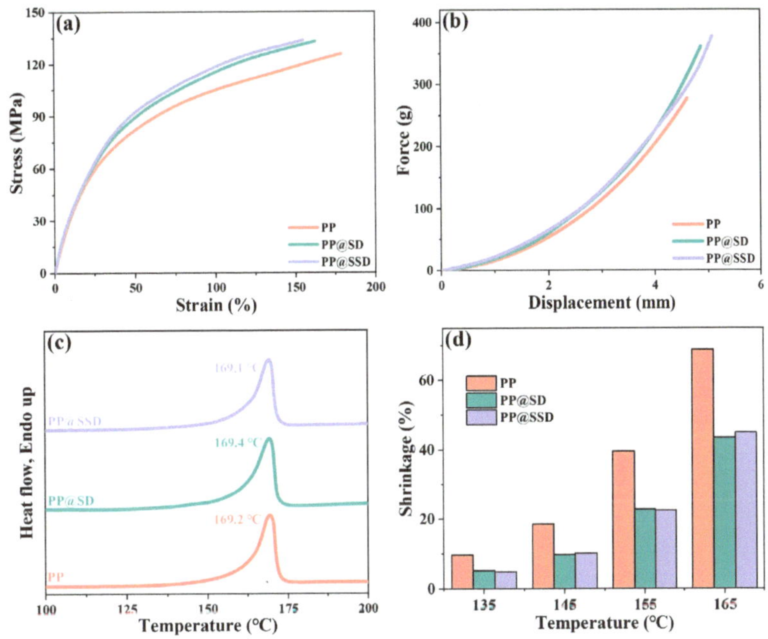

Figure 4. (**a**) Tensile and (**b**) puncture plots of separators. (**c**) DSC heating curves of separators. (**d**) Thermal shrinkage conditions at various temperatures.

3.4. Electrochemical Performance

The LSV curves of Li/SS cells containing separators are presented in Figure 5a to evaluate separator electrochemical stability windows. Three separators show placid voltage platforms within 4.6 V, followed by the subsequent current increases owing to electrolyte decompositions. The PP separator maintains stable voltages until 4.73 V. PP@SD shows a stable voltage limit of 4.93 V. The highest stable voltage of 5.08 V and minimum current uncover the superior electrolyte stabilizing for PP@SSD, which can endow separators with more adaptable capacity to high power density LIBs and even the expectant lithium-metal batteries.

Figure 5b shows the AC impedance curves of SS symmetric cells, accompanied by the calculated ionic conductivity (σ) in Figure 5d. The ions can only be compelled to transfer by using electrolytes as carriers owing to the insulating polypropylene substrate. σ is thus mainly governed by the intrinsic pore channel linearity of separators and electrolyte filling conditions within the porous structure. The PP separator exhibits an impedance of 3.54 Ω and σ of 0.47 mS/cm. The SD coating on the porous skeleton optimizes electrolyte affinity and permeability, which endow PP@SD with an impedance of 2.39 Ω and a higher σ of 0.72 mS/cm. Sulfonated SSD coating with an abundant sulfonic acid group owns the optimal hydrophily and thus improves σ to 0.94 mS/cm. Furthermore, in order to assess the ion removability facilitated by the separators, ion transfer activation energy (E_a) is further calculated by the Arrhenius formula in Figure 5c,d [49]. The high E_a of 9.21 kJ/mol for the PP separator signifies the high ion migration barriers. The PP@SD separator with SD coating gives a lower E_a of 7.18 kJ/mol, while the lowest E_a of 6.45 kJ/mol reveals the highest ion mobility within the PP@SSD separator with SSD coating.

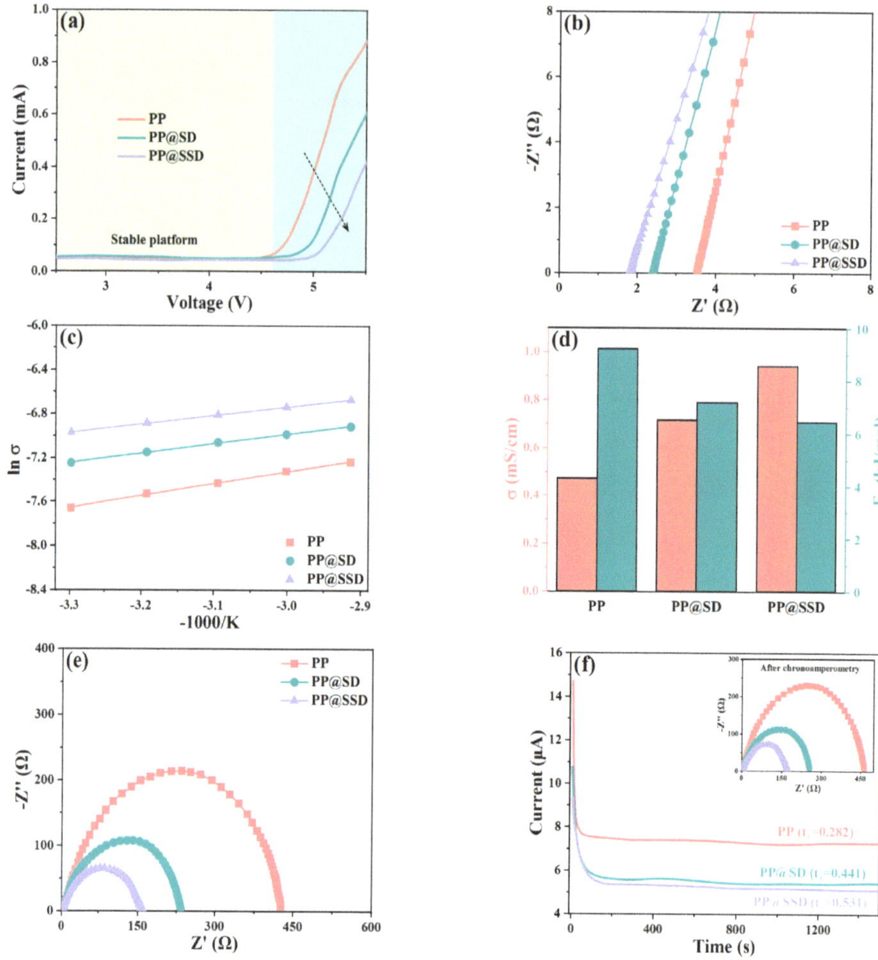

Figure 5. (**a**) LSV curves of Li/SS cells with separators. (**b**) AC impedance plots of SS symmetric cells. (**c**) Arrhenius plots as the temperature rises. (**d**) Ionic conductivity (σ) and activation energy (E_a). (**e**) AC impedance plots of Li symmetric cells. (**f**) Chronoamperometry curves and calculated Li^+ transfer number (t_+) of three separators.

The σ of separators represents the overall ion migration capacity within separators owing to the dual-ion conductions of LIBs, whereas the migration of anion inevitably generates electrode polarizations and sets off chain side reactions. Hence, the Li^+ transfer number (t_+) is obtained based on the chronoamperometry to evaluate the current proportion contributed by only Li^+ migrations [50]. The weak electrolyte affinity of polyolefin separators generates scant contact between porous channels and electrolytes and thus elongates ion migration routes in reverse. The PP separator with inferior wettability thus gives the low-level t_+ of 0.282. While superior electrolyte affinity can improve electrolyte filling and contact conditions within pore channels, the luxuriant hydroxyl groups on SD surfaces optimize electrolyte compatibility. Meanwhile, the lone pairs in hydroxyl electrostatically interact with Li^+, which accelerates Li^+ desolvation and raises free ion concentrations [51]. Furthermore, hydrogen bonds form between PF_6^- and hydroxyl, which hinders anion transfer within separators. Consequently, PP@SD exhibits the t_+ of 0.441. The highest t_+ of PP@SSD (0.531) indicates that sulfonic acid groups are significantly more advantageous in promoting Li^+ transport compared with hydroxyl, which is consistent with the calculated E_a.

The AC impedance curves of Li/Li cells are displayed in Figure 5e to compare the interfacial impedances (R_i) of three separators, which state the compatibility between the separator and Li electrode [52]. The high R_i value of 426 Ω for the PP separator reflects poor compatibility with Li electrodes. Better electrolyte uptake and retention enable sufficient contact with the Li electrode and improve more suitable interfaces, leading to the much lower R_i of 233 and 158 Ω for PP@SD and PP@SSD.

3.5. Battery Performance

The C-rate capacities of $LiCoO_2$/Li cells with three separators are shown in Figure 6a to testify new-type separator-determined battery combination performance. Cells containing the PP separator show a discharge capacity of about 139–140 mA h/g when cycling at 0.2 C for the first five charge–discharge processes. The limited ion migrations caused by ohmic polarization significantly reduce discharge capacities at the higher current densities, which decrease to 40.9 mA h/g when cycling once at 8 C and reach the minimum of 31.6 mA h/g only after the 5th cycle. Cells including PP@SD give discharge capacities of 140.3 and 139.9 mA h/g during the 1st and 5th cycles at 0.2 C but maintain higher capacities as the current density rises, dropping to 63.5 mA h/g for the 1st cycle at 8C and 57.1 mA h/g after the 5th cycle. The electrode active materials routinely determine the battery capacity. Separators between two electrodes can also affect Li^+ transfer channels and the separator/electrode interfacial process, which alters battery dynamics and comprehensive performance. The enhanced electrolyte affinity ensures superior uptake and retention conditions, thus wetting the electrode effectively and facilitating Li^+ insertions and removals. The optimized σ, E_a, and t_+ state the accelerated migration rate of Li^+ within separators and alleviative anti-anion polarization effects. Also, the superior compatibilities between separators and electrodes promote Li^+ diffusions through the electrode/separator interfaces. Consequently, PP@SSD cells show the optimal capacity retentions of 77.9 and 70.4 mA h/g for the 1st and 5th cycles at 8 C.

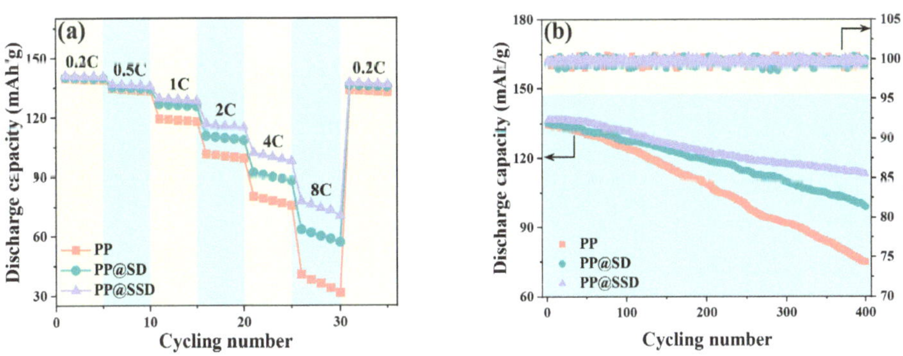

Figure 6. (**a**) C-rate capacity and (**b**) cycling stability of assembled $LiCoO_2$/Li cells with three separators.

Figure 6b displays the capacity retentions and corresponding coulomb efficiencies of $LiCoO_2$/Li cells when cycled at 0.5/0.5 C for 400 cycles to assess the effects of a coated separator on long-term charge–discharge behaviors of LIBs. Similar capacities of 134.5, 135.2, and 136.8 mA h/g of PP, PP@SD, and PP@SSD can be detected for the first cycle, whereas the discharge capacity gaps are highlighted as the cycle increases. The capacities of cells with PP achieve 124.7 mA h/g after the 100th cycle and quickly lower to 75.1 mA h/g (capacity retention: 55.8%) after the 400th cycle. Improved cycling stability emerges for the PP@SD cell, which retains a sluggish capacity decay of 127.5 mA h/g and 99.0 mA h/g after the 100th and 400th cycles. Especially, PP@SSD with sulfonated SSD coating can expedite Li^+ within cells to where it should migrate, leading to the high capacity during the cycling test and ultimately maintaining a capacity of 113.1 mA h/g (retention: 82.7%).

4. Conclusions

In this research, the novel PP@SSD composite separator SSD composited separators were designed based on the DPBD. This approach simultaneously homogenized pore size distributions and fixed to porous skeleton surfaces during PP@SSD separator fabrications, which facilely integrated the ceramic coating during separator fabrications and feature low production difficulty and cost. In addition, the robust SSD coating on porous skeleton surfaces provided superior thermal dimensional stability and adequate security at elevated temperatures. Numerous sulfonic acid groups of the SSD coating also endowed the PP@SSD composite separator with better electrolyte affinity, which lowered barriers for Li^+ transfer and optimized overall battery performance. This research focused on offering a facile separator manufacturing process, which combines the characteristics of low cost, high security, and high performance for the next generation of LIBs and the expected lithium-metal batteries.

Author Contributions: Conceptualization: L.D.; software: L.D., D.L., Y.Z. and F.D.; writing—original draft: L.D.; writing—reviewing and editing: D.L., S.Z. (Sihang Zhang) and F.Y.; data curation: D.L., Y.Z. and F.D.; formal analysis: S.Z. (Shuyue Zhao) and F.D.; funding acquisition: L.D.; supervision: S.Z. (Sihang Zhang) and F.Y.; project administration: S.Z. (Sihang Zhang). All authors have read and agreed to the published version of the manuscript.

Funding: We sincerely thank the Natural Science Foundation of Shandong Province (ZR2022QB050) and the Liaocheng University Doctoral Initial Fund (318052137) for financial support.

Institutional Review Board Statement: Not applicable.

Data Availability Statement: The original contributions presented in the study are included in the article, further inquiries can be directed to the corresponding author.

Conflicts of Interest: The authors declare no conflicts of interest.

References

1. Jiang, L.; Deng, Y.; Luo, T.; Xie, R.; Ju, X.; Wang, W.; Pan, D.; Liu, Z.; Chu, L. A smart membrane with negative thermo-responsiveness in battery electrolyte solution. *J. Membr. Sci.* **2024**, *692*, 122266. [CrossRef]
2. Guo, J.X.; Gao, C.; Liu, H.; Jiang, F.; Liu, Z.; Wang, T.; Ma, Y.; Zhong, Y.; He, J.; Zhu, Z.; et al. Inherent thermal-responsive strategies for safe lithium batteries. *J. Energy Chem.* **2023**, *89*, 519–534. [CrossRef]
3. Wei, Z.; Yu, L.; Lu, S.; Zhao, Y. Reversibly thermo-responsive materials applied in lithium batteries. *Energy Storage Mater.* **2023**, *61*, 102901. [CrossRef]
4. Chen, S.; Li, Y.; Feng, Y.; Feng, W. Thermally responsive polymers for overcoming thermal runaway in high-safety electrochemical storage devices. *Mater. Chem. Front.* **2023**, *7*, 1562–1590. [CrossRef]
5. Zhang, X.; Zhang, C.; Li, H.; Cao, Y.; Yang, H.; Ai, X. Reversible Temperature-Responsive Cathode for Thermal Protection of Lithium-Ion Batteries. *ACS Appl. Energy Mater.* **2022**, *5*, 5236–5244. [CrossRef]
6. Hu, X.; Li, Y.; Chen, Z.; Duan, C.; Yuan, B. Anchoring porous F-TiO_2 particles by directed-assembly on PMIA separators for enhancing safety and electrochemical performances of Li-ion batteries. *Electrochim. Acta* **2023**, *443*, 141926. [CrossRef]
7. Yang, Y.; Bai, Z.; Liu, S.; Zhu, Y.; Zheng, J.; Chen, G.; Huang, B. Self-Protecting Aqueous Lithium-Ion Batteries. *Small* **2022**, *18*, 2203035. [CrossRef]
8. Francis CF, J.; Kyratzis, I.L.; Best, A.S. Lithium-Ion Battery Separators for Ionic-Liquid Electrolytes: A Review. *Adv. Mater.* **2020**, *32*, 1904205. [CrossRef]
9. Saal, A.; Hagemann, T.; Schubert, U.S. Polymers for Battery Applications-Active Materials, Membranes, and Binders. *Adv. Energy Mater.* **2020**, *11*, 2001984. [CrossRef]
10. Liu, Z.; Jiang, Y.; Hu, Q.; Guo, S.; Yu, L.; Li, Q.; Liu, Q.; Hu, X. Safer Lithium-Ion Batteries from the Separator Aspect: Development and Future Perspectives. *Energy Environ. Mater.* **2020**, *4*, 336–362. [CrossRef]
11. Jang, J.; Oh, J.; Jeong, H.; Kang, W.; Jo, C. A Review of Functional Separators for Lithium Metal Battery Applications. *Materials* **2020**, *13*, 4625. [CrossRef] [PubMed]
12. Tang, C.; Bruening, M.L. Ion separations with membranes. *J. Polym. Sci.* **2020**, *58*, 2831–2856. [CrossRef]
13. Wang, Q.; Mao, B.; Stoliarov, S.I.; Sun, J. A review of lithium ion battery failure mechanisms and fire prevention strategies. *Prog. Energy Combust. Sci.* **2019**, *73*, 95–131. [CrossRef]

14. Costa, C.M.; Lee, Y.; Kim, J.; Lee, S.; Lanceros-Méndez, S. Recent advances on separator membranes for lithium-ion battery applications: From porous membranes to solid electrolytes. *Energy Storage Mater.* **2019**, *22*, 346–375.
15. Zhang, T.; Tian, T.; Shen, B.; Song, Y.; Yao, H. Recent advances on biopolymer fiber based membranes for lithium-ion battery separators. *Compos. Commun.* **2019**, *14*, 7–14. [CrossRef]
16. Rana, M.; Li, M.; Huang, X.; Luo, B.; Gentle, I.; Knibbe, R. Recent advances in separators to mitigate technical challenges associated with re-chargeable lithium sulfur batteries. *J. Mater. Chem. A* **2019**, *7*, 6596–6615. [CrossRef]
17. Asghar, M.R.; Anwar, M.T.; Naveed, A.; Zhang, J. A Review on Inorganic Nanoparticles Modified Composite Membranes for Lithium-Ion Batteries: Recent Progress and Prospects. *Membranes* **2019**, *9*, 78. [CrossRef]
18. Mauger, A.; Julien, C.M. Critical review on lithium-ion batteries: Are they safe? Sustainable? *Ionics* **2017**, *23*, 1933–1947. [CrossRef]
19. Zhang, S.S. A review on the separators of liquid electrolyte Li-ion batteries. *J. Power Sources* **2007**, *164*, 351–364. [CrossRef]
20. Dong, H.; Wang, P.; Yan, S.; Xia, Y.; Wang, B.; Wang, X.; Liu, K. A thermoresponsive composite separator loaded with paraffin@SiO_2 microparticles for safe and stable lithium batteries. *J. Energy Chem.* **2021**, *62*, 423–430. [CrossRef]
21. Liu, J.; Liu, Y.; Yang, W.; Ren, Q.; Li, F.; Huang, Z. Lithium ion battery separator with high performance and high safety enabled by tri-layered SiO_2@PI/m-PE/SiO_2@PI nanofiber composite membrane. *J. Power Sources* **2018**, *396*, 265–275. [CrossRef]
22. Chen, W.; Shi, L.; Wang, Z.; Zhu, J.; Yang, H.; Mao, X.; Chi, M.; Sun, L.; Yuan, S. Porous cellulose diacetate-SiO_2 composite coating on polyethylene separator for high-performance lithium-ion battery. *Carbohydr. Polym.* **2016**, *147*, 517–524. [CrossRef] [PubMed]
23. Jeon, K.S.; Nirmala, R.; Navamathavan, R.; Kim, K.J.; Chae, S.H.; Kim, T.W.; Kim, H.Y.; Park, S.J. The study of efficiency of Al_2O_3 drop coated electrospun meta-aramid nanofibers as separating membrane in lithium-ion secondary batteries. *Mater. Lett.* **2014**, *132*, 384–388. [CrossRef]
24. Ding, L.; Yan, N.; Zhang, S.; Xu, R.; Wu, T.; Yang, F.; Cao, Y.; Xiang, M. Low-Cost Mass Manufacturing Technique for the Shutdown-Functionalized Lithium-Ion Battery Separator Based on Al_2O_3 Coating Online Construction during the β-iPP Cavitation Process. *ACS Appl. Mater. Interfaces* **2022**, *14*, 6714–6728. [CrossRef]
25. Liu, L.; Wang, Y.; Gao, C.; Yang, C.; Wang, K.; Li, H.; Gu, H. Ultrathin ZrO_2-coated separators based on surface sol-gel process for advanced lithium ion batteries. *J. Membr. Sci.* **2019**, *592*, 117368. [CrossRef]
26. Luo, X.; Liao, Y.; Zhu, Y.; Li, M.; Chen, F.; Huang, Q.; Li, W. Investigation of nano-CeO_2 contents on the properties of polymer ceramic separator for high voltage lithium ion batteries. *J. Power Sources* **2017**, *348*, 229–238. [CrossRef]
27. Ding, L.; Yan, N.; Zhang, S.; Xu, R.; Wu, T.; Yang, F.; Cao, Y.; Xiang, M. Low-Cost and Large-Scale Fabricating Technology for High-Performance Lithium-Ion Battery Composite Separators with Connected Nano-Al_2O_3 Coating. *ACS Appl. Energy Mater.* **2022**, *5*, 615–626. [CrossRef]
28. Ding, L.; Yan, N.; Zhang, S.; Xu, R.; Wu, T.; Yang, F.; Cao, Y.; Xiang, M. Facile manufacture technique for lithium-ion batteries composite separator via online construction of fumed SiO_2 coating. *Mater. Des.* **2022**, *215*, 110476. [CrossRef]
29. Ding, L.; Li, D.; Liu, L.; Zhang, P.; Du, F.; Wang, C.; Zhang, D.; Zhang, S.; Zhang, S.; Yang, F. Dependence of lithium metal battery performances on inherent separator porous structure regulation. *J. Energy Chem.* **2023**, *84*, 436–447. [CrossRef]
30. Xu, H.; Usseglio-Viretta, F.; Kench, S.; Cooper, S.J.; Finegan, D.P. Microstructure reconstruction of battery polymer separators by fusing 2D and 3D image data for transport property analysis. *J. Power Sources* **2020**, *480*, 229101. [CrossRef]
31. Zhong, Y.; Zhong, L.; Wang, S.; Qin, J.; Han, D.; Ren, S.; Xiao, M.; Sun, L.; Meng, Y. Ultrahigh Li-ion conductive single-ion polymer electrolyte containing fluorinated polysulfonamide for quasi-solid-state Li-ion batteries. *J. Mater. Chem. A* **2019**, *7*, 24251–24261. [CrossRef]
32. Suharto, Y.; Lee, Y.; Yu, J.; Choi, W.; Kim, K.J. Microporous ceramic coated separators with superior wettability for enhancing the electrochemical performance of sodium-ion batteries. *J. Power Sources* **2018**, *376*, 184–190. [CrossRef]
33. Lin, Y.; Li, X.; Meng, L.; Chen, X.; Lv, F.; Zhang, Q.; Zhang, R.; Li, L. Structural Evolution of Hard-Elastic Isotactic Polypropylene Film during Uniaxial Tensile Deformation: The Effect of Temperature. *Macromolecules* **2018**, *51*, 2690–2705. [CrossRef]
34. Lin, Y.; Meng, L.; Wu, L.; Li, X.; Chen, X.; Zhang, Q.; Zhang, R.; Zhang, W.; Li, L. A semi-quantitative deformation model for pore formation in isotactic polypropylene microporous membrane. *Polymer* **2015**, *80*, 214–227. [CrossRef]
35. Yoneda, H.; Nishimura, Y.; Doi, Y.; Fukuda, M.; Kohno, M. Development of microporous PE films to improve lithium ion batteries. *Polym. J.* **2010**, *42*, 425–437. [CrossRef]
36. Ihm, D.; Noh, J.; Kim, J. Effect of polymer blending and drawing conditions on properties of polyethylene separator prepared for Li-ion secondary battery. *J. Power Sources* **2002**, *109*, 388–393. [CrossRef]
37. Ding, L.; Yan, N.; Zhang, S.; Xu, R.; Wu, T.; Yang, F.; Cao, Y.; Xiang, M. Separator impregnated with polyvinyl alcohol to simultaneously improve electrochemical performances and compression resistance. *Electrochim. Acta* **2022**, *403*, 139568. [CrossRef]
38. Yang, F.; Yang, F.; Xiang, M.; Wu, T. Preparation of highly oriented β polypropylene and its pore formation mechanism during stretching. *Polym. Cryst.* **2021**, *4*, e10183. [CrossRef]
39. Habumugisha, J.C.; Feng, S.; Iqbal, O.; Lin, Y.; An, M.; Meng, L.; Wang, D.; Chen, W.; Li, L. Stretch-induced structural evolution of pre-oriented isotactic polypropylene films: An in-situ synchrotron radiation SAXS/WAXS study. *Polymer* **2021**, *214*, 123234. [CrossRef]
40. Yang, Y.; Wu, G.G.; Ding, C.; Zhang, Y.; Yang, W.; Yang, M.B. Formation of oriented β-transcrystals induced by self-assembly of nucleating agent and its micropores formation during uniaxial stretching. *Polym. Cryst.* **2020**, *3*, e10129. [CrossRef]
41. Yang, X.; Tuinea Bobe, C.; Whiteside, B.; Coates, P.; Lu, Y.; Men, Y. Molecular weight dependency of β phase formation in injection-molded isotactic polypropylene. *J. Appl. Polym. Sci.* **2020**, *137*, 48555. [CrossRef]

42. Zhang, D.; Ding, L.; Yang, F.; Lan, F.; Cao, Y.; Xiang, M. Effect of annealing on the microvoid formation and evolution during biaxial stretching of β nucleated isotactic polypropylene. *Polym. Plast. Technol. Mater.* **2020**, *59*, 1595–1607. [CrossRef]
43. Ding, L.; Xu, R.; Pu, L.; Yang, F.; Wu, T.; Xiang, M. Pore formation and evolution mechanism during biaxial stretching of β-iPP used for lithium-ion batteries separator. *Mater. Des.* **2019**, *179*, 107880. [CrossRef]
44. Wu, D.; He, J.; Zhang, M.; Ni, P.; Li, X.; Hu, J. Fabrication of a novel sandwich-like composite separator with enhanced physical and electrochemical performances for lithium-ion battery. *J. Power Sources* **2015**, *290*, 53–60. [CrossRef]
45. Wang, Y.; Wang, S.; Fang, J.; Ding, L.; Wang, H. A nano-silica modified polyimide nanofiber separator with enhanced thermal and wetting properties for high safety lithium-ion batteries. *J. Membr. Sci.* **2017**, *537*, 248–254. [CrossRef]
46. Shi, J.; Fang, L.; Li, H.; Zhang, H.; Zhu, B.; Zhu, L. Improved thermal and electrochemical performances of PMMA modified PE separator skeleton prepared via dopamine-initiated ATRP for lithium ion batteries. *J. Membr. Sci.* **2013**, *437*, 160–168. [CrossRef]
47. Guo, Z.; Wu, L.; Wang, Y.; Zhu, Y.; Wan, G.; Li, R.; Zhang, Y.; Qian, D.; Wang, Y.; Zhou, X.; et al. Design of Dendritic Large-Pore Mesoporous Silica Nanoparticles with Controlled Structure and Formation Mechanism in Dual-Templating Strategy. *ACS Appl. Mater. Interfaces* **2020**, *12*, 18823–18832. [CrossRef]
48. Lagadec, M.F.; Zahn, R.; Wood, V. Designing Polyolefin Separators to Minimize the Impact of Local Compressive Stresses on Lithium Ion Battery Performance. *J. Electrochem. Soc.* **2018**, *165*, A1829–A1836. [CrossRef]
49. Liao, H.; Zhang, H.; Hong, H.; Li, Z.; Qin, G.; Zhu, H.; Lin, Y. Novel cellulose aerogel coated on polypropylene separators as gel polymer electrolyte with high ionic conductivity for lithium-ion batteries. *J. Membr. Sci.* **2016**, *514*, 332–339. [CrossRef]
50. Kim, J.; Lee, D.; Jung, H.; Sun, Y.; Hassoun, J.; Scrosati, B. Lithium-Sulfur Batteries: An Advanced Lithium-Sulfur Battery. *Adv. Funct. Mater.* **2013**, *23*, 1092. [CrossRef]
51. Wang, Z.; Guo, F.; Chen, C.; Shi, L.; Yuan, S.; Sun, L.; Zhu, J. Self-Assembly of PEI/SiO_2 on Polyethylene Separators for Li-Ion Batteries with Enhanced Rate Capability. *ACS Appl. Mater. Interfaces* **2015**, *7*, 3314–3322. [CrossRef] [PubMed]
52. Schaefer, J.L.; Yanga, D.A.; Archer, L.A. High Lithium Transference Number Electrolytes via Creation of 3-Dimensional, Charged, Nanoporous Networks from Dense Functionalized Nanoparticle Composites. *Chem. Mater.* **2013**, *25*, 834–839. [CrossRef]

polymers

Article

Characterization and Performance Evaluation of Liquid Biodegradable Mulch Films and Its Effects on Peanut Cultivation

Jie Shi [1], Shaoli Wang [1], Zhongxue Yang [2], Baoyan Li [1], Ruijue Chen [2], Fanzhi Bu [2], Binghui Luan [1], Baoyou Liu [1,*] and Peiqiang Li [2,*]

1 Yantai Academy of Agricultural Sciences, Yantai 265500, China; zhibaoshijie@163.com (J.S.); shaoliwang123@126.com (S.W.); byli1314@163.com (B.L.); lbh815@163.com (B.L.)
2 College of Chemistry and Material Science, Shandong Agricultural University, 61 Daizong Road, Taian 271018, China; zxyang@sdau.edu.cn (Z.Y.); rjchen6310@163.com (R.C.); 17662540427@163.com (F.B.)
* Correspondence: ytnkyzbs@yt.shandong.cn (B.L.); pqli@sdau.edu.cn (P.L.)

Abstract: With the development of material science and increasing awareness of ecological environmental protection, liquid biodegradable mulch films (LBDMs) have garnered significant public interest. In this research, new LBDMs were developed using hydrophobically modified polymer materials, surfactants, and photosensitive catalysts. Characterization by scanning electron microscopy (SEM) revealed good material compatibility. LBDMs exhibited excellent wettability and degradability, effectively covering soil surfaces and enhancing soil moisture conservation, with a degradation rate of 76.09% after 80 days of burial. The field performance experiment was conducted over two consecutive years, 2021 and 2022, to assess differences in soil temperature and moisture, peanut agronomic traits, pod traits, and yield under four treatments: non-mulching (CK), LBDMs, clear polyethylene mulch films (CPEMs), and black polyethylene mulch films (BPEMs). LBDMs increased soil temperature by 0.56 °C and soil moisture by 19.25%, accelerated the seedling stage by 4-to-6 days, and improved the average emergence rate by 15.91%. Furthermore, LBDMs significantly promoted peanut growth, and it increased yield by 14.34% compared to CK. LBDMs performed comparably to the two types of PE films in maintaining soil conditions and different crop phenotype traits, including plant height, branch number, yield, and quality, and they even outperformed PE films in productivity per plant and 100-kernel weight. These findings suggest that LBDMs are a promising eco-friendly alternative to traditional PE films.

Keywords: liquid biodegradable mulch films; characterization; performance; soil conditions; peanut growth; peanut yield

check for updates

Citation: Shi, J.; Wang, S.; Yang, Z.; Li, B.; Chen, R.; Bu, F.; Luan, B.; Liu, B.; Li, P. Characterization and Performance Evaluation of Liquid Biodegradable Mulch Films and Its Effects on Peanut Cultivation. *Polymers* **2024**, *16*, 2487. https://doi.org/10.3390/polym16172487

Academic Editor: Beom Soo Kim

Received: 15 July 2024
Revised: 19 August 2024
Accepted: 27 August 2024
Published: 31 August 2024

1. Introduction

The peanut (*Arachis hypogaea* L.) is a significant grain and oil crop in China, with its yield ranking first globally [1]. For over half a century, polyethylene (PE) mulch films have been widely used in agriculture to enhance peanut productivity [2]. Film mulching can elevate soil temperature, maintain soil moisture, suppress weed growth, reduce the occurrence and spread of phytopathogens, and enhance crop growth and yield [2–5]. However, the extensive use of PE mulch films has led to severe agricultural non-point source pollution with long-lasting ecological impacts [6,7]. Residual plastic fragments in the soil degrade the structure of the plow layer, impede water and fertilizer transport, hinder soil microorganism activity, and eventually cause soil compaction, which negatively affects crop growth [8,9]. To promote sustainable agricultural practices, biodegradable films have emerged as a research focus. These films offer similar warming and moisture conservation benefits as conventional PE films and often surpass PE films in improving soil properties and crop growth [10,11]. Biodegradable films naturally degrade through microbial action and finally break down into CO_2 and H_2O [12,13]. Therefore, biodegradable films represent a promising alternative to traditional PE films in agricultural ecosystems [14,15].

In recent years, researchers have proposed the feasibility of using LBDMs as substitutes for traditional plastic films. Various research institutions, both in China and abroad, have studied and applied LBDMs [16–19]. These films are emulsion suspensions with organic polymers as the main carbon skeleton, significantly reducing labor intensity and improving efficiency compared to the manual application of plastic films [17]. LBDMs exhibit excellent wettability, forming a multi-molecular network of gelatinous film after spraying on the soil surface [20]. Spraying LBDMs binds soil particles together, forming an aggregate structure that effectively preserves soil temperature and moisture, while reducing water evaporation without impeding water infiltration [21]. And then, it promotes the growth and yield of crops [22,23]. Additionally, integrating water-soluble fertilizers and pesticides beneficial to crops into the LBDM system creates a multifunctional film, further enhancing its value by reducing labor intensity through combined applications [4,7]. However, existing LBDMs are typically composed of chemical polymers with poor degradability or highly hydrophilic materials such as humic acid, starch, cellulose, ethyl cellulose, and polyglutamic acid, or simply mixed with these materials [5]. These compositions are either not environmentally friendly or fail to form a durable film, resulting in a poor performance characterized by inelasticity, fragility, and susceptibility to rainfall erosion, with a relatively short effective duration [24–27]. Currently, the research and development of LBDMs are still in the small-scale experimental stage, with significant challenges remaining before large-scale adoption can be achieved. Therefore, developing a high-performance, production-suitable liquid biodegradable mulch film has become an urgent need in this field.

In this study, newly developed LBDMs were prepared using hydrophobically modified polymer materials, surfactants, and photosensitive catalysts. These films exist in liquid form and can be directly sprayed onto the soil to form a biodegradable film. The use of hydrophobically modified polymer materials enhances the film's erosion resistance and improves soil water retention. To determine whether LBDMs can meet the varying soil condition requirements at different crop growth stages, we characterized their performance and selected peanuts as a model crop to evaluate their effects on crop growth and yield. The objective of this study is to verify the superior performance of LBDMs, including their wettability, degradability, temperature regulation, water retention capabilities, etc. Additionally, we aim to explore the impact of LBDMs on the physiological traits, quality, and yield of peanuts in the Yantai area, guiding the application of LBDMs in Yantai and surrounding regions.

2. Materials and Methods

2.1. Experimental Site

The field experiments were conducted at the experimental plot of Yantai Academy of Agricultural Sciences (37°29′ N, 121°16′ E), Yantai, Shandong Province, China, in 2021 and 2022. The soil was loam, and the plot was flat with medium and uniform fertility. The contents of hydrolyzable nitrogen, available potassium, and available phosphorus at 0~20 cm depth were 32.61~39.03 mg·kg^{-1}, 183.69~188.53 mg·kg^{-1}, and 111.34~126.96 mg·kg^{-1}, respectively. The content of organic matter was 1.68~1.83%, and the soil pH value was 6.40~6.66. In the study area, the climate was the temperate monsoon with four distinctive seasons, sufficient sunshine, and moderate rainfall with a mean annual temperature ranging from 12.7 °C to 13.0 °C. The annual rainfall was from 830.6 mm to 989.9 mm, of which 70~90% fell in a major part of the growing season between June and September. The rainfall and the air temperature during the experimental period were measured using an automatic weather station (RS-ECTH-N01-TR temperature and humidity sensors, Jinan, China; ZQZ-A automatic weather station, Beijing, China) at the experimental site.

2.2. Field Experimental Design and Treatments

The big-fruit-type peanut variety 'Huayu 22', provided by Shandong Peanut Research Institute and approved by the Shandong Province Crop Variety Approval Committee in February 2003, was selected in this study. This variety is an early-maturing ordinary peanut, with high quality, high yield potential, high stress resistance, moderate disease resistance, and a 130-day growth period [28].

Four treatments were designed and applied: (1) peanut cultivation mulched with liquid biodegradable mulch films (LBDMs), (2) peanut cultivation mulched with black polyethylene mulch films (BPEMs), (3) peanut cultivation mulched with clear polyethylene mulch films (CPEMs), and (4) peanut cultivation with no mulching as the control (CK). Each treatment was replicated three times, a total of 12 plots, and each plot area measured 42 m^2 (21 m × 2 m) in a randomized block arrangement. Each plot contained 2 ridges, and 2 rows were planted in each ridge. The plant spacing was 20 cm, the ridge length was 21 m, and 2 seeds were sown in each hole. A 0.8 m wide border was set between each plot for field management and sampling activities. All the polyethylene (PE) film mulches were 0.01 mm thick and 130 cm wide (Yantai Changsheng Plastic Factory, Yantai, China). With PE film mulching, the films were used flat to cover the surface of the ridges, where the film edges were covered carefully and compacted with soil. LBDMs were sprayed evenly on the soil surface, with the help of the knapsack sprayer, and the spraying range was 1 m. The seed cultivar 'Huayu 22' was sown at a rate of 420 per plot using a handheld hole-sowing machine, with a sowing depth of 4~5 cm. The seeding and film mulching were conducted on 3 May and 8 May, and the peanuts were harvested on 13 September and 16 September, in 2021 and 2022, respectively.

Herbicides were applied before sowing, and weeds were manually controlled during the crop growth period. Before the experiment, deep plowing to a depth of approximately 20 cm was performed using a tractor-mounted moldboard plow. No additional tillage or irrigation was conducted during the entire experimental period.

2.3. Preparation and Characterization of LBDMs Mulching Films

2.3.1. Preparation of LBDMs

A certain amount of polymer materials chitin (0.72 wt%), polycarbonate (1.52 wt%), alkyl glucoside (0.06 wt%), coconut oil based glucoside (0.06 wt%), cellulose acetate (0.85 wt%), and carboxymethyl cellulose (0.85 wt%) were weighed and mixed with water at a mass ratio (solid/liquid) of 1:2 [18,29,30]. They were heated in water to 50~60 °C and dispersed. Carboxylic acids, epoxy compounds, halogenated hydrocarbons, aliphatic acyl chloride, and isocyanate were added as hydrophobic reagents (0.68 wt%) for hydrophobic modification, and amine compounds (0.34 wt%) were used as an end capping reagent to obtain hydrophobically modified polymer materials [31–33]. Then, the above-mentioned hydrophobically modified polymer material was weighed, and the surfactant and photosensitive catalyst were added and stirred with water to obtain a gelatinous viscous transparent liquid film. The composition ratio of surfactant, photo-sensitive catalyst, and hydrophobically modified polymer material was 0.05:0.02:1. The mass ratio of the above surfactants, including calcium dodecyl benzene sulfonate and styrene phenol polyoxyvinyl ether, is 1.5:1. The photosensitive catalyst was metal porphyrin [34]. The prepared mixture was poured into culture plates and dried under natural conditions, and then the films were uncovered and reserved as spares.

2.3.2. Characterization of LBDMs

The surface morphology and microstructure of the LBDMs sample films were observed using a scanning electron microscope (SEM) (Phenom Pure, The Netherlands) at various magnifications (Mag = 500×, 1500×, 3000×, and 13,000×). Before testing, the LBDM samples were dried in a 60 °C drying oven for 12 h to form thin-film samples. These samples were then affixed to the test bench with a conductive tape. The samples were sputter-coated with gold and subsequently imaged using the SEM with an accelerating voltage of 10 kV.

2.3.3. LBDMs Wettability

To evaluate the wettability of LBDMs, the contact angle between LBDMs and soil surface was measured by a simplified sessile drop method at room environment (25 ± 2 °C, 50% RH) [35,36]. Four treatments were set up according to the soil granularity size in this experiment, and an LBDMs droplet with a volume of 15 μL was deposited onto the soil surface. The contact angle and the complete infiltration time of each treatment were recorded, and subsequently the wettability of LBDMs was characterized under a microscope. In addition, the soil collected from the field was placed in a petri dish, and the liquid film was sprayed evenly on the surface of the soil. The treatment without spraying liquid film was used as the control. After the film was formed by natural drying, a certain amount of water was sprayed on the surface of the soil, so that the soil was completely immersed in water to simulate the natural rainwater soaking process. After seven days, the effect of water immersion on the stability of the liquid film was observed.

2.3.4. LBDMs Degradation

The degradable property of LBDMs was investigated by the soil burial test method [37]. After drying the LBDMs to form solid films, three kinds of films (LBDMs, CPEMs, and BPEMs) were cut into a square of 3 cm × 3 cm, buried in the soil at a depth of 10 cm, and maintained by 30–40% soil moisture. Within 80 days after burial, the films were taken out every 10 days, and the degradation situation was observed. The surface morphology of the sample films after soil burial degradation was analyzed using the SEM characterization method described in Section 2.3.2. In addition, to determine the degradation rate of LBDMs, the film samples, before being buried with soil, were weighed, and then they were taken out every 10 days, washed with distilled water, dried at 40 °C for 6 h, and weighed again. Each treatment was repeated three times to take the average value, and the mass loss was obtained. The degradation rate of LBDMs was calculated according to the following formula:

$$\text{Degradation rate (\%)} = (\text{mass before degradation} - \text{mass after degradation})/\text{mass before degradation} \times 100$$

2.4. Determination of Soil Temperature and Moisture

After sowing, the soil temperature and moisture of the different treatments at 10 cm depth were measured with the 485-type soil temperature and moisture sensor (RS-ECTH-N01-TR; Shandong Renke Measurement and Control Technology Co., Ltd., Jinan, China), and the temperature and moisture were recorded every 2 h to see the detailed changes until the end of the whole growth period.

2.5. Peanut Growth and Developmental Progress and Yield

2.5.1. Seedling Emergence

The seedling stage and emergence rates for the different treatments were recorded after sowing. During this period, the number of successfully emerged seedlings in all test plots was recorded daily. The seedling emergence rate was calculated by the formula as follows. When the seeding emergence rate reached 50%, that day was determined as the seeding stage.

$$\text{Seedling emergence rate (\%)} = (\text{the number of successful seedling emergence}/\text{the number of total seeds}) \times 100$$

2.5.2. Growth Parameters of Peanuts

Each growth stage of the peanut was recorded. During the seedling stage, anthesis stage, pod bearing stage, and harvesting stage, the values of plant height were recorded every 10 days until they were seriously lodging and could not be measured. Fifteen peanut plants were continuously selected from the middle ridge of each test plot (a total of forty-five plants for every treatment) to record the total branch number and lateral branch length.

2.5.3. Determination of Peanut Pod Traits

To determine peanut pod traits, fifteen peanut plants were continuously selected from the middle ridge of each replicate (repeat three times, a total of forty-five plants for every treatment), and all peanut pods were collected from selected plants. Total pod number, full pod number, immature pod number, total kernel number, germinated kernel number, single-kernel pod number, and double-kernel pod number were recorded. The full pod rate, immature pod rate, kernel rate, single kernel rate, double kernel rate, and germinated kernel rate were calculated for each treatment.

2.5.4. Peanut Yield

During the harvest period, 2 sample points were randomly selected from each plot, in which both ends were removed, and a 1 m^2 area with uniform growth of each sample point was harvested to determine the final pod yield. The pod yield for each experimental plot was determined by fresh weight and was used to calculate yield per hectare. 'Huayu 22' groundnut is usually harvested at about 130 days in the eastern part of Shandong, and the final harvest date was determined based on visual observations of leaf senescence and peanut kernel maturity. At the same time, fifteen peanut plants were continuously selected from the middle ridge of each replicate (repeat three times, a total of forty-five plants for every treatment) to count the productivity per plant, 100-kernel weight, and 100-pod weight. Among them, the determination of single plant productivity required the mature pods to be fully sundried and weighed to calculate the average weight of the pods per plant.

2.6. Statistical Analyses

The effects of the treatments on the measured parameters were evaluated using one-way ANOVA from the SAS package, and the least significant difference (LSD) was used to compare means [38]. In all cases, differences were deemed to be significant if $p < 0.05$. Data graphs were generated using GraphPad Prism 8.0.2 software.

3. Results and Discussion

3.1. Weather Conditions

The weather conditions, including rainfall and air temperature, varied between the two growing seasons of peanuts (Figure 1). From May to September in 2021, air temperatures ranged from 9.9 °C to 28.8 °C, with a total rainfall of 530.2 mm and a mean monthly rainfall of 106.0 mm. The highest rainfall occurred in August. In contrast, during the same period in 2022, temperatures ranged from 11.8 °C to 31.5 °C, with a total rainfall of 788.2 mm and a mean monthly rainfall of 157.6 mm. The maximum rainfall was recorded in September. These differences in rainfall and air temperature between the two years were expected to influence peanut development and result in yield variations.

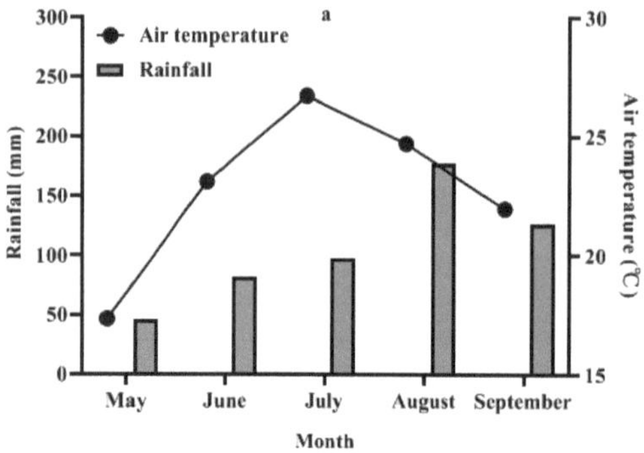

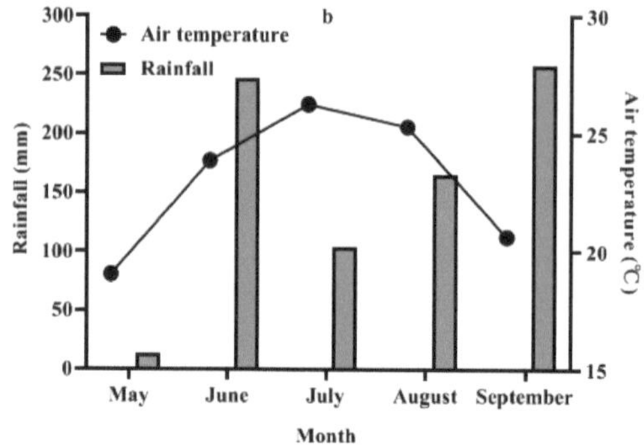

Figure 1. Monthly air temperature and rainfall distribution were monitored throughout the experimental periods in 2021 (**a**) and 2022 (**b**).

3.2. LBDMs Surface Morphologies, Wettability, and Degradability

3.2.1. LBDMs Surface Morphologies

To visually assess the surface morphology and dispersion of LBDMs, the dried films were observed using scanning electron microscopy (SEM) at different magnifications (500×, 1500×, 3000×, and 13,000×), as shown in Figure 2. The low-magnification images (Figure 2A) revealed a smooth, compact, and uniform surface with the transparency of plastic, free from noticeable holes, folds, or cracks, indicating good compatibility among the materials used in the preparation of LBDMs. The higher-magnification images (Figure 2B–D) displayed small particles on the film surface, likely resulting from the agglomeration of polymer materials forming small, aggregated structures. Additionally, during the drying process, small molecular substances may have volatilized first, with macromolecular substances continuously precipitating and depositing on the film surface.

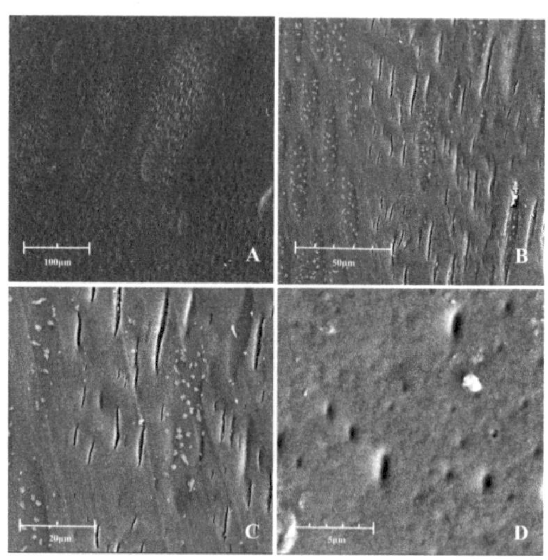

Figure 2. SEM images of the LBDMs surface morphology at different magnifications. (**A**) Mag = 500×. (**B**) Mag = 1500×. (**C**) Mag = 3000×. (**D**) Mag = 13,000×.

3.2.2. Wettability Analysis of LBDMs

To analyze the wettability of LBDMs, contact angle tests were conducted. The contact angles of LBDM droplets on soil surfaces with different particle sizes are shown in Figure 3.

As the soil particle size increased, the contact angle also increased ($\theta e_I = 37.5°$, $\theta e_{II} = 52.0°$, $\theta e_{III} = 90.0°$, $\theta e_{IV} = 116.2°$), and the complete infiltration time of the film lengthened ($T_I = 24$ s, $T_{II} = 6$ s, $T_{III} = 4$ s, $T_{IV} = 3$ s). These results indicated a good wettability effect on soil surfaces of varying particle sizes. It was also observed in the experiment that, as the soil particle size became smaller, the soil surface layer after spraying LBDMs was more likely to form a film layer. This phenomenon confirmed that, under uniform soil flatness, reducing soil granularity and enhancing the binding force between soil aggregates made it easier to form a continuous film on the soil surface, significantly improving the film-forming effect.

Figure 3. The wettability of LBDMs on soil surfaces with different particle sizes. (I) Grinding soil after passing through a 100-mesh sieve. (II) Soil passing through a 100-mesh sieve. (III) Conventional soil with large particles removed. (IV) Untreated conventional soil. (SL) Microscopic images of the soil surface sprayed with LBDMs (mag = 40×). (NO) Microscopic images of the soil surface without LBDMs application (mag = 40×).

LBDMs demonstrated excellent wettability, effectively covering soil surfaces with various particle sizes, bonding soil particles, and forming film-covered soil surface layers (Figure 4C). This good film-forming property is crucial for its warming and moisturizing effects. When examining the impact of simulated rainwater immersion on LBDMs' stability (Figure 4B), many dry cracks of varying depths were observed on the surface of soil without LBDMs mulching. In contrast, the soil surface with LBDM mulching had a relatively uniform structure with only a few small cracks (Figure 4A). Soil cracks result from rapid water evaporation. Compared to the untreated soil sample, fewer cracks appeared on the LBDM-covered soil surface after rain immersion due to the protective layer formed by LBDMs. This layer slows water evaporation, reduces the agglomeration and shrinkage of soil aggregates, and thus minimizes the occurrence of surface cracks. These results further indicate that LBDMs effectively enhance the bonding force between surface soil aggregates, maintain soil granular structure stability, prevent soil moisture evaporation, and improve soil water stability and moisture conservation.

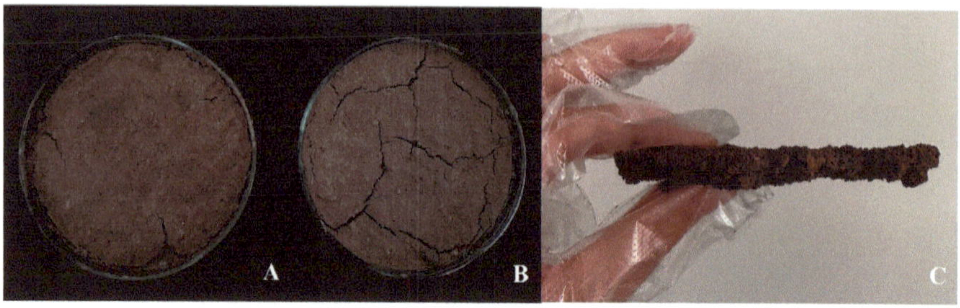

Figure 4. The film-forming effect of LBDMs on the soil surface. (**A**) Comparison of the soil surface with LBDM mulching after 7 days of water soaking. (**B**) Comparison of the soil surface without LBDM mulching after 7 days of water soaking. (**C**) Digital images of the soil cross-section after spraying LBDM mulching film.

3.2.3. Degradation Performance of LBDMs

The degradation of the film in the natural environment is influenced by seasonal and location-specific factors, such as humidity, temperature, sunlight, and microorganisms [39]. These factors compromise the film's internal structure, leading to a loss of rigidity and toughness. Some researchers consider the weight loss method as a quantitative indicator of the degradation performance of films, with an increased degradation rate over time proving the degradability of LBDMs [40]. As shown in Figure 5, LBDMs softened and thinned after ten days of burial, with observable rupture and degradation. The LBDMs lost their original appearance and structural integrity during the degradation process, developed many holes and cracks, and gradually decomposed into small fragments. SEM images revealed that the surface morphology of the degraded film sample became wrinkled and uneven, with visible fragmented structures protruding from the surface (Figure 6). We also found that the degradation rate increased over time (Figure 6). After 50 days of burial in the soil, the film's degradation rate exceeded 50%, classifying it as a biodegradable material [41]. Specifically, the degradation rate of LBDMs was 61.33% after 50 days and reached 76.09% after 80 days of burial, indicating that LBDMs have good degradability. These results demonstrated that LBDMs were one kind of biodegradable and environmentally friendly material.

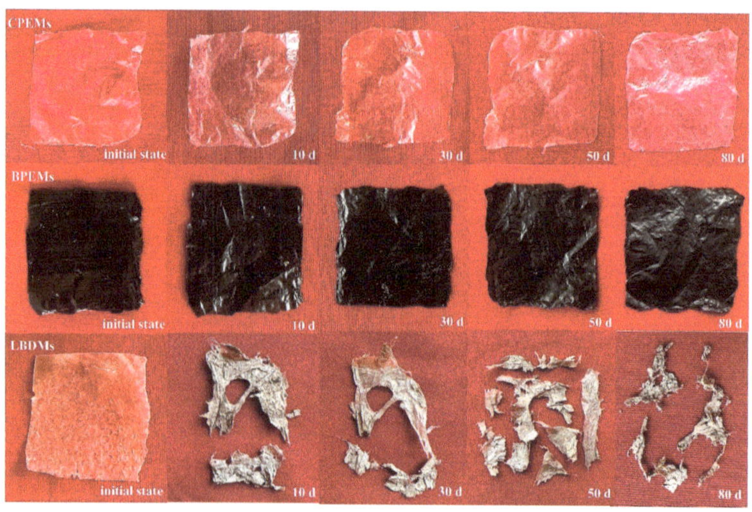

Figure 5. The surface morphology of CPEMs, BPEMs, and LBDMs after degradation at different soil burial times.

Figure 6. SEM images of the LBDMs' surface morphology (mag = 13,000×) and the correlation between the degradation rate of LBDMs and soil burial duration. BD = before degradation (0 days after burial); AD = after degradation (80 days after burial).

3.3. Soil Temperature and Moisture

Film mulching significantly impacts the soil microenvironment, with soil temperature and moisture levels affecting crop yield [3]. Warmer soil and higher soil moisture content are positively related to peanut-seedling emergence and phenological development, canopy formation, radiation use efficiency, and pod yield. From sowing to harvest, the trend of soil temperature and moisture during the experimental period is shown in Figure 7. Compared to the control (CK), LBDMs demonstrated a more remarkable warming effect, raising the average temperature over two years by 0.56 °C. However, this warming effect was less pronounced than that of CPEMs and BPEMs, which increased temperatures by 1.21 °C and 0.72 °C, respectively. These findings align with the results of Sun et al. [42] and Sartore et al. [43]. In the later stages of crop growth, the difference in soil temperature among the treatments diminished compared to the earlier stages. This may be attributed to the crop canopy affecting the soil's heat absorption from solar radiation, thus influencing soil temperature. Several studies have shown that the warming effect of mulching decreased in the later growth stages, as the plant canopy became fully established, narrowing the soil temperature gap between mulching and non-mulching treatments [44,45].

Soil moisture is a critical physical property of soil, playing a vital role in crop growth. Film mulching forms a barrier between the soil and the atmosphere, preventing soil moisture loss [10,46]. During the 2021 and 2022 growth stages, significant differences in soil moisture at a 10 cm depth were observed under different mulching conditions. LBDMs exhibited better soil moisture retention than CK, increasing average humidity by 19.25%. In comparison, CPEMs increased average humidity by 20.09%, while BPEMs showed the highest moisture retention, with a 35.75% increase over CK. This difference from soil temperature trends can be explained by the fact that increased soil temperature not only accelerates surface moisture loss but also promotes root development, enhances crop aboveground growth, and increases leaf transpiration. Excessive transpiration is not conducive to soil moisture retention [47]. Additionally, Chen et al. [48] found that higher soil moisture increased heat capacity and slowed temperature rise. This study could confirm the above point that LBDMs and BPEMs retained soil moisture better than CPEMs, though their soil temperature preservation was not as effective as that of CPEMs.

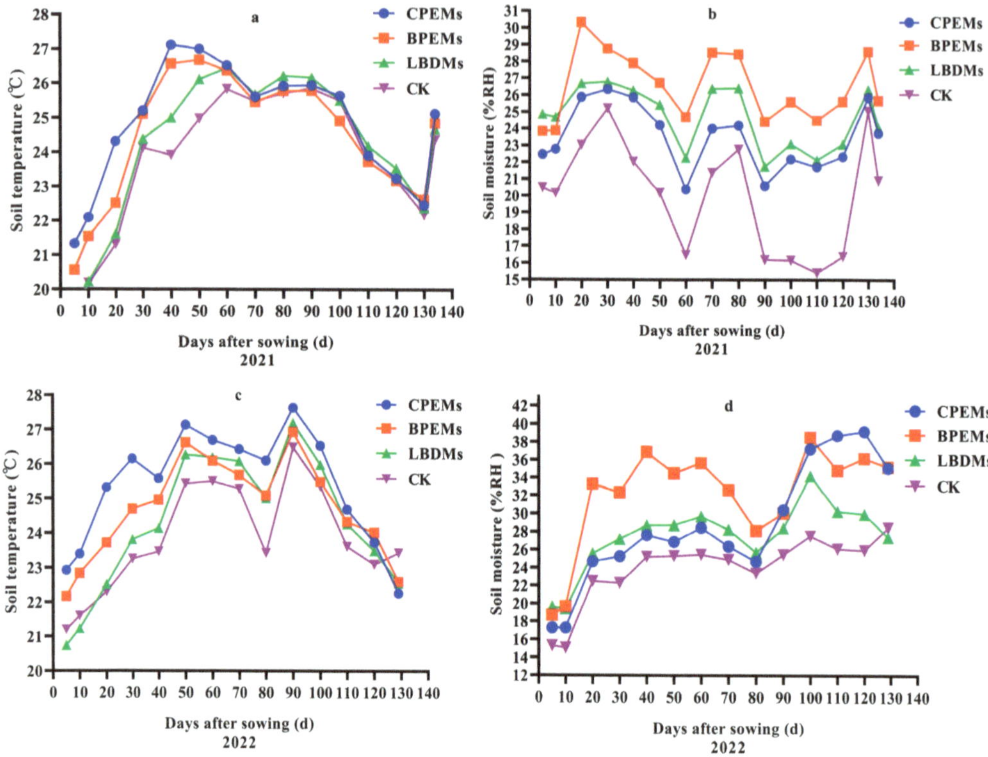

Figure 7. Soil temperature (**a**,**c**) and moisture (**b**,**d**) in the 0–10 cm layer under CPEMs, BPEMs, LBDMs, and CK treatments during the growing seasons of 2021 and 2022.

3.4. Effects of mulching films on Peanut Seedling Emergence

The seedling emergence rate is a crucial factor in determining yield. We recorded daily seedling emergence data. As shown in Table 1, all mulching treatments promoted peanut emergence and increased the emergence rate. Among them, BPEMs and CPEMs had the most significant effect, followed by LBDMs. In both 2021 and 2022, compared to CK, the LBDMs treatment advanced the peanut seedling stage by 4-to-6 days and effectively increased the emergence rate by 16.88% and 14.93%, respectively. It was found that increased soil temperature could improve seed germination and emergence [49]. At the seedling stage, the plant canopy was small, which allowed most of the film-mulched area to receive solar energy and the soil temperature to warm up. In addition, the water underneath the film could reduce the longwave radiation, which reduces the rate of decrease in soil temperature at night. Therefore, the diurnal temperature fluctuation in this stage involved faster warming up of mulched than un-mulched soil during the day and slower cooling at night, producing a mini-greenhouse effect.

Table 1. Seedling emergence analysis in 2021 and 2022.

Year	2021			2022		
Treatments	Seedling Stage		Emergence Rate (%)	Seedling Stage		Emergence Rate (%)
	Date	DAS		Date	DAS	
CPEMs	15 May	12	87.66 ± 3.52 b	19 May	11	89.52 ± 3.29 a
BPEMs	15 May	12	91.35 ± 2.66 a	18 May	10	92.17 ± 4.29 a

Table 1. *Cont.*

Year	2021			2022		
Treatments	Seedling Stage		Emergence Rate (%)	Seedling Stage		Emergence Rate (%)
	Date	DAS		Date	DAS	
LBDMs	19 May	16	82.59 ± 2.28 c	23 May	15	81.08 ± 3.10 b
CK	25 May	22	65.71 ± 2.36 d	27 May	19	66.15 ± 3.69 c

Note: Data are presented as mean ± SE, $n = 3$. Values followed by different lowercase letters in the same column are significantly different among treatments at 0.05 level for the same factor. Seedling stage: the time required for 50% of seedlings to emerge and expand their first true leaf; DAS, days after sowing.

3.5. Effects of Mulching Films on Peanut Growth Parameters

To analyze the agronomic traits of peanuts, we recorded the plant height, lateral branch length, and number of branches. Based on two years of field data, the plant height of peanuts under different treatments showed a gradual increase over time (Figure 8). Compared to CK, mulching with LBDMs significantly increased plant height throughout the growth period. Additionally, the lateral branch length and number of branches mulched with LBDMs resulted in longer lateral branch lengths and more branches than CK and comparable to or even better than CPEMs and BPEMs treatments (Figure 9).

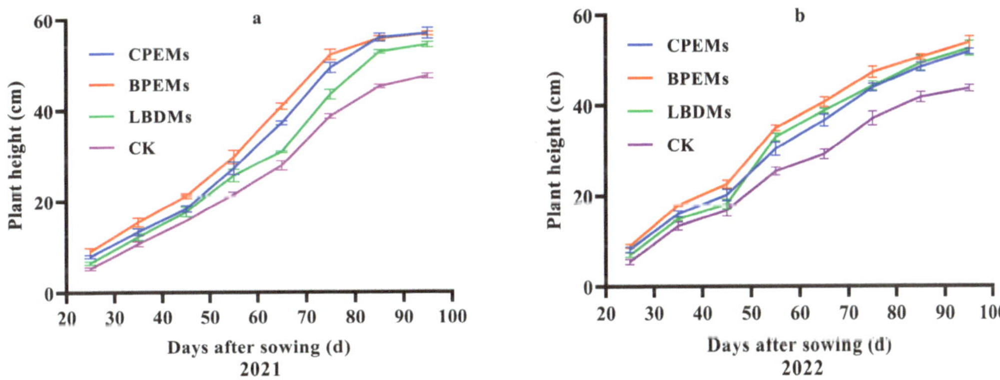

Figure 8. The trend in peanut plant height under CPEM, BPEM, LBDM, and CK treatments in 2021 (**a**) and 2022 (**b**).

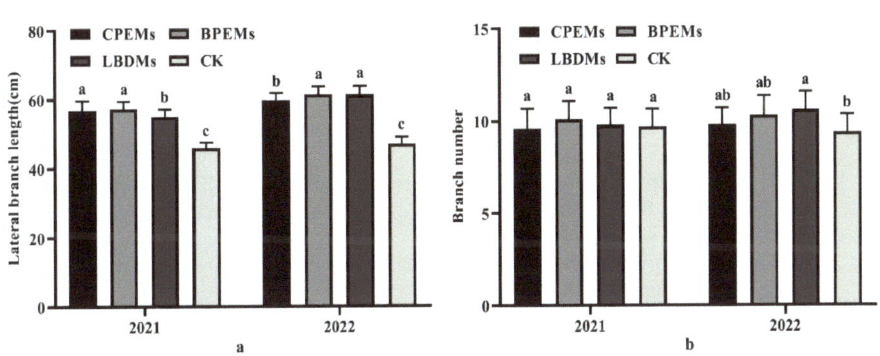

Figure 9. Lateral branch length (**a**) and branch number (**b**) of peanut plants at harvest period under CPEM, BPEM, LBDM, and CK treatments in 2021 and 2022. Values followed by different lowercase letters in the same column are significantly different among treatments at 0.05 level for the same factor.

In 2021, the average length of lateral branches mulched with LBDMs was 55.15 cm, which was 9.2 cm longer than CK, but 1.8 cm and 2.2 cm shorter than CPEMs and BPEMs, respectively (Figure 9a). The mean number of branches for LBDM treatment was 9.8, which is not significantly different from CK, CPEMs, and BPEMs ($p < 0.05$, Figure 9b). Similar results were observed in 2022, where the average length of lateral branches mulched with LBDMs was 61.35 cm, 14.3 cm longer than CK, and 1.55 cm longer than CPEMs, with no significant differences from BPEMs ($p < 0.05$, Figure 9a). The mean number of branches for LBDMs treatment in 2022 was 10.6, which was 1.2 more than that of CK and not significantly different from that of CPEMs and BPEMs ($p < 0.05$, Figure 9b). Favorable soil temperature plays a crucial role in optimal plant growth. These findings confirm that LBDM mulching promotes the early growth of peanut plants, resulting in faster seedling emergence and better overall plant growth. This phenomenon can be attributed to film mulching, which alters soil temperature and moisture to conditions more suitable for seedling emergence and plant growth. This conclusion aligns with the reports of Wang et al. [50] and Sun et al. [42].

3.6. Effects of Mulching Films on Peanut Pod Traits

To assess the effects of LBDM mulching treatments on peanut pod quality, we recorded the full pod rate, immature pod rate, kernel rate, single kernel rate, double kernel rate, and germinated kernel rate. The results in Table 2 show that the full pod rate, kernel rate, and germinated kernel rate for LBDM mulching treatments were significantly higher than those for CK ($p < 0.05$), while the immature pod rate was lower. Compared to the CPEM and BPEM treatments, the germination rate of LBDMs was lower, and the full pod rate and kernel rate were equivalent to that of CPEMs, equivalent to or slightly worse than that of BPEMs. In addition, there were no significant differences among the four treatments regarding the single kernel rate and double kernel rate.

Table 2. Effect of mulching on peanut pod quality in 2021 and 2022.

	Full Pod Rate (%)	Immature Pod Rate (%)	Kernel Rate (%)	Single Kernel Rate (%)	Double Kernel Rate (%)	Germinated Kernel Rate (%)
Year				2021		
CPEMs	72.31 ± 0.96 b	25.24 ± 1.18 b	74.02 ± 1.42 ab	21.27 ± 1.40 a	77.59 ± 1.85 a	6.96 ± 0.92 b
BPEMs	76.66 ± 1.75 a	21.06 ± 0.92 c	75.37 ± 0.86 a	20.87 ± 0.83 a	78.29 ± 0.97 a	10.88 ± 1.30 a
LBDMs	72.73 ± 1.13 b	24.97 ± 0.64 b	73.40 ± 1.00 b	21.37 ± 0.89 a	77.06 ± 1.38 a	4.02 ± 0.58 c
CK	66.69 ± 1.37 c	30.57 ± 1.59 a	73.32 ± 1.06 b	21.68 ± 1.22 a	77.74 ± 1.25 a	1.18 ± 0.62 d
Year				2022		
CPEMs	66.02 ± 2.23 bc	31.13 ± 2.65 ab	81.32 ± 1.54 a	24.91 ± 1.72 a	72.99 ± 1.82 a	3.59 ± 0.28 ab
BPEMs	73.41 ± 3.36 a	25.11 ± 2.68 b	80.59 ± 4.67 a	20.98 ± 4.18 a	74.48 ± 4.74 a	8.38 ± 2.63 a
LBDMs	67.67 ± 2.58 b	29.37 ± 3.91 ab	83.09 ± 7.86 a	22.61 ± 1.01 a	75.93 ± 0.29 a	0.74 ± 0.82 c
CK	60.44 ± 3.20 c	36.52 ± 4.23 a	70.21 ± 2.76 b	26.34 ± 3.68 a	71.59 ± 4.85 a	1.25 ± 1.44 bc

Data are presented as mean ± SE, $n = 3$. Values followed by different lowercase letters in the same column indicate significant differences among treatments at the 0.05 level for the same factor.

In 2021, the full pod rate of peanuts under the LBDM treatment was 72.73%, which was 6.04% higher than that of CK, 3.93% lower than that of BPEMs, and not significantly different from that of CPEMs ($p < 0.05$). The immature pod rate for LBDMs was 24.97%, 5.60% lower than that of CK, 3.91% higher than that of BPEMs, and not significantly different from that of CPEMs ($p < 0.05$). The kernel rate for LBDMs was 73.40%, 1.97% lower than that of BPEMs, and not significantly different from that of CK and CPEMs ($p < 0.05$). The single and double kernel rates for LBDMs were 21.37% and 77.06%, respectively, with no significant differences from the other treatments ($p < 0.05$). The germinated kernel rate for LBDMs was 4.02%, 2.84% higher than that of CK, and 2.97% and 6.86% lower than that of CPEMs and BPEMs, respectively. The results for 2022 showed some differences from 2021. The full pod rate under LBDM treatment was 67.67%, 7.23% higher than that of

CK, 5.74% lower than that of BPEMs, and not significantly different from that of CPEMs ($p < 0.05$). The immature pod rate for LBDMs was 29.37%, with no significant differences from the other treatments ($p < 0.05$). The kernel rate for LBDMs was 83.09%, 12.88% higher than that of CK, and not significantly different from that of CPEMs and BPEMs ($p < 0.05$). The single and double kernel rates for LBDMs were 22.61% and 75.93%, respectively, with no significant differences from the other treatments ($p < 0.05$). The germinated kernel rate for LBDMs was 0.74%, 7.64%, and 2.85% lower than that if BPEMs and CPEMs, respectively, and not significantly different from CK ($p < 0.05$).

3.7. Effects of Mulching Films on Peanut Yield

To explore the effect of LBDMs treatment on yield, we studied productivity per plant, 100-pod weight, and 100-kernel weight. Over two years, we recorded peanut yield and confirmed that mulching with LBDMs could significantly affect yield. The data in Table 3 indicate that mulching with LBDMs, CPEMs, and BPEMs significantly increased pod yield compared to CK.

Table 3. Effect of mulching on peanut yield and yield components in 2021 and 2022.

	Pod Yield (kg·ha^{-1})	Yield Increase Compared to CK (%)	Productivity per Plant (g)	100-Pod Weight (g)	100-Kernel Weight (g)
Year			2021		
CPEMs	4975.00 ± 131.05 ab	10.19	18.43 ± 0.49 ab	238.63 ± 5.47 ab	97.24 ± 2.55 a
BPEMs	5175.00 ± 119.06 a	14.62	20.37 ± 0.97 a	247.33 ± 2.25 a	97.15 ± 1.82 a
LBDMs	4895.00 ± 129.33 bc	8.42	19.38 ± 1.04 a	245.00 ± 1.80 a	96.80 ± 1.04 a
CK	4515.00 ± 83.52 c	-	17.08 ± 0.96 b	234.00 ± 5.00 b	96.00 ± 0.98 a
Year			2022		
CPEMs	4965.00 ± 70.53 b	14.27	18.22 ± 1.07 b	239.99 ± 2.54 bc	89.10 ± 2.32 b
BPEMs	5360.00 ± 106.42 a	23.36	21.24 ± 0.41 a	246.45 ± 3.65 a	96.23 ± 3.29 a
LBDMs	5225.00 ± 92.60 a	20.25	20.67 ± 0.61 a	242.08 ± 0.68 ab	95.47 ± 1.14 a
CK	4345.00 ± 102.10 c	-	16.93 ± 0.60 b	235.53 ± 1.60 c	88.92 ± 4.20 b

Data are presented as mean ± SE, $n = 3$. Values followed by different lowercase letters in the same column indicate significant differences among treatments at the 0.05 level for the same factor.

In 2021, the yield increase rate for the LBDM mulching treatment was 8.42%, which was 2.49% and 6.2% lower than that of CPEMs and BPEMs, respectively. In 2022, the yield increase rate for LBDMs was 20.25%, 5.98% higher than that of CPEMs and 3.11% lower than that of BPEMs. Overall, the average yield increase rate over the two years of LBDMs mulching treatment was significantly better than that of CPEMs but not as good as that of BPEMs. In 2021, the productivity per plant and 100-pod weight for LBDMs were 19.38 g and 245.00 g, respectively, which were 2.30 g and 11.00 g higher than CK's values but not significantly different from the values of CPEMs and BPEMs ($p < 0.05$). There was no significant difference in 100-kernel weight among the four treatments ($p < 0.05$). In 2022, the productivity per plant for LBDMs was 20.67 g, which was 3.74 g and 2.45 g higher than that of CK and CPEMs, respectively, but not significantly different from that of BPEMs ($p < 0.05$). The 100-pod weight for LBDMs was 242.08 g, which was 6.55 g higher than that of CK, with no significant differences from that of CPEMs and BPEMs ($p < 0.05$). The 100-kernel weight for LBDMs was 95.47 g, which was 6.55 g and 6.37 g higher than that of CK and CPEMs, respectively, but not significantly different from that of BPEMs ($p < 0.05$).

The pod yield in the mulched treatments, including the LBDM-mulched treatments, was higher than that in CK. Similar results were reported by Waterer [51]. Film mulching increases soil temperature by several degrees, promoting better growth during the early growth period and more water absorption in the later period. Notably, LBDM mulching improved plant height, lateral branch length, number of branches, and dry matter accumulation in individual plants and increased productivity per plant, full pod rate, kernel rate, 100-pod weight, and 100-kernel weight, leading to an increased pod yield. This result aligns with Song's report [22]. Compared with CK, the pod yield, productivity per plant, 100-pod weight, 100-kernel weight, full pod rate, and kernel rate of LBDMs were better, but it did not affect pod traits such as the single kernel rate. This may be due to film mulching promoting peanut flower bud differentiation, increasing the number of effective flowers and needles, thereby promoting an increase in the number of pods per plant and increasing peanut yield [52]. Additionally, film mulching prevents pegs developing during later growth stages from entering the soil, thus conserving nutrients for developing pods set earlier, increasing the number of full pods, and reducing the number of immature pods [53,54].

The productivity per plant, 100-pod weight, and 100-kernel weight of LBDMs treatment was significantly better than that of CK and equivalent to or even better than PE mulching treatments. However, in terms of pod yield, the LBDM treatment's result was significantly higher than that of the CK but lower than that of the BPEM, possibly due to the emergence rate. This finding is consistent with the research of Kunzova et al. [55], which highlighted that seedling emergence and establishment are key processes in grain yield determination.

4. Conclusions

Overall, this study underscores the potential of LBDMs as environmentally friendly alternatives to traditional PE films. LBDMs demonstrated good degradability and wettability and can be simply sprayed onto the soil to form a film layer to serve as agriculture mulching, which is highly facile and efficient as compared with traditional PE films. It optimizes the soil environment, promoting peanut seed germination and root development, supporting healthy plant growth, and laying the foundation for increased peanut yield. These improvements were comparable to those achieved with ordinary PE films and even outperformed in regard to some aspects of crop growth and yield. In general, LBDMs not only provide the heat- and moisture-retention benefits of PE films, advancing the seedling stage, increasing the emergence rate, promoting crop growth and development, accelerating the growth process, and increasing yield, but also reduce labor input and residual soil pollution due to their simple application and degradation performance. Therefore, promoting the use of LBDMs as a substitute for ordinary PE films in agricultural production is of great significance. However, further studies on economic cost, field application characteristics, and supporting spraying equipment are necessary.

Author Contributions: Conceptualization, J.S., B.L. (Baoyou Liu) and P.L.; Methodology, J.S., Z.Y. and B.L.(Baoyou Liu); Software, J.S., S.W., Z.Y. and F.B.; Validation, J.S., B.L. (Baoyou Liu) and P.L.; Formal analysis, J.S. and B.L. (Baoyan Li); Investigation, J.S., S.W., B.L. (Baoyan Li) and B.L. (Binghui Luan); Data curation, J.S. and S.W.; Writing—original draft, J.S.; Writing—review & editing, B.L. (Baoyou Liu) and P.L.; Visualization, R.C. and P.L.; Supervision, B.L. (Baoyou Liu) and P.L.; Project administration, B.L. (Baoyou Liu) and P.L.; Funding acquisition, B.L. (Baoyou Liu). All authors have read and agreed to the published version of the manuscript.

Funding: This research was funded by the Natural Science Foundation of Shandong Province (ZR2022MD045); Shandong Province Key R&D Program (2021CXGC010602, 2021CXGC010802); Shandong Province Fruit Industry Technology System (SDAIT-06-11); 2023 Tai'an Science and Technology Innovation "Double Ten Project" (Major Technological Research) (23JSGG01); and Shandong Province Ecological Agriculture Technology System Construction Task (SDAIT-30-04).

Institutional Review Board Statement: Not applicable

Data Availability Statement: Data are contained within the article.

Conflicts of Interest: The authors declare no conflict of interest.

References

1. Li, H.D.; Li, W.J.; Kang, T.; Zhang, L.M.; Chen, J.S.; Zhang, Y.Y.; Ma, W.Y. Effects of different harvest time on yield and its components of peanut under plastic film mulching and open field cultivation. *J. Peanut Sci.* **2021**, *50*, 75–79.

2. Sun, Z.H.; Ning, R.X.; Qin, M.H.; Liang, J.; Jiang, J.X.; Sun, W.; Liu, X.; Zi, M. Sustainable and hydrophobic polysaccharide-based mulch film with thermally stable and ultraviolet resistance performance. *Carbohydr. Polym.* **2022**, *295*, 119865. [CrossRef] [PubMed]
3. Ramakrishna, A.; Hoang, M.T.; Suhas, P.W.; Tranh, D.L. Effect of mulch on soil temperature, moisture, weed infestation and yield of groundnut in northern Vietnam. *Field Crops Res.* **2006**, *95*, 115–125. [CrossRef]
4. Li, X.L.; Zheng, G.F.; Li, Z.Y.; Fu, P. Formulation, performance and environmental/agricultural benefit analysis of biomass-based biodegradable mulch films: A review. *Eur. Polym. J.* **2024**, *203*, 112663. [CrossRef]
5. Yang, M.; Yan, L.; Li, Y.C.; Huang, P.; Han, W.J.; Dang, X.G. An environment-friendly leather waste-based liquid film mulching and its application for facilitating the growth of maize crops. *Process Saf. Environ. Prot.* **2022**, *159*, 1236–1244. [CrossRef]
6. Moreno, M.M.; Moreno, A. Effect of different biodegradable and polyethylene mulches on soil properties and production in a tomato crop. *Sci. Hortic.* **2008**, *116*, 256–263. [CrossRef]
7. Tian, D.; Zhang, J.; Hu, J.G.; Huang, M.; Zhao, L.; Lei, Y.J.; Zou, J.M.; Zhang, S.B.; Shen, F. A new water-soluble lignin incorporation enhanced the barrier performance of liquid mulching film. *Chem. Eng. J.* **2023**, *452*, 139383. [CrossRef]
8. Kumar, M.K.; Sheela, M.A. Effect of plastic film mulching on the distribution of plastic residues in agricultural fields. *Chemosphere* **2021**, *273*, 128590. [CrossRef]
9. Wang, K. Study on the effects of agricultural residual film on soil physicochemical properties and crop yield. *Mod. Agric. Res.* **2021**, *27*, 27–28.
10. Qin, J.; Song, L.; Li, X.; Wu, X.Y.; Zhao, B. Effects of liquid mulching film on plant community characteristics and soil water use efficiency in degraded land. *Soil Water Conserv. China* **2022**, *8*, 21–24+17.
11. Yin, G.H.; Tong, N.; Hao, L.; Gu, J.; Liu, Z.X. Effects of soil moisture and yield under different material film mulching on peanut. *J. Soil Water Conserv.* **2012**, *26*, 204–207+250.
12. Mei, L.; Dong, W.Y.; Zhou, J.H.; Liu, Q.; Song, H.X.; Wang, J.Y. Performance of biodegradable films and their application on fresh corn and sweet potato production in Beijing. *J. China Agric. Univ.* **2021**, *26*, 54–65.
13. Li, C.; Moore-Kucera, J.; Lee, J.; Corbin, A.; Brodhagen, M.; Miles, C.; Inglis, D. Effects of biodegradable mulch on soil quality. *Appl. Soil Ecol.* **2014**, *79*, 59–69. [CrossRef]
14. Kyrikou, I.; Briassoulis, D. Biodegradation of agricultural plastic films: A critical review. *J. Polym. Environ.* **2007**, *15*, 125–150. [CrossRef]
15. Yang, X.J.; Li, L.; Zhao, W.Y.; Li, X.; Mu, Y.S.; Chen, M.H.; Wu, X.Q. Substitute for polyethylene (PE) films: A novel cow dung-based liquid mulch on silage cornfields. *PLoS ONE* **2022**, *17*, e0271273. [CrossRef] [PubMed]
16. Chen, L.W. Fabrication and Application of One Biodegradable Starch-Based Liquid Mulching Materials. Master's Thesis, Sichuan University, Chengdu, China, 2021.
17. Zhang, J. Lignin-Reinforced Functionality of Biomass-Based Liquid Mulching Film. Master's Thesis, Sichuan Agricultural University, Chengdu, China, 2023.
18. Xu, Y.; Li, Q.; Man, L. Bamboo-derived carboxymethyl cellulose for liquid film as renewable and biodegradable agriculture mulching. *Int. J. Biol. Macromol.* **2021**, *192*, 611–617. [CrossRef]
19. Harada, J.; Souza, A.G.; Macedo, J.R.N. Soil culture: Influence of different natural fillers incorporated in biodegradable mulching film. *J. Mol. Liq.* **2019**, *273*, 33–36. [CrossRef]
20. Ning, R.X.; Liu, C.J.; Cheng, X.C.; Lei, F.H.; Zhang, F.L.; Xu, W.; Zhu, L.W.; Jiang, J.X. Fabrication of multi-functional biodegradable liquid mulch utilizing xyloglucan derived from tamarind waste for agricultural application. *Int. J. Biol. Macromol.* **2024**, *257*, 128627. [CrossRef]
21. Bao, H.H.; Cao, L.K.; Wei, C.H. Field characteristics of potato pulp based multifunctional and biodegradable liquid mulching film. *Acad. Period. Farm Prod. Process.* **2014**, *16*, 1–3.
22. Song, L.; Qin, J.; Li, X.; Wu, X.Y. Effects of liquid mulching film on agronomic traits, yield of peanut and soil fertility. *Chin. Agric. Sci. Bull.* **2023**, *39*, 33–37.
23. Liu, X.G.; Fan, Y.; Zhao, X.F.; Chen, Z.; Miao, B.Y.; Chen, S.P. Effects of different film mulching treatments on physiological traits and yield of peanut in Tangshan area. *J. Peanut Sci.* **2021**, *50*, 80–84.
24. Gao, X.; Fu, C.X.; Li, M.X.; Qi, X.J.; Jia, X. Effects of biodegradation of corn-starch–sodium-alginate-based liquid mulch film on soil microbial functions. *Int. J. Environ. Res. Public Health* **2022**, *19*, 8631. [CrossRef] [PubMed]
25. Li, M.N.; Cheng, T.; Sun, C.K.; Gao, H.Y.; Yan, Y.J.; Yang, Y.Q. Research prospect of environmentally friendly liquid mulch film prepared by biomass materials. *Henan Chem. Ind.* **2024**, *41*, 33–38.
26. Zainul, A.N.A.; Zulkifli, A.; Gunaseelan, S.; Palanivelu, S.D.; Salleh, K.M.; Che Othman, M.H.; Zakaria, S. Regenerated cellulose products for agricultural and their potential: A review. *Polymers* **2021**, *13*, 3586. [CrossRef] [PubMed]
27. Chiara, M.M.; Luigi, B.; Paolo, F.M.L.; Antonino, D.F.; Marco, C. Film Blowing of biodegradable polymer nanocomposites for agricultural applications. *Macromol. Mater. Eng.* **2021**, *306*, 2100177.
28. Zheng, Y.; Zheng, Y.; Wu, Z.; Sun, X.; Wang, C. Study on physiological characteristics and supporting techniques of Huayu 22 peanut. *Asian Agric. Res.* **2015**, *7*, 74–82.
29. Santos, M.; Carlo, D.O.; Hong, J.; Liu, Z.; Jiang, S.; Hrapovic, S.; Lam, E.; Jin, T.; Moores, A. Effect of surface functionality on the rheological and self-assembly properties of chitin and chitosan nanocrystals and use in biopolymer films. *Biomacromolecules* **2023**, *24*, 4180–4189. [CrossRef]

30. Huang, Z.; Wang, Y.; Zhang, N.; Zhang, L.; Darensbourg, D.J. One-Pot Synthesis of ion-containing CO_2-based polycarbonates using protic ionic liquids as chain transfer agents. *Macromolecules* **2018**, *51*, 9122–9130. [CrossRef]
31. Lin, Z.; Cheng, H.; He, K.; McClements, D.J.; Jin, Z.; Xu, Z.; Meng, M.; Peng, X.; Chen, L. Recent progress in the hydrophobic modification of starch-based films. *Food Hydrocoll.* **2024**, *151*, 109860. [CrossRef]
32. AlSaleem, S.S.; Zahid, W.M.; Alnashef, I.M.; Haider, H. Extraction of halogenated hydrocarbons using hydrophobic ionic liquids. *Sep. Purif. Technol.* **2017**, *184*, 231–239. [CrossRef]
33. Marczak, J.; Kargol, M.; Psarski, M.; Celichowski, G. Modification of epoxy resin, silicon and glass surfaces with alkyl- or fluoroalkylsilanes for hydrophobic properties. *Appl. Surf. Sci.* **2016**, *380*, 91–100. [CrossRef]
34. Deng, Z.A.; Wu, M.; Shen, C.; Yang, X.; Wang, D.; Li, J.; Wu, D.; Chen, K. Microfluidic-blow-spinning of carvacrol-loaded porphyrin metal-organic framework nanofiber films with synergistic antibacterial capabilities for food packaging. *Food Chem.* **2024**, *460*, 140707. [CrossRef]
35. Sun, Z.H.; Chen, F.S. Hydrophilicity and antifouling property of membrane materials from cellulose acetate/polyethersulfone in DMAc. *Int. J. Biol. Macromol.* **2016**, *91*, 143–150. [CrossRef]
36. Crouvisier-Urion, K.; Lagorce-Tachon, A.; Lauquin, C.; Winckler, P.; Tongdeesoontorn, W.; Domenek, S.; Debeaufort, F.; Karbowiak, T. Impact of the homogenization process on the structure and antioxidant properties of chitosan-lignin composite films. *Food Chem.* **2017**, *236*, 120–126. [CrossRef] [PubMed]
37. Yu, H.X.; Guo, W.L.; Le, Z.P. A study on preparation and performance of yam residues biodegradable liquid mulching film. *Acta Agric. Univ. Jiangxiensis* **2017**, *39*, 581–586.
38. Steel, R.G.D.; Torrie, J.H. *Principles and Procedures of Statistics: A Biometrical Approach*, 2nd ed.; McGraw-Hill: New York, NY, USA, 1980.
39. Yadav, N.; Hakkarainen, M. Degradable or not? Cellulose acetate as a model for complicated interplay between structure, environment and degradation. *Chemosphere* **2021**, *265*, 128731. [CrossRef] [PubMed]
40. Wen, P.; Wu, Z.S.; Han, Y.J.; Cravotto, G.; Wang, J.; Ye, B.C. Microwave assisted synthesis of a novel biochar based slow release nitrogen fertilizer with enhanced water retention capacity. *ACS Sustain. Chem. Eng.* **2017**, *5*, 7374–7382. [CrossRef]
41. Deshmukh, R.A.; Aloui, H.; Khomlaem, C.; Negi, A.; Yun, J.H.; Kim, H.S.; Kim, B.S. Biodegradable films based on chitosan and defatted *Chlorella* biomass: Functional and physical characterization. *Food Chem.* **2021**, *337*, 127777. [CrossRef]
42. Sun, T.; Zhang, Z.; Ning, T.; Mi, Q.; Liu, Z. Colored polyethylene film mulches on weed control, soil conditions and peanut yield. *Plant Soil Environ.* **2015**, *61*, 79–85. [CrossRef]
43. Sartore, L.; Vox, G.; Schettini, E. Preparation and performance of novel biodegradable polymeric materials based on hydrolyzed proteins for agricultural application. *J. Polym. Environ.* **2013**, *21*, 718–725. [CrossRef]
44. Brault, D.; Stewart, K.A.; Jenni, S. Growth, development, and yield of head lettuce cultivated on paper and polyethylene mulch. *HortScience* **2002**, *37*, 92–94. [CrossRef]
45. Zhao, H.; Xiong, Y.C.; Li, F.M.; Wang, R.Y.; Qiang, S.C.; Yao, T.F.; Mo, F. Plastic film mulch for half growing-season maximized WUE and yield of potato via moisture-temperature improvement in a semi-arid agroecosystem. *Agric. Water Manag.* **2012**, *104*, 68–78. [CrossRef]
46. Mulumba, L.N.; Lal, R. Mulching effects on selected soil physical properties. *Soil Tillage Res.* **2008**, *98*, 106–111. [CrossRef]
47. Anikwe, M.A.N.; Mbah, C.N.; Ezeaku, P.I.; Onyia, V.N. Tillage and plastic mulch effect on soil properties and growth and yield of cocoyam (*Colocasia esculenta*) on an ultisol in southeastern Nigeria. *Soil Tillage Res.* **2007**, *93*, 264–272. [CrossRef]
48. Chen, J.K.; Zhang, Y.; Chen, J.S.; Chen, F.; Zhang, H.L. Characteristics of soil temperature and its response to air temperature under different tillage systems characteristics of soil temperature and thermal properties. *Sci. Agric. Sin.* **2009**, *42*, 2747–2753.
49. Zhou, L.M.; Li, F.M.; Jin, S.L.; Song, Y.J. How two ridges and the furrow mulched with plastic film affect soil water, soil temperature and yield of maize on the semiarid Loess Plateau of China. *Field Crops Res.* **2009**, *113*, 41–47. [CrossRef]
50. Wang, F.X.; Feng, S.Y.; Hou, X.Y.; Kang, S.Z.; Han, J.J. Potato growth with and without plastic mulch in two typical regions of Northern China. *Field Crop Res.* **2009**, *110*, 123–129. [CrossRef]
51. Waterer, D. Evaluation of biodegradable mulches for production of warm season vegetable crops. *Can. J. Plant Sci.* **2010**, *90*, 737–743. [CrossRef]
52. Zhang, W.F.; Wu, H.F.; Wang, R.J.; Liu, Y.F.; Zheng, L. Effects of different tillage and plastic film mulching practices on physiological characteristics, yield and quality of peanut. *J. Peanut Sci.* **2024**, *53*, 1002–4093.
53. Ghosh, P.K.; Dayal, D.; Bandyopadhyay, K.; Mohanty, M. Evaluation of straw and polythene mulch for enhancing productivity of irrigated summer groundnut. *Field Crops Res.* **2006**, *99*, 76–86. [CrossRef]
54. Zhu, G.D.; Xia, N.N.; Xue, M.; Li, Z.P.; Zhao, H.P.; Qu, C. Effects of colored polyethylene film mulch on pest populations, plant growth and yield of peanut in Northern China. *J. Asia-Pac. Entomol.* **2022**, *25*, 101944. [CrossRef]
55. Kunzova, E.; Hejcman, M. Yield development of winter wheat over 50 years of FYM, N, P and K fertilizer application on black earth soil in the Czech Republic. *Field Crops Res.* **2009**, *111*, 226–234. [CrossRef]

polymers MDPI

Article

Highly Sensitive and Flexible Capacitive Pressure Sensors Combined with Porous Structure and Hole Array Using Sacrificial Templates and Laser Ablation

Yibin Zhao [1,2], Jingyu Zhou [1,2], Chenkai Jiang [1,2], Tianlong Xu [1,2], Kaixin Li [1,2], Dawei Zhang [1,2] and Bin Sheng [1,2,*]

1 School of Optical Electrical and Computer Engineering, University of Shanghai for Science and Technology, Shanghai 200093, China; 2135050829@st.usst.edu.cn (Y.Z.); 213330579@st.usst.edu.cn (J.Z.); 15003438539@163.com (C.J.); 223330649@st.usst.edu.cn (T.X.); 233350670@st.usst.edu.cn (K.L.); dwzhang@usst.edu.cn (D.Z.)
2 Shanghai Key Laboratory of Modern Optical Systems, Engineering Research Center of Optical Instruments and Systems, Shanghai 200093, China
* Correspondence: bsheng@usst.edu.cn

Abstract: Flexible, wearable pressure sensors offer numerous benefits, including superior sensing capabilities, a lightweight and compact design, and exceptional conformal properties, making them highly sought after in various applications including medical monitoring, human–computer interactions, and electronic skins. Because of their excellent characteristics, such as simple fabrication, low power consumption, and short response time, capacitive pressure sensors have received widespread attention. As a flexible polymer material, polydimethylsiloxane (PDMS) is widely used in the preparation of dielectric layers for capacitive pressure sensors. The Young's modulus of the flexible polymer can be effectively decreased through the synergistic application of sacrificial template and laser ablation techniques, thereby improving the functionality of capacitive pressure sensors. In this study, a novel sensor was introduced. Its dielectric layer was developed through a series of processes, including the use of a sacrificial template method using NaCl microparticles and subsequent CO_2 laser ablation. This porous PDMS dielectric layer, featuring an array of holes, was then sandwiched between two flexible electrodes to create a capacitive pressure sensor. The sensor demonstrates a sensitivity of 0.694 kPa^{-1} within the pressure range of 0–1 kPa and can effectively detect pressures ranging from 3 Pa to 200 kPa. The sensor demonstrates stability for up to 500 cycles, with a rapid response time of 96 ms and a recovery time of 118 ms, coupled with a low hysteresis of 6.8%. Furthermore, our testing indicates that the sensor possesses limitless potential for use in detecting human physiological activities and delivering signals.

Keywords: flexible capacitive pressure sensor; porous structure; array of holes; polymer; laser ablation

check for updates

Citation: Zhao, Y.; Zhou, J.; Jiang, C.; Xu, T.; Li, K.; Zhang, D.; Sheng, B. Highly Sensitive and Flexible Capacitive Pressure Sensors Combined with Porous Structure and Hole Array Using Sacrificial Templates and Laser Ablation. *Polymers* **2024**, *16*, 2369. https://doi.org/10.3390/polym16162369

Academic Editors: Jiangtao Xu and Sihang Zhang

Received: 20 July 2024
Revised: 20 August 2024
Accepted: 20 August 2024
Published: 21 August 2024

1. Introduction

As the complexity of wearable systems [1–3] continues to advance, scholars have shown a growing interest in the advancement of flexible and wearable pressure sensors [4]. The utilization of these sensors has experienced a notable increase in various applications, including medical monitoring [5–8], human–computer interactions [9–12], electronic skins [13–15], and other fields, owing to their exceptional sensing capabilities, compact dimensions, and robust shape retention properties, among other benefits. Wearable pressure sensors can be classified based on their operating principles as piezoresistive [16,17], capacitive [18–20], friction electric [21,22], and piezoelectric [23] pressure sensors. Capacitive pressure sensors are distinguished among various sensor types for their straightforward manufacturing process, minimal energy consumption, superior stability, and rapid response time [20,24–26]. Previous studies have indicated that conventional capacitive sensors primarily utilize solid silicone rubbers such as polydimethylsiloxane (PDMS) [27,28] and ecoflex [29]. However, despite the flexibility of these silicone rubbers, their low compressive strain and high Young's modulus [30] result in reduced sensitivity [31], thereby failing to meet the requirements for high sensitivity sensors. To improve the sensitivity of capacitive pressure sensors within a certain pressure range, scholars have suggested the incorporation of microstructures within the dielectric layer [32]. Drawing from previous research, scholars have developed microstructures in the forms of micropyramidal [33–35], microcylindrical [36–39], and microconical [19,40] shapes. For example, Luo et al. present a capacitive pressure sensor with a tilted micropillar array structure in its dielectric layer prepared by a photolithographic method. This sensor exhibits high pressure sensitivity (0.42 kPa^{-1}) and a very small detection limit (1 Pa) [39].

Moreover, incorporating a porous structure into the dielectric layer is a method that can enhance sensitivity [41–44]. This structural modification reduces the Young's modulus of the dielectric layer, facilitating compression of the flexible sensor and ultimately increasing sensitivity [45]. Previous research has explored different techniques for creating polymer-based porous dielectric layers. For example, sacrificial templates [46–48], gas foaming [49–51], 3D printing [52,53], and other methods [54,55] have been used in various studies. The sacrificial template method is a popular choice due to its ease of use and high effectiveness. The preparation process entails blending removable particles with silicone rubber prior to its curing, allowing for the embedding of particles within the cured silicone rubber. These particles facilitate the formation of interconnected air exchange channels [41]. Pores can be generated in the silicone rubber by extracting the particles. Salt and sugar [47,48] are commonly used as sacrificial templates due to their high water solubility, facilitating their removal. In addition to these conventional templates, researchers are actively investigating alternative sacrificial templates such as polystyrene (PS) beads [56,57]. Yang et al. obtained porous micropyramidal structured dielectric layers by pressing PDMS into a micropyramidal silicon mold filled with PS beads and curing it, followed by dissolving away the PS beads using toluene. Capacitive pressure sensors which were based on this dielectric layer exhibit an extremely high sensitivity of up to 44.5 kPa^{-1} in the pressure range of 0–100 Pa [33].

The susceptibility of the dielectric layer to compression at low pressures, resulting in rapid saturation of the sensor and a reduced operating range, hinders its ability to accurately detect higher pressures. Additionally, repeated compressions may lead to destruction of the microstructures due to interaction forces between them. In order to combine high sensitivity with a wide pressure detection range and to maintain good recovery properties of the dielectric layer, Li et al. obtained a highly porous dielectric layer by the sacrificial NaCl template method, while the insertion of multiple metal pins introduced a through-hole array in the porous dielectric layer, thus further improving the porosity. This working design of the sensor achieves a sensitivity of 1.15 kPa^{-1} within 0–1 kPa and the device has a very wide operating ranging from 5 Pa to 1 MPa [45]. However, damage to the dielectric layer during metal pin demolding can compromise sensor performance. Jiang et al. demonstrate the fabrication of microstructures with uniform geometry and adjustable size through laser ablation on a dielectric substrate [40]. This cost-effective and efficient method offers a practical approach for designing arrays.

This research presents the fabrication of an innovative flexible capacitive sensor that boasts a distinctive design. Notably, this novel sensor incorporates a dielectric layer characterized by a porous architecture and an array of holes, setting it apart from conventional sensors. The porous dielectric layer was fabricated using the sacrificial NaCl template method, while the hole array was created through laser ablation of the porous dielectric layer. By sandwiching this porous dielectric layer with the hole array between flexible electrodes made of polyimide tape and copper foil, a flexible capacitive pressure sensor was developed. The sensor exhibited a notable sensitivity of 0.694 kPa^{-1} within the 0–1 kPa range, while demonstrating a broad pressure response spanning from 0 to 200 kPa. Experimental evaluations revealed its capability to discern pressures as minute as 3 Pa, accompanied by remarkable consistency across 500 high-pressure cycles, underscoring its reliability and durability. Furthermore, the sensor's utility is showcased in monitoring human physiological activities, such as swallowing and elbow flexion, as well as in Morse code communication.

2. Materials and Methods

2.1. Materials

Polydimethylsiloxane (PDMS) and a silicone elastomer curing agent were purchased from Dow Corning in Midland, MI, USA. Sodium chloride (NaCl microparticles, ≥99.5%, with a particle size distribution in the range of 5–70 µm) was purchased from China Salt Shanghai Salt Industry Co., Ltd., in Shanghai, China. The flexible electrodes were manufactured using copper foil purchased from Anhui Zhengying Company in Fuyang, Anhui, China, and polyimide (PI) tape from Hangzhou Ubisoft Company in Hangzhou, Zhejiang, China.

2.2. Preparation of Porous PDMS Dielectric

The sacrificial template method was employed to fabricate porous dielectric layers using salt microparticles with a particle size distribution ranging from 5 to 70 µm. The procedure for creating the porous PDMS dielectric layer is illustrated in Figure 1a. Specifically, 5.521 g of salt was introduced into a mold measuring 60 mm in diameter and 2 mm in height and then compressed to conform to the mold shape. Subsequently, 3.110 g of PDMS prepolymer was introduced into the templates, and the molds containing the PDMS prepolymer were subjected to a vacuum treatment in a vacuum machine (LC-DZF-6050AB, Lichen Instrument Technology Co., Shanghai, China) at a temperature of 25 °C and an air pressure of 0.07 MPa for 20 min. This process facilitated the thorough infiltration of the PDMS prepolymer into the molded salt templates. The PDMS prepolymer utilized in this study was prepared by blending PDMS with a curing agent through homogeneous stirring for a period of 20 min, with a mass ratio of PDMS to curing agent set at 10:1. After complete infiltration of the PDMS, the sample was subjected to atmospheric pressure and heated to a temperature of 50 °C for 3 h to allow the PDMS prepolymer to undergo full curing. Subsequently, the demolded sample was transferred to a water bath heating unit (DF-101S5L, Lichen Instrument Technology Co., Shanghai, China) and immersed in hot water at 80 °C for 12 h in order to eliminate salt residues. Finally, the sample was dried in a drying oven at 100 °C for 1 h, resulting in the formation of a porous PDMS dielectric layer with a porosity of approximately 46%.

2.3. Preparation of Porous PDMS Dielectric Layers with Hole Array

Following the preparation of the porous PDMS dielectric layer, the porous PDMS material was subjected to ablation using a carbon dioxide laser (K3020, Julong Laser Co., Ltd., Liaocheng, Shandong, China) with a wavelength of 10.6 µm, leading to the formation of the hole array. The specific procedures for this process are outlined in Figure 1b. Utilizing AutoCAD2013^{TM} software (G. 55. 0. 0), a 6 × 6 matrix of circular holes with a diameter of 1 mm was designed. The spacing (*d*) between two adjacent holes was identified as a key parameter for optimization purposes in our research. Subsequently, the designed arrays were fed into the laser machine, where the samples underwent ablation with precise adjustments of laser power and scanning speed, resulting in the creation of a PDMS dielectric layer containing an array of holes. The ablation depth of porous PDMS using varying laser powers at a consistent scanning speed is illustrated in Figure S1.

It is important to acknowledge that achieving greater hole depth at lower power levels necessitates multiple ablations to meet the specified criteria, resulting in a more time-intensive process. Elevated power levels raise the ambient temperature of the laser output beam, potentially causing "over-burning" of the hole arrays and impacting their morphology [58]. In this study, a laser power of 30 W and a scanning speed of 200 mm/s was utilized to process the porous PDMS samples.

Furthermore, the depth of the hole array can be controlled by adjusting the number of ablations. When utilizing a power of 30 W and a scanning speed of 200 mm/s, it was observed that two ablations of the porous PDMS film resulted in a hole depth of approximately 2 mm, which corresponds to the thickness of the sample films prepared.

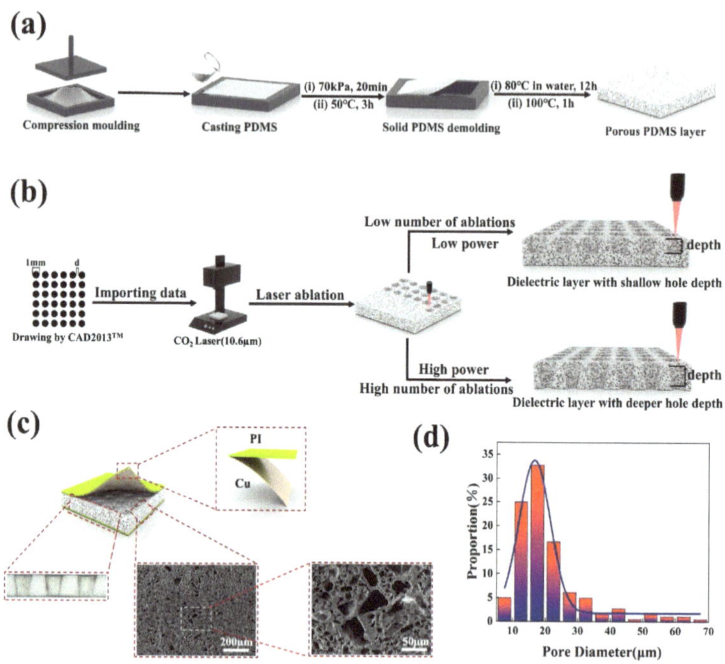

Figure 1. (**a**) Schematic illustration of the process for preparing a porous PDMS dielectric layer. (**b**) A detailed schematic representation of the laser ablation process. (**c**) A visual representation of the flexible capacitive pressure sensor, including an optical microscope view of the hole array profile in the bottom left panel and scanning electron microscopy images of microporous structures in the bottom two panels on the right. (**d**) Graph depicting the distribution of micropore sizes within the porous structures of the dielectric layers.

2.4. Experimental Setup

The morphology of the dielectric layer was analyzed using optical microscopy (RY001, Ksgaopin, Kunshan, Jiangsu, China) and scanning electron microscopy (JSM-IT500HR, Japan Electronics, Tokyo, Japan). The porous PDMS dielectric layer, measuring 8 mm × 8 mm × 2 mm (2 mm thick), containing an array of holes, was positioned between two flexible electrodes, each measuring 8 mm × 8 mm × 0.025 mm. These electrodes were composed of PI tape (0.02 mm thick) and copper foil (5 μm thick). Due to the pressure sensitive adhesive (PSA) on the purchased copper sheet, it possessed some adhesive properties. The PSA on the electrodes was bonded to thin (50 μm thick) layers of PDMS precursor (with a 10:1 mass ratio of PDMS to curing agent), which had been pre-coated (using the scratch-coating method) on the upper and bottom surfaces of porous PDMS dielectric layers with the hole array. After the PDMS precursor is cured, the flexible electrodes can be firmly connected to the dielectric layer, ensuring no detachment occurs during compression and bending. The sample was mounted on a 5 cm diameter disk and secured to the testing apparatus using PI tape to maintain flatness and optimal contact during testing. A manual press (HLD, Handpi, Yueqing, Zhejiang, China) was utilized to exert pressure on the sensor, while a digital force gauge (HP-20, Handpi, Yueqing, Zhejiang, China) was employed to measure the real-time pressure value. The digital force gauge boasts a reading accuracy of 0.001 N and a range of 0–20 N. Copper wire was employed to connect the two electrodes of the sensor with the LCR bridge (TH2822D, Tonghui, Changzhou, Jiangsu, China). The real-time capacitance data of the sensor were recorded using the LCR bridge at a temperature of 25 °C. The data acquisition frequency of the bridge was set at 100 kHz. Afterwards, the collected data were compiled and analyzed on a computer system. The circuit connection diagram for the test of the capacitive flexible pressure sensor developed in this study is illustrated in Figure S3.

3. Results and Discussion

3.1. Measurement of Porosity in Porous Dielectric Layers

The mass of multiple prepared porous PDMS samples was individually recorded in order to improve the accuracy of porosity estimation. Subsequently, these samples were immersed in separate beakers filled with water for 8 h to ensure complete penetration of water into the porous PDMS samples. After taking these samples out of the water, the mass of each sample post-water absorption was measured and documented. The porosity of each sample was then calculated using Equation (1) [59].

$$P = \frac{(m_w - m_d)/\rho_w}{(m_w - m_d)/\rho_w + m_d/\rho_d} \times 100\% \quad (1)$$

where m_d is the mass of the initial porous dielectric layer, m_w is the mass of the porous dielectric layer after sufficient water absorption, ρ_d is the density of the PDMS, and ρ_w is the density of water. The porosity of the porous PDMS dielectric layer, which was fabricated using the sacrificial NaCl template method, was determined to fall within the range of 46% $\pm$ 0.7% based on Equation (1).

3.2. Characterization of Flexible Capacitive Sensors

The capacitive flexible pressure sensor developed in this study comprises two flexible electrodes composed of flexible PI tape (0.02 mm thick) and copper foil (5 µm thick), as well as porous PDMS dielectric layer which incorporates an array of holes. Figure S2 illustrates the external appearance of the sensor and highlights its remarkable flexibility. Meanwhile, Figure 1c provides a comprehensive illustration of the flexible capacitive pressure sensor's composition, complemented by optical and scanning electron microscope images that reveal the cross-section of the dielectric layer in detail. Analysis of the sensor dielectric layer's cross-section reveals that the laser-ablated holes exhibit a slight tilt angle on their sidewalls, deviating from perfect verticality. The analysis of the cross-sectional intensity distribution of the laser output power reveals a Gaussian function pattern [58,60], indicating that the highest energy concentration is located at the center of the beam. Consequently, materials positioned near the beam's center experience complete ablation during the ablation process, whereas those situated at the periphery of the beam exhibit lower energy levels and, consequently, reduced ablation efficiency. This discrepancy in energy distribution results in a slightly inclined sidewall formation. Scanning electron micrographs of the cross sections provide visual representation of the dimensions and spatial arrangement of micropores. It is observed that, with the exception of a few larger pores resulting from larger salt particles, the variation in pore size among the remaining pores is minimal. Analysis of the particle size distribution shown in Figure 1d reveals that the majority of pores fall within the range of 8–32 µm.

3.3. Sensing Mechanism of Capacitive Pressure Sensors

A capacitive pressure sensor's capacitance is determined by the effective overlap area of the electrodes and the dielectric layer (A), the relative permittivity of the dielectric layer (ε_r), and the distance between the electrode plates (d). Specifically, it is determined according to the conventional capacitance formula, Equation (2), for a parallel flat capacitor.

$$C = \frac{\varepsilon_0 \varepsilon_r A}{d} \tag{2}$$

In this equation, ε_0 represents the dielectric constant of air, while ε_r stands for the relative dielectric constant of the dielectric layer, and d denotes the distance between the two electrode plates. In the case of a porous dielectric layer containing an array of holes, a fraction of the dielectric layer's volume is initially filled with air. Upon application of force, the air-filled pores within the dielectric layer gradually collapse and are substituted with solid PDMS material. This process alters the sensor's geometry, leading to variations in the dielectric constant (ε_r) and thickness (d), consequently impacting the capacitance. The sensor's response to pressure is determined by monitoring the relative changes in capacitance. Additionally, the close and secure attachment of the flexible electrodes to the dielectric layer serves to mitigate noise interference. The electrodes employed in this study exhibit superior conductivity and flexibility, thereby enhancing the precision of data collection within narrow pressure ranges. The sensing mechanism of the sensor we designed is illustrated in Figure 2a. The dielectric layer's relative dielectric constant is influenced by both air and PDMS due to the presence of microporous pores and the hole array. This relationship can be quantitatively determined using Equation (3).

$$\varepsilon_r = \varepsilon_a v_a + \varepsilon_c v_c \tag{3}$$

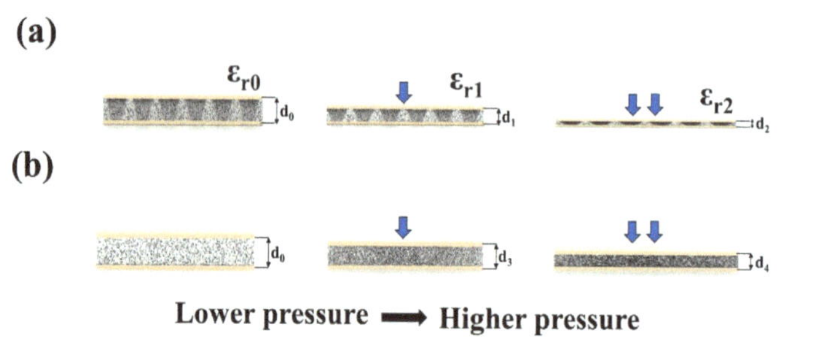

Figure 2. (**a**) Illustration of the workflow for a porous flexible capacitive pressure sensor, featuring an array of holes structure. (**b**) A schematic diagram elucidating the sensing mechanism of a flexible capacitive pressure sensor, featuring a porous dielectric layer without an array of holes structure.

The relative permittivity of air, denoted as ε_a and approximately equal to 1, and the relative permittivity of PDMS, denoted as ε_c, are key parameters in the analysis of the dielectric layer. The volumes occupied by air (v_a) and PDMS (v_c) within the dielectric layer play a crucial role in the compression sensing mechanism of the sensor. When the dielectric layer undergoes compression, a portion of the air is displaced by PDMS, resulting in changes to both v_a and v_c, ultimately leading to variations in the relative permittivity (ε_r).

The diagram in Figure 2a illustrates the workflow of a porous PDMS dielectric layer containing an array of holes. The initial thickness of the sensor's dielectric layer is denoted as d_0, with a relative dielectric constant of ε_{r0}. The pressure response of the sensor can be categorized into two distinct phases as pressure increases from 0. During the initial phase, the sensor's high density of micropores and the hole array results in a significant amount of air being trapped within the sensor, leading to the low Young's modulus of the dielectric layer, which facilitates easy compression of the sensor. During the low-pressure phase, there is a rapid increase in $\Delta d_1 = d_1 - d_0$, as the relative permittivity of the dielectric layer transitions from ε_{r0} to ε_{r1} due to the partial replacement of air with solid PDMS. This enhanced sensitivity of the sensor is particularly pronounced in the low-pressure regime, contributing to its improved performance in detecting subtle pressure variations. In the subsequent stage, as the pressure increases to a higher level, the microporous pores of the PDMS layer become densified, leading to a less significant change in $\Delta d_2 = d_2 - d_1$. The alteration in the relative dielectric constant ε_r, transitioning from ε_{r1} to ε_{r2}, significantly influences the variation in capacitance value. This shift in relative permittivity primarily correlates with the saturation of microporous pores and laser-ablated holes. In comparison to the porous structure-only sensor depicted in Figure 2b, our analysis revealed that the porous dielectric layer with an array of holes illustrated in Figure 2a exhibited greater compression than the porous structure-only dielectric layer shown in Figure 2b when subjected to identical pressure levels; that is, $d_0 - d_1 > d_0 - d_3$ and $d_1 - d_2 > d_3 - d_4$. The notable increase in sensor sensitivity, particularly evident in low-pressure conditions, stems directly from the incorporation of a hole matrix within the porous dielectric layer. This design feature results in a reduced Young's modulus, enabling greater compressibility and subsequently enhancing the sensor's ability to detect even minute pressure fluctuations. The uniformly distributed micropores within the dielectric layer, in conjunction with the array of holes extending to the base, function synergistically to enhance pressure sensing across the entire operational range. During the initial phase, when the pressure was applied, the majority of the microporous pores were filled by solid PDMS, whereas only a minor proportion of the air introduced through the array of holes was filled with solid PDMS. At this stage, the microporous pores assume a primary function, while the array of holes assumes a secondary role. In the second stage, when the pressure gradually increases, the air introduced through the array of holes assumes a primary role because it still has a large volume fraction, whereas the residual microporous pores contribute a secondary function because most of the micropores were densified. The synergistic interaction between the microporous pores and the hole array is crucial in ensuring a broad operational range and high sensitivity at low pressures for the sensors. As the pressure on the sensor is gradually released, the densified pores within the dielectric layer are re-established, resulting in the restoration of the hole array to its initial height and the morphology of the dielectric layer. This observation serves as evidence of the sensor's favorable recoverability.

3.4. Sensor Performance Optimization and Improvement

The specific test steps for the performance of capacitive pressure sensors are described in Section 2.4. Pressure is applied using the HLD pressure testing machine, pressure data are recorded with the HP-200 dynamometer, and capacitance data are collected using the TH2830 LCR meter. The sensitivity equation for the sensor is presented as Equation (2).

$$S = \frac{\delta\left(\frac{\Delta C}{C_0}\right)}{\delta p} \tag{4}$$

where ΔC is the capacitance value C of the sensor after it has been compressed minus the initial capacitance value C_0, and p is the pressure loaded on the sensor. It is recognized that the larger the value of $\Delta C/C_0$ per unit pressure, the higher the sensitivity of the sensor. Based on this theory, five different sample designs were created for comparative analysis. Specifically, the samples can be categorized into five types: bulk PDMS (bPDMS), bulk PDMS with an array of holes (bPDMS-h_2) with a hole depth of 2 mm, porous PDMS (pPDMS), porous PDMS with an array of holes depth of 1 mm (pPDMS-h_1), and porous PDMS with an array of holes depth of 2 mm (pPDMS-h_2). Each sample measures 8 mm × 8 mm × 2 mm (2 mm thick) and features a 6 × 6 round hole array. The manufacturing procedure entailed the repetitive ablation of bulk PDMS material to a depth of 2 mm, accomplished through five sequential cycles using a laser power setting of 30 W and a scanning velocity of 200 mm/s. To produce porous PDMS with a uniform array of 1 mm deep holes, a singular ablation step was employed, utilizing a laser intensity of 30 W and a scanning velocity of 200 mm/s. Furthermore, to create porous PDMS featuring 2 mm deep holes, a two-step ablation process was conducted under the same laser conditions, precisely reaching the bottom of the dielectric layer during the second ablation iteration. Figure S4 illustrates the variation in hole depths resulting from ablating the porous PDMS dielectric layer once (Figure S4a), twice (Figure S4b), and ablating the bulk PDMS dielectric layer three (Figure S4d) and five times (Figure S4c) using a CO_2 laser, as per the specified laser parameters. The data indicate that a greater number of ablations are required for the bulk PDMS in comparison to the porous PDMS to achieve equivalent hole depths. This discrepancy can be attributed to the poor light absorption of the transparent bulk PDMS [61], necessitating an increased number of ablations for effective ablation.

The performance curves illustrating the relative capacitance versus pressure for the five tested sensors are presented in Figure 3. Specifically, Figure 3a displays the curves within the pressure range of 0–200 kPa. Analysis of the results indicates that the sensitivity of the sensors, namely pPDMS, pPDMS-h_1, and pPDMS-h_2, can be significantly improved through the incorporation of multiple micropores. This enhancement is attributed to the lower Young's modulus of the porous samples, rendering them more compressible compared to the bulk PDMS. Despite the introduction of an array of holes in the dielectric layer of the bPDMS-h_2 sensor, the lack of a porous structure in the dielectric layer results in all other areas being occupied by PDMS. This limits the presence of air gaps, thereby hindering the reduction of the Young's modulus of the dielectric layer. Consequently, the dielectric layer is difficult to compress, leading to a relatively low sensitivity. Regarding the two sensors, pPDMS-h_1 and pPDMS-h_2, which have been modified with an array of holes in addition to the existing porous structure, their porosity is greater and their Young's modulus is lower compared to the pPDMS sensor with only a porous structure. This results in increased compressibility and higher sensitivity of the two sensors compared to the pPDMS sensor.

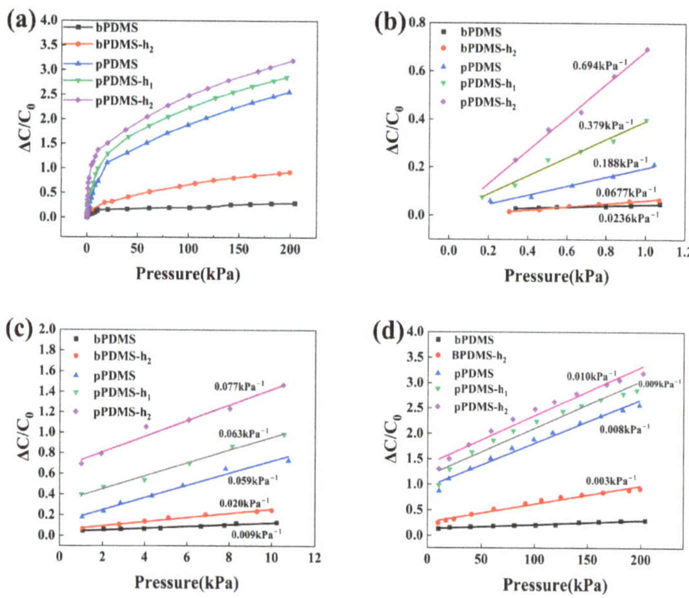

Figure 3. (**a**) The variation curves depicting the relationship between relative capacitance and pressure within the range of 0 to 200 kPa for five test samples. (**b**) A linear fit curve illustrating the correlation between relative capacitance change values and pressure values within the pressure range of 0 to 1 kPa. (**c**) Linear fit curves demonstrating the relationship between relative capacitance change values and pressure values within the pressure range of 1 kPa to 10 kPa. (**d**) Fit curves representing the relationship between relative capacitance change values and pressure values within the pressure range of 10 to 200 kPa.

To better elucidate the operational principles of the sensor, three distinct pressure intervals were selected for individual analysis of the capacitive response. In particular, the sensitivities of pPDMS-h_2, pPDMS-h_1, and pPDMS were determined to be 0.694 kPa^{-1}, 0.379 kPa^{-1}, and 0.188 kPa^{-1}, respectively, within the range of 0–1 kPa, as illustrated in Figure 3b. When the applied pressure falls within the range of 1–10 kPa, the sensitivity of various sensor types diminishes, as shown in Figure 3c. However, sensors utilizing a dielectric layer with a composite structure of micropores and an array of holes still exhibit a level of sensitivity. Furthermore, the sensitivity of these sensors increases with the depth of the holes within the array. Even under extremely high pressures, pPDMS-h_2 continues to exhibit a certain degree of sensitivity, as depicted in Figure 3d. In addition, the tests conducted yielded the relationship curves between pressure and strain in the dielectric layer for the five samples within the range of 0–200 kPa, as illustrated in Figure S5. It is evident that the sample pPDMS-h_2, characterized by high compressibility, exhibited the highest strain of 72% at a pressure of 200 kPa, whereas bulk PDMS displayed the lowest strain of approximately 26% at 200 kPa. Additionally, all samples demonstrated strain–pressure curves that followed an exponential function increase.

The dielectric layer of a capacitive pressure sensor exhibits a capacitive response that is tied to both the material's relative permittivity and the inter-electrode distance between its two conductive plates. Under elevated pressure, the porous structure of the dielectric layer undergoes compression, leading to the infiltration of solid PDMS into the majority of its pores. This mechanism entails a deceleration in the shrinking rate of the gap between the electrode plates, coupled with a diminished rate of growth in the relative permittivity of the dielectric layer, collectively contributing to a decrease in sensor sensitivity as the applied pressure escalates. Introducing an array of holes into the dielectric layer enhances the air content, enabling a more effortless compression response to applied pressure in contrast to a purely porous dielectric layer. This modification leads to an improved compressibility characteristic under pressure. Consequently, the sensitivity is heightened. As pressure levels escalate, the air gap within the porous dielectric layer becomes nearly solidified by the PDMS material, leading to an increase in Young's modulus and rendering compression more challenging. Therefore, the sensitivity of dielectric layers with only microporous structures is relatively low at high pressures. Nevertheless, in the case of a porous dielectric layer containing an array of holes, the presence of a significant volume of air at the location of the hole array allows the dielectric layer to maintain an air gap even when subjected to high levels of compression. This characteristic provides the dielectric layer with the ability to respond to the higher pressures, thereby enabling sensors utilizing such porous dielectric layers to maintain sensitivity even under high-pressure conditions.

The impact of the sparsity of an array of holes in porous dielectric layers on the Young's modulus of a sensor, and subsequently on its sensitivity, was investigated by maintaining a constant individual hole area (1 mm in diameter) and adjusting the laser ablation spacing between neighboring holes in the array. The laser ablation process results in concentrated thermal energy at the location of the holes, leading to over-burning. When the spacing between two holes is too small, the high laser energy causes an increase in material temperature outside the ablation area. This can result in the destruction of the PDMS between neighboring holes, leading to the cross-linking of the holes and compromising the intended structure of the dielectric layer. Considering the above issues and the ablation accuracy (0.06 mm) of the CO_2 laser we used, we set the minimum pitch to 0.4 mm. The spacing was sequentially adjusted to 0.8 mm, 1.0 mm, and 1.2 mm. The depth of the holes was all 2 mm. The samples of the various classes mentioned above were named pPDMS-$w_{0.4}$ (which is the same sample as pPDMS-h_2), pPDMS-$w_{0.8}$, pPDMS-$w_{1.0}$, and pPDMS-$w_{1.2}$. Figure S6a–c shows the physical diagrams of the pPDMS-$w_{0.8}$, pPDMS-$w_{1.0}$, and pPDMS-$w_{1.2}$ dielectric layers, respectively.

The results of their performance are illustrated in Figure 4. Our investigation revealed that, under identical pressure conditions, sensitivity increases as the spacing of the hole array decreases. Figure 4a displays the pressure response curves of the four samples within the pressure range of 0–200 kPa. Specifically, the sensor pPDMS-$w_{0.4}$ demonstrated the highest sensitivity of 0.694 kPa^{-1}, within the pressure range of 0–1 kPa, as shown in Figure 4b. As the spacing between the hole array of various samples increases in a sequential manner, the sensitivity of the sensors correspondingly decreases across the same pressure range. The sensitivity of the sensor pPDMS-$w_{0.8}$ is 0.396 kPa^{-1} within the pressure range of 0–1 kPa. The sensor pPDMS-$w_{1.0}$ exhibits a sensitivity of 0.248 kPa^{-1}, while the sensor pPDMS-$w_{1.2}$ demonstrates a sensitivity of merely 0.197 kPa^{-1}. It is important to highlight that despite the decrease in sensitivity of a sensor utilizing a porous dielectric layer with an array of holes, as the spacing of the array increases over the same pressure range, they remain superior to pPDMS. This is precisely the result of the increased sensitivity due to the introduction of a hole array in the porous dielectric layer, thus leading to a further reduction of the Young's modulus of the dielectric layer. In the pressure range of 20–200 kPa (Figure 4c), the decrease in sensitivity of all four sensors is attributed to the increased density and reduced compressibility of the dielectric layer at higher pressures.

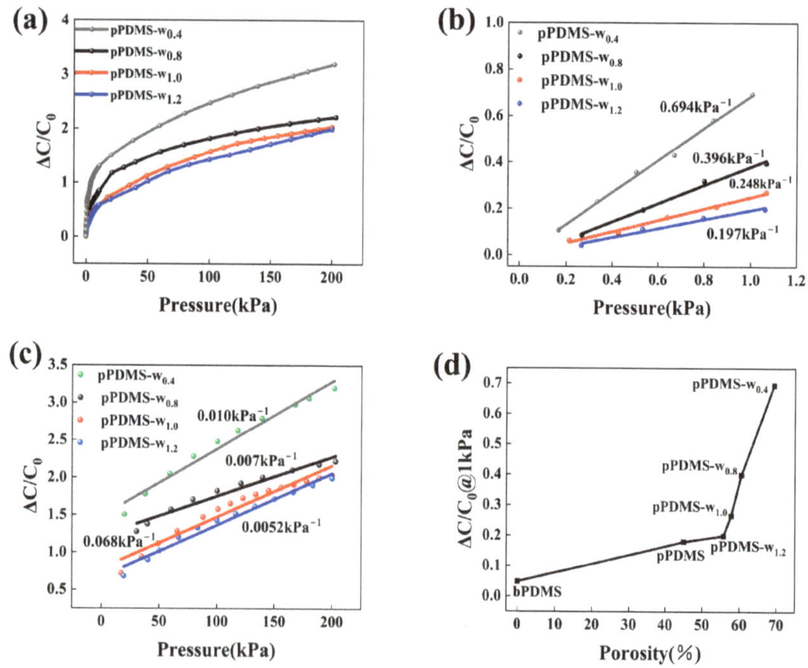

Figure 4. (**a**) Variation curves of relative capacitance values as a function of pressure ranging from 0 to 200 kPa for four sensors with different hole array spacings. (**b**) Linear fitting curves of relative capacitance change values versus pressure values for the four tested sensors within a pressure range of 0–1 kPa. (**c**) Linear fit curves of relative capacitance change versus pressure values for the four tested transducers for pressures ranging from 1 to 200 kPa. (**d**) Variation curve of relative capacitance with porosity at a pressure of 1 kPa.

Laser ablation creates a series of perforations that introduce additional air into the porous dielectric layer, which has a porosity of approximately 46%. This process lowers the Young's modulus of the material, facilitating compression and ultimately enhancing the sensitivity of the flexible sensor. In a specification-consistent porous dielectric layer, the volume fraction of air introduced by the hole array increases as the holes are closer together. This enhanced porosity facilitates greater compressibility of the dielectric layer, thereby improving the sensitivity of the sensor. Theoretically, the porosity of the samples can be determined by combining the air volume introduced by the sacrificial NaCl template method with the air volume introduced by the pore array created through laser ablation [45], as shown in Figure 4d. According to this theory, the porosity values for pPDMS, pPDMS-$w_{1.2}$, pPDMS-$w_{1.0}$, pPDMS-$w_{0.8}$, and pPDMS-$w_{0.4}$ are approximately 46%, 56.8%, 58.8%, 61.5%, and 69.3%, respectively. It is important to highlight that the porosity calculated through this method closely aligns with the results obtained from the porosity measurement method for porous dielectric layers discussed in Section 3.1, with a maximum discrepancy of 1%. Figure 4d illustrates the variation in relative capacitance of the sensor in relation to the porosity of the dielectric layer under a pressure of 1 kPa. The curve clearly exhibits an exponential transformation as the porosity increases. Therefore, we believe that the hole array provides an effective way to reduce the Young's modulus of the dielectric layer and improve the responsiveness of capacitive sensors in a certain pressure range.

3.5. The Comprehensive Performance of Flexible Capacitive Sensors

Based on the optimization process outlined above, the flexible capacitive sensor pPDMS-h_2 was chosen for further comprehensive performance evaluations in this study. To demonstrate the sensor's minimum detection limit, incremental pressure was applied starting from 1 Pa, resulting in a notable change in capacitance value at 3 Pa, as illustrated in Figure 5a. The sensor exhibited a significant response at 3 Pa, indicating that its minimum pressure detection threshold is approximately 3 Pa. The graphical representation in Figure 5b showcases three distinct traces, each portraying the sensor's relative capacitance variation $\Delta C/C_0$ in response to static loads of 5 g, 25 g, and 40 g, respectively. It is evident that $\Delta C/C_0$ exhibits rapid variation during the loading and unloading, demonstrating consistent responsiveness, effective recovery, and high sensitivity. The sensor's response and recovery times were assessed by swiftly applying and releasing a pressure of 800 Pa. Rapid application and release of pressure can be crucial for ensuring the accuracy of the measurement data. Analysis of Figure 5c reveals that the sensor achieves a response time of 96 ms when the capacitance value change rate reaches 0.58, and a recovery time of 118 ms when pressure is rapidly released. Based on this observation, the sensor's response time aligns closely with that of human skin's sensitivity to pressure stimuli [62], suggesting its potential applicability in monitoring human physiological activities, thereby expanding its utilization domain. To evaluate the reproducibility of the sensor's performance, the sensor was subjected to repetitive loading and unloading cycles at a constant pressure of 150 kPa on a testing platform, totaling 500 iterations. This methodology allowed for an assessment of the sensor's consistency over multiple cycles. The results, depicted in Figure 5d, indicate that the maximum change in samples exhibited a margin of error of ±14% throughout the cycling process. The observed maximum error value of 14% can be attributed to external interference affecting the testing machine at some point in time. Analysis of the graphical results indicates that the sensor exhibits excellent repeatability across each cycle under stable operating conditions. Consequently, it can be inferred that the sensor demonstrates enhanced durability, an extended service life, consistent responsiveness after multiple uses, and the capacity to endure high pressures. Small errors due to interference from the test equipment do not affect the overall sensing performance. The phenomenon of hysteresis, arising from the cyclic loading and unloading of the sensor within a pressure spectrum spanning 0 to 200 kPa, is visually presented in Figure 5e. The analysis reveals a maximum delay of approximately 6.8%. This lag level is nearly identical to that reported in some previous work [63,64]. This unavoidable hysteresis is due to the fact that polymers show viscoelastic behavior [65]. Figure 5f demonstrates the sensor's exceptional dynamic pressure response within the 0–70 kPa range.

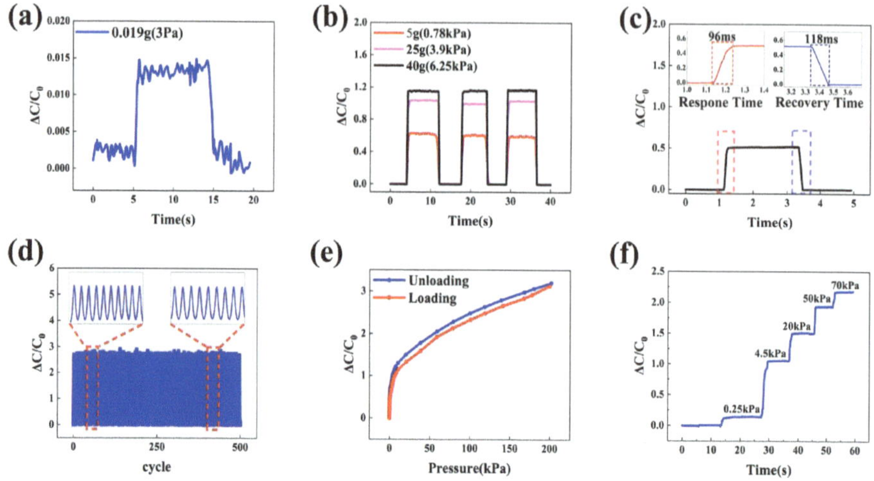

Figure 5. Comprehensive performance testing of capacitive pressure sensors utilizing the sample pPDMS-h_2. (**a**) The sensor's minimum pressure detection limit. (**b**) The sensor's pressure response to the loading and unloading of 5 g, 25 g, and 40 g weights. (**c**) The sensor's response time at a pressure of 0.85 kPa. (**d**) The stability test of the sensor's responsiveness at a pressure of 105 kPa for 500 cycles. (**e**) The test of the hysteresis in the capacitive response of pressure sensors between 0 and 200 kPa during loading and unloading. (**f**) The test of the sensor's response to stepped pressure changed between 0 and 70 kPa.

3.6. Performance Comparison of Different Porous Capacitive Pressure Sensors

Table 1 demonstrates some research conducted in recent years to investigate the performance of porous capacitive pressure sensors. It can be noted that the vast majority of work is developed using templates, or by dissolving away the soluble material after the elastomer layer has cured. The introduction of air gap increases the compressibility of the dielectric layer, so that the sensitivity increases when the porosity is higher. Therefore, by introducing an array of holes into the porous dielectric layer, we have obtained a higher porosity and thus improved the sensitivity of our sensors compared to other sensors with only micropores. In addition, due to the synergistic effect of micropores and a hole array, the sensor obtains a wide pressure detection range.

Table 1. A review of the performance specifications of several types of porous capacitive pressure sensors (NR = Not Reported).

Electrodes/Dielectric Layer	Key Materials to Fabricate the Dielectric	Pressure Range	Sensitivity	Response Time	Reference
AgNPs-SBS/ Microporous PDMS	PDMS/Glucose particles	0–2 kPa	0.278 kPa^{-1}	340 ms	[63]
AgNWs and CFs-PDMS/ Microporous ecoflex	Ecoflex/Sugar	0–10 kPa	0.161 kPa^{-1}	NR	[48]
ITO coated flexible PET/ Porous PDMS	PDMS/Sugar/Salt particles	0–5 kPa	0.171 kPa^{-1}	162 ms	[66]
Ag-TPU/ microporous PDMS	PDMS/$NaHCO_3$/ HNO_3	0–50 Pa 0.2–1 MPa	0.3 kPa^{-1} 3.2 MPa^{-1}	116 ms	[49]
ITO coated flexible PET/porous PDMS	PDMS/Deionized water	0.1–0.5 kPa	0.095 kPa^{-1}	110 ms	[67]
CB-PDMS/ porous PDMS	PDMS/Citric acid monohydrate	0–4 kPa 4–14 kPa	0.1 kPa^{-1} 0.049 kPa^{-1}	80 ms	[68]
PI-Cu/ microporous PDMS with hole array	PDMS/Salt microparticles	0–1 kPa 1–10 kPa 10–200 kPa	0.694 kPa^{-1} 0.077 kPa^{-1} 0.01 kPa^{-1}	96 ms	This work

3.7. Applications Related to Flexible Capacitive Pressure Sensors

This study showcased the potential applications of the sensor developed for monitoring human physiological activities. The fluorescent point in Figure 6a indicates the specific site where the sensor was deployed to monitor the volunteers' physiological activities. By affixing the sensor to the elbow joint of the volunteers, it was feasible to track the degree of flexion in their arms. In Figure 6b, it is evident that the capacitive output is optimized when the arm assumes a 90° flexion, followed by a decrease in this response as the arm bends to a 45° angle. Even at minimal bending, a response is still generated. Swallowing, a critical physiological function, was examined by affixing the sensor to the volunteer's throat during normal swallowing activity. Real-time capacitance data collected from the sensor revealed its responsiveness to the activity, including differences in swallowing speed and strength (Figure 6c). The sensor was affixed to the volunteer's knee prior to engaging in leg-raising movements. As shown in Figure 6d, subsequent repetitive stretching and bending activities elicited periodic changes in the sensor's response. To quantify the pressure applied by each finger on the water cup, sensors were affixed to the glove worn by the participant. The volunteer subsequently lifted the water cup, allowing for the measurement of pressure through the detection of capacitance value changes in the sensors on each finger. Analysis of the data, as depicted in Figure 6e, reveals that the thumb exerts the highest pressure on the water cup, while the little finger exerts the least pressure. Morse Code, a signaling code characterized by alternating signals, is utilized to convey various alphabets, numbers, and punctuation marks through unique sequences. This code serves a crucial function in radio communication, navigation, emergency signaling, and other domains. Volunteers attempted to transmit Morse code through the actuation of sensors, and the resulting Morse code signal was successfully detected on the testing equipment. The results of this experiment are depicted in Figure 6f,g, showcasing the successful transmission of the messages "USST" and "OECE" by the volunteers. The aforementioned applications were effectively demonstrated through the utilization of the sensor developed in this study, thereby showcasing the significant potential applications of this sensor in the realm of human body monitoring and signal transmission.

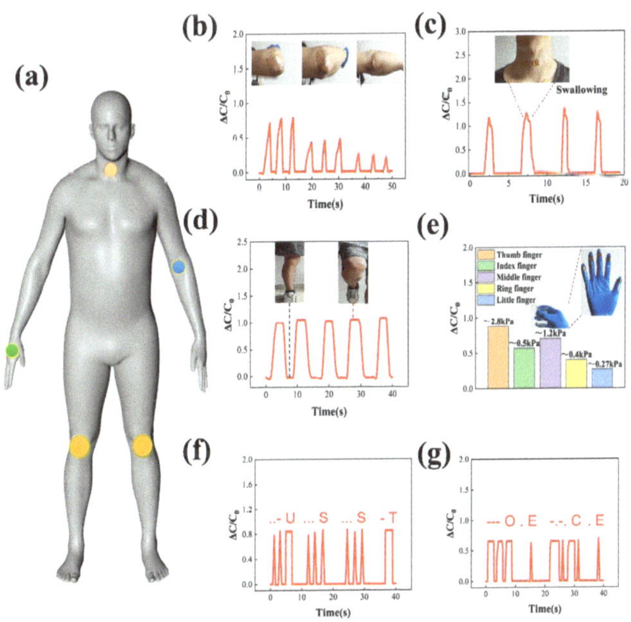

Figure 6. (**a**) Detection of human physiological activities at various locations on the body through the utilization of capacitive sensor pPDMS-h$_2$. (**b**) Variations in capacitance levels during the flexion of

the elbow. (**c**) Tracking changes in relative capacitance during the act of swallowing. (**d**) The response of the sensor from a volunteer engaging in leg lifts. (**e**) The pressure applied by individual fingers on a cup of water while being held is measured. (**f,g**) Generating specific Morse code signals through the act of pressing the sensor with the fingers.

4. Conclusions

The objective of this study is to introduce a new flexible capacitive pressure sensor fabricated through the utilization of laser ablation and sacrificial templates. Initially, a PDMS film with approximately 46% porosity was produced using the sacrificial template technique. Subsequently, a 6 × 6 array of perforations was created on the porous PDMS film via a CO_2 laser, leading to the development of the dielectric layer employed in this investigation. The preparation method described is cost-effective, environmentally sustainable, and easily manageable, allowing for the customization of array patterns and demonstrating significant potential for various applications. Through optimization of the ablation power of the CO_2 laser and careful regulation of the number of ablations, we were able to identify the specific ablation parameters necessary to achieve the desired hole depth. Additionally, our research delved into the impact of varying spacing between adjacent holes within the array on sensor sensitivity, attributing differences in sensitivity to variations in Young's modulus. Our research revealed a positive correlation between the proximity of neighboring cavities and sensor sensitivity, within the limitations of the machining process. The top-performing sensor in our study demonstrated a sensitivity of 0.694 kPa^{-1} within the range of 0–1 kPa. Additionally, the sensor exhibited a broad pressure detection range, remaining responsive even at pressures as high as 200 kPa and detecting pressures as low as 3 Pa. Furthermore, the device maintained consistent performance after undergoing 500 consecutive pressure loadings and unloadings. In order to showcase the practical utility of the sensor, we utilized it to effectively identify signals of human physiological activity and explored its capacity to convey pertinent information. These findings underscore the promising application prospects of the capacitive sensor developed by our team, thus holding significant implications for its broader implementation in the future.

Supplementary Materials: The following supporting information can be downloaded at: https://www.mdpi.com/article/10.3390/polym16162369/s1, Figure S1: Depth of ablation in porous PDMS dielectric layer as a function of CO_2 laser power, at a scanning rate of 200 mm/s; Figure S2: (a) Physical illustration of a flexible capacitive sensor. (b–d) The exhibition of the mechanical flexibility of the sensor designed in this work, the upper right image representing the overall flexibility of the sensor. The two figures below show the flexibility of the electrodes and the dielectric layer, respectively; Figure S3: The circuit connection diagram of each experimental device during the performance test of the capacitive pressure sensor; Figure S4: Optical microscopy image of the cross-section of the dielectric layer of the sensors we designed during the hole depth optimization. (a) Optical microscopy image of the cross-section of the dielectric layer of the sensor pPDMS-h_1. (b) Optical microscopy image of the cross-section of the dielectric layer of the sensor pPDMS-h_2. (c) Optical microscopy image of the cross-section of the dielectric layer of the sensor PDMS-h_2. (d) Hole depths obtained by laser ablation of bulk PDMS, performed three times at a laser power of 30 W and a scanning speed of 200 mm/s.; Figure S5: Relationship between pressure and compressive strain for capacitive sensors based on PDMS, pPDMS, pPDMS-h_1, and pPDMS-h_2 dielectric layers. Figure S6: The show of the actual drawings of three dielectric layers designed in the process of hole spacing optimization. (a–c) The three figures show, from left to right, the top views of the dielectric layer of the sensors pPDMS-$w_{0.8}$, pPDMS-$w_{1.0}$, and pPDMS-$w_{1.2}$, respectively.

Author Contributions: Conceptualization, Y.Z. and B.S.; methodology, Y.Z.; software, J.Z.; validation, Y.Z., B.S., and C.J.; formal analysis, Y.Z.; investigation, Y.Z.; resources, B.S.; data curation, Y.Z.; writing—original draft preparation, Y.Z. and B.S.; writing—review and editing, Y.Z. and B.S.; visualization, T.X. and K.L.; supervision, D.Z. and B.S.; project administration, B.S.; funding acquisition, B.S. All authors have read and agreed to the published version of the manuscript.

Funding: This research was funded by the Natural Science Foundation of Shanghai, grant number 19ZR1436100, the National Natural Science Foundation of China, grant number 11105149, and the Shanghai Municipal Innovation and Entrepreneurship Training Program for College Students, grant number SH2024054.

Institutional Review Board Statement: Not applicable.

Data Availability Statement: All data from this work have been included in the article.

Acknowledgments: The authors express gratitude to the editors and the reviewers for their constructive and helpful review comments.

Conflicts of Interest: The authors declare no conflicts of interest.

References

1. Hu, T.; Sheng, B. A Highly Sensitive Strain Sensor with Wide Linear Sensing Range Prepared on a Hybrid-Structured CNT/Ecoflex Film via Local Regulation of Strain Distribution. *ACS Appl. Mater. Interfaces* **2024**, *16*, 21061–21072. [CrossRef]
2. Zhou, J.Y.; Zhao, S.S.; Tang, L.; Zhang, D.W.; Sheng, B. Programmable and Weldable Superelastic EGaIn/TPU Composite Fiber by Wet Spinning for Flexible Electronics. *ACS Appl. Mater. Interfaces* **2023**, *15*, 57533–57544. [CrossRef]
3. Si, R.; Chen, W.; Chen, J.; Yang, Y.; Zhou, W.; Zhang, Q.; Chen, C.; Han, B. Green chemistry fabrication of durable antimicrobial peptide- immobilized silk fibroin films for accelerated full-thickness wound healing. *Mater. Today Chem.* **2023**, *29*, 12. [CrossRef]
4. Zhong, W.; Ming, X.; Jiang, H.; Ke, Y.; Ding, X.; Li, M.; Jia, K.; Wang, D. Full-Textile Human Motion Detection Systems Integrated by Facile Weaving with Hierarchical Core-Shell Piezoresistive Yarns. *ACS Appl. Mater. Interfaces* **2021**, *13*, 52901–52911. [CrossRef]
5. Lin, Q.; Huang, J.; Yang, J.; Huang, Y.; Zhang, Y.; Wang, Y.; Zhang, J.; Wang, Y.; Yuan, L.; Cai, M.; et al. Highly Sensitive Flexible Iontronic Pressure Sensor for Fingertip Pulse Monitoring. *Adv. Healthc. Mater.* **2020**, *9*, 2001023. [CrossRef]
6. Yang, L.; Wang, H.; Yuan, W.; Li, Y.; Gao, P.; Tiwari, N.; Chen, X.; Wang, Z.; Niu, G.; Cheng, H. Wearable Pressure Sensors Based on MXene/Tissue Papers for Wireless Human Health Monitoring. *ACS Appl. Mater. Interfaces* **2021**, *13*, 60531–60543. [CrossRef] [PubMed]
7. Zhu, B.; Xu, Z.; Liu, X.; Wang, Z.; Zhang, Y.; Chen, Q.; Teh, K.S.; Zheng, J.; Du, X.; Wu, D. High-Linearity Flexible Pressure Sensor Based on the Gaussian-Curve-Shaped Microstructure for Human Physiological Signal Monitoring. *ACS Sens.* **2023**, *8*, 3127–3135. [CrossRef]
8. Vinh Van, T.; Lee, S.; Lee, D.; Thanh-Hai, L. Recent Developments and Implementations of Conductive Polymer-Based Flexible Devices in Sensing Applications. *Polymers* **2022**, *14*, 3730. [CrossRef] [PubMed]
9. Sakhuja, N.; Kumar, R.; Katare, P.; Bhat, N. Structure-Driven, Flexible, Multilayered, Paper-Based Pressure Sensor for Human-Machine Interfacing. *ACS Sustain. Chem. Eng.* **2022**, *10*, 9697–9706. [CrossRef]
10. Wang, Y.; Duan, S.; Liu, J.; Zhao, F.; Chen, P.; Shi, Q.; Wu, J. Highly-sensitive expandable microsphere-based flexible pressure sensor for human-machine interaction. *J. Micromech. Microeng.* **2023**, *33*, 115009. [CrossRef]
11. Yuan, Y.; Xu, H.; Zheng, W.; Liu, M.; Li, S.; Yan, J.; Wang, D.; Liu, K.; Zhang, H.; Chen, G.; et al. Bending and Stretching-Insensitive, Crosstalk-Free, Flexible Pressure Sensor Arrays for Human-Machine Interactions. *Adv. Mater. Technol.* **2024**, *9*, 2301615. [CrossRef]
12. Franco, M.; Correia, V.; Marques, P.; Sousa, F.; Silva, R.; Figueiredo, B.R.; Bernardes, A.; Silva, R.P.; Lanceros-Mendez, S.; Costa, P. Environmentally Friendly Graphene-Based Conductive Inks for Multitouch Capacitive Sensing Surfaces. *Adv. Mater. Interfaces* **2021**, *8*, 2100578. [CrossRef]
13. Lei, M.; Feng, K.; Ding, S.; Wang, M.; Dai, Z.; Liu, R.; Gao, Y.; Zhou, Y.; Xu, Q.; Zhou, B. Breathable and Waterproof Electronic Skin with Three-Dimensional Architecture for Pressure and Strain Sensing in Nonoverlapping Mode. *ACS Nano* **2022**, *16*, 12620–12634. [CrossRef]
14. Yang, C.; Wang, W.; Zhang, B.; Liu, W.; Zhang, H.; Zhang, D. High sensitivity $SnSe_2$/MWCNTs flexible pressure sensors based on a lotus leaf biomimetic microstructure for electronic skin. *J. Mater. Chem. C* **2024**, *12*, 10669–10677. [CrossRef]
15. Veeramuthu, L.; Cho, C.-J.; Liang, F.-C.; Venkatesan, M.; Kumar, G.R.; Hsu, H.-Y.; Chung, R.-J.; Lee, C.-H.; Lee, W.-Y.; Kuo, C.-C. Human Skin-Inspired Electrospun Patterned Robust Strain-Insensitive Pressure Sensors and Wearable Flexible Light-Emitting Diodes. *ACS Appl. Mater. Interfaces* **2022**, *14*, 30160–30173. [CrossRef]
16. Cao, M.; Su, J.; Fan, S.; Qiu, H.; Su, D.; Li, L. Wearable piezoresistive pressure sensors based on 3D graphene. *Chem. Eng. J.* **2021**, *406*, 126777. [CrossRef]
17. Huang, X.; Zhang, X. Investigating the advanced characteristics of SiC based piezoresistive pressure sensors. *Mater. Today Commun.* **2020**, *25*, 101493. [CrossRef]
18. Kim, H.; Kim, G.; Kim, T.; Lee, S.; Kang, D.; Hwang, M.-S.; Chae, Y.; Kang, S.; Lee, H.; Park, H.-G.; et al. Transparent, Flexible, Conformal Capacitive Pressure Sensors with Nanoparticles. *Small* **2018**, *14*, 1703432. [CrossRef]
19. Li, J.; Chen, S.; Zhou, J.; Tang, L.; Jiang, C.; Zhang, D.; Sheng, B. Flexible $BaTiO_3$-PDMS Capacitive Pressure Sensor of High Sensitivity with Gradient Micro-Structure by Laser Engraving and Molding. *Polymers* **2023**, *15*, 3292. [CrossRef]
20. Wang, H.; Li, Z.; Liu, Z.; Fu, J.; Shan, T.; Yang, X.; Lei, Q.; Yang, Y.; Li, D. Flexible capacitive pressure sensors for wearable electronics. *J. Mater. Chem. C* **2022**, *10*, 1594–1605. [CrossRef]

21. Lei, Y.; Yang, J.; Xiong, Y.; Wu, S.; Guo, W.; Liu, G.-S.; Sun, Q.; Wang, Z.L. Surface engineering AgNW transparent conductive films for triboelectric nanogenerator and self-powered pressure sensor. *Chem. Eng. J.* **2023**, *462*, 142170. [CrossRef]
22. Zhong, Y.; Wang, J.; Wu, L.; Liu, K.; Dai, S.; Hua, J.; Cheng, G.; Ding, J. Dome-Conformal Electrode Strategy for Enhancing the Sensitivity of $BaTiO_3$-Doped Flexible Self-powered Triboelectric Pressure Sensor. *ACS Appl. Mater. Interfaces* **2023**, *16*, 1727–1736. [CrossRef]
23. Wang, B.; Liu, C.; Xiao, Y.; Zhong, J.; Li, W.; Cheng, Y.; Hu, B.; Huang, L.; Zhou, J. Ultrasensitive cellular fluorocarbon piezoelectret pressure sensor for self-powered human physiological monitoring. *Nano Energy* **2017**, *32*, 42–49. [CrossRef]
24. Cao, S.; Li, R.; Panahi-Sarmad, M.; Chen, T.; Xiao, X. A Flexible and Highly Sensitive Capacitive Pressure Sensor with Microstructured Dielectric TPU Layer Based on Mesh Fabric as Template. *IEEE Sens. J.* **2022**, *22*, 20276–20284. [CrossRef]
25. Huang, Y.; Fan, X.; Chen, S.-C.; Zhao, N. Emerging Technologies of Flexible Pressure Sensors: Materials, Modeling, Devices, and Manufacturing. *Adv. Funct. Mater.* **2019**, *29*, 1808509. [CrossRef]
26. Ye, J.; Chen, K.; Chen, L.; You, Z.; Jiang, J.; Wu, H. Highly linear capacitive tactile sensor with elastic dome-shaped electrodes. *Smart Mater. Struct.* **2022**, *31*, 075002. [CrossRef]
27. Lipomi, D.J.; Vosgueritchian, M.; Tee, B.C.K.; Hellstrom, S.L.; Lee, J.A.; Fox, C.H.; Bao, Z. Skin-like pressure and strain sensors based on transparent elastic films of carbon nanotubes. *Nat. Nanotechnol.* **2011**, *6*, 788–792. [CrossRef]
28. Viry, L.; Levi, A.; Totaro, M.; Mondini, A.; Mattoli, V.; Mazzolai, B.; Beccai, L. Flexible Three-Axial Force Sensor for Soft and Highly Sensitive Artificial Touch. *Adv. Mater.* **2014**, *26*, 2659–2664. [CrossRef]
29. Zhang, H.; Wang, M.Y.; Li, J.; Zhu, J. A soft compressive sensor using dielectric elastomers. *Smart Mater. Struct.* **2016**, *25*, 035045. [CrossRef]
30. Ariati, R.; Sales, F.; Souza, A.; Lima, R.A.; Ribeiro, J. Polydimethylsiloxane Composites Characterization and Its Applications: A Review. *Polymers* **2021**, *13*, 4258. [CrossRef]
31. Li, R.; Zhou, Q.; Bi, Y.; Cao, S.; Xia, X.; Yang, A.; Li, S.; Xiao, X. Research progress of flexible capacitive pressure sensor for sensitivity enhancement approaches. *Sens. Actuators A-Phys.* **2021**, *321*, 112425. [CrossRef]
32. Mannsfeld, S.C.B.; Tee, B.C.K.; Stoltenberg, R.M.; Chen, C.; Barman, S.; Muir, B.V.O.; Sokolov, A.N.; Reese, C.; Bao, Z.N. Highly sensitive flexible pressure sensors with microstructured rubber dielectric layers. *Nat. Mater.* **2010**, *9*, 859–864. [CrossRef] [PubMed]
33. Yang, J.C.; Kim, J.-O.; Oh, J.; Kwon, S.Y.; Sim, J.Y.; Kim, D.W.; Choi, H.B.; Park, S. Microstructured Porous Pyramid-Based Ultrahigh Sensitive Pressure Sensor Insensitive to Strain and Temperature. *ACS Appl. Mater. Interfaces* **2019**, *11*, 19472–19480. [CrossRef] [PubMed]
34. Wang, S.; Fan, X.; Zhang, Z.; Su, Z.; Ding, Y.; Yang, H.; Zhang, X.; Wang, J.; Zhang, J.; Hu, P. A Skin-Inspired High-Performance Tactile Sensor for Accurate Recognition of Object Softness. *ACS Nano* **2024**, *18*, 17175–17184. [CrossRef] [PubMed]
35. Javidi, R.; Zand, M.M.; Majd, S.A. Numerical simulation analysis of flexible capacitive pressure sensors based on porous pyramidal microstructures. *J. Comput. Electron.* **2024**, *23*, 108–121. [CrossRef]
36. Li, X.; Liu, Y.; Ding, Y.; Zhang, M.; Lin, Z.; Hao, Y.; Li, Y.; Chang, J. Capacitive Pressure Sensor Combining Dual Dielectric Layers with Integrated Composite Electrode for Wearable Healthcare Monitoring. *ACS Appl. Mater. Interfaces* **2024**, *16*, 12974–12985. [CrossRef] [PubMed]
37. Rawal, G.; Ghatak, A. Highly sensitive flexible capacitive pressure sensor with structured elastomeric dielectric layers. *J. Micromech. Microeng.* **2024**, *34*, 025012. [CrossRef]
38. Sumer, B.; Koc, I.M. Fabrication of a Flexible Tactile Sensor with Micro-Pillar Array. In Proceedings of the Conference on EUROSENSORS, Freiberg, Germany, 6–9 September 2015; pp. 134–137.
39. Luo, Y.; Shao, J.; Chen, S.; Chen, X.; Tian, H.; Li, X.; Wang, L.; Wang, D.; Lu, B. Flexible Capacitive Pressure Sensor Enhanced by Tilted Micropillar Arrays. *ACS Appl. Mater. Interfaces* **2019**, *11*, 17796–17803. [CrossRef]
40. Jiang, C.; Sheng, B. Linear Capacitive Pressure Sensor with Gradient Architecture through Laser Ablation on MWCNT/Ecoflex Film. *Polymers* **2024**, *16*, 962. [CrossRef]
41. Lee, S.; Liang, X.; Kim, J.S.; Yokota, T.; Fukuda, K.; Someya, T. Permeable Bioelectronics toward Biointegrated Systems. *Chem. Rev.* **2024**, *124*, 6543–6591. [CrossRef]
42. Ha, K.H.; Huh, H.; Li, Z.J.; Lu, N.S. Soft Capacitive Pressure Sensors: Trends, Challenges, and Perspectives. *ACS Nano* **2022**, *16*, 3442–3448. [CrossRef] [PubMed]
43. Ma, Z.Y.; Zhang, Y.; Zhang, K.Y.; Deng, H.; Fu, Q. Recent progress in flexible capacitive sensors: Structures and properties. *Nano Mater. Sci.* **2023**, *5*, 265–277. [CrossRef]
44. Zhu, D.; Handschuh-Wang, S.; Zhou, X. Recent progress in fabrication and application of polydimethylsiloxane sponges. *J. Mater. Chem. A* **2017**, *5*, 16467–16497. [CrossRef]
45. Li, W.; Jin, X.; Zheng, Y.; Chang, X.; Wang, W.; Lin, T.; Zheng, F.; Onyilagha, O.; Zhu, Z. A porous and air gap elastomeric dielectric layer for wearable capacitive pressure sensor with high sensitivity and a wide detection range. *J. Mater. Chem. C* **2020**, *8*, 11468–11476. [CrossRef]
46. Jung, Y.; Lee, W.; Jung, K.; Park, B.; Park, J.; Ko, J.; Cho, H. A Highly Sensitive and Flexible Capacitive Pressure Sensor Based on a Porous Three-Dimensional PDMS/Microsphere Composite. *Polymers* **2020**, *12*, 1412. [CrossRef]
47. Atalay, O.; Atalay, A.; Gafford, J.; Walsh, C. A Highly Sensitive Capacitive-Based Soft Pressure Sensor Based on a Conductive Fabric and a Microporous Dielectric Layer. *Adv. Mater. Technol.* **2018**, *3*, 1700237. [CrossRef]

48. Park, S.W.; Das, P.S.; Chhetry, A.; Park, J.Y. A Flexible Capacitive Pressure Sensor for Wearable Respiration Monitoring System. *IEEE Sens. J.* **2017**, *17*, 6558–6564. [CrossRef]
49. Masihi, S.; Panahi, M.; Maddipatla, D.; Hanson, A.J.; Bose, A.K.; Hajian, S.; Palaniappan, V.; Narakathu, B.B.; Bazuin, B.J.; Atashbar, M.Z. Highly Sensitive Porous PDMS-Based Capacitive Pressure Sensors Fabricated on Fabric Platform for Wearable Applications. *ACS Sens.* **2021**, *6*, 938–949. [CrossRef]
50. Chen, S.; Zhuo, B.; Guo, X. Large Area One-Step Facile Processing of Microstructured Elastomeric Dielectric Film for High Sensitivity and Durable Sensing over Wide Pressure Range. *ACS Appl. Mater. Interfaces* **2016**, *8*, 20364–20370. [CrossRef]
51. Jung, S.; Kim, J.H.; Kim, J.; Choi, S.; Lee, J.; Park, I.; Hyeon, T.; Kim, D.-H. Reverse-Micelle-Induced Porous Pressure-Sensitive Rubber for Wearable Human-Machine Interfaces. *Adv. Mater.* **2014**, *26*, 4825–4830. [CrossRef]
52. Karagiorgis, X.; Khandelwal, G.; Beniwal, A.; Chirila, R.; Skabara, P.J.; Dahiya, R. Polydimethylsiloxane Foam-Based Fully 3D Printed Soft Pressure Sensors. *Adv. Intell. Syst.* **2023**. *early view*. [CrossRef]
53. Wang, Y.; Li, Y.; Song, Y.; Gao, Z. Flexible Capacitive Pressure Sensors and Their Applications in Electronic Skin. In Proceedings of the 2023 3rd International Conference on Electronic Information Engineering and Computer Science, Changchun, China, 22–24 September 2024.
54. Zhong, Y.; Wu, L.; Gu, F.; Wang, J.; Dai, S.; Zhu, H.; Cheng, G.; Ding, J. Negative pressure-assisted porous structure with gradient dielectrics design for linearity enhancement of flexible capacitance pressure sensor. *Colloids Surf. A-Physicochem. Eng. Asp.* **2023**, *676*, 132306. [CrossRef]
55. Giustiniani, A.; Guegan, P.; Marchand, M.; Poulard, C.; Drenckhan, W. Generation of Silicone Poly-HIPEs with Controlled Pore Sizes via Reactive Emulsion Stabilization. *Macromol. Rapid Commun.* **2016**, *37*, 1527–1532. [CrossRef] [PubMed]
56. Kim, Y.; Yang, H.; Oh, J.H. Simple fabrication of highly sensitive capacitive pressure sensors using a porous dielectric layer with cone-shaped patterns. *Mater. Des.* **2021**, *197*, 109203. [CrossRef]
57. Yoon, S.; Seok, M.; Kim, M.; Cho, Y.-H. Wearable porous PDMS layer of high moisture permeability for skin trouble reduction. *Sci. Rep.* **2021**, *11*, 938. [CrossRef]
58. Pan, S.; Li, Q.; Xian, Z.; Su, N.; Zeng, F. The Effects of Laser Parameters and the Ablation Mechanism in Laser Ablation of C/SiC Composite. *Materials* **2019**, *12*, 3076. [CrossRef]
59. Liu, Q.; Liu, Y.; Shi, J.; Liu, Z.; Wang, Q.; Guo, C.F. High-Porosity Foam-Based Iontronic Pressure Sensor with Superhigh Sensitivity of 9280 kPa^{-1}. *Nano-Micro Lett.* **2022**, *14*, 21. [CrossRef]
60. Qu, C.; Lu, M.; Zhang, Z.; Chen, S.; Liu, D.; Zhang, D.; Wang, J.; Sheng, B. Flexible Microstructured Capacitive Pressure Sensors Using Laser Engraving and Graphitization from Natural Wood. *Molecules* **2023**, *28*, 5339. [CrossRef]
61. Shin, J.; Ko, J.; Jeong, S.; Won, P.; Lee, Y.; Kim, J.; Hong, S.; Li Jeon, N.; Ko, S.H. Monolithic digital patterning of polydimethylsiloxane with successive laser pyrolysis. *Nat. Mater.* **2021**, *20*, 100–107. [CrossRef]
62. Chortos, A.; Bao, Z.N. Skin-inspired electronic devices. *Mater. Today* **2014**, *17*, 321–331. [CrossRef]
63. Chhetry, A.; Yoon, H.; Park, J.Y. A flexible and highly sensitive capacitive pressure sensor based on conductive fibers with a microporous dielectric for wearable electronics. *J. Mater. Chem. C* **2017**, *5*, 10068–10076. [CrossRef]
64. Chen, Y.; Zhang, P.; Li, Y.; Zhang, K.; Su, J.; Huang, L. Flexible capacitive pressure sensor based on multi-walled carbon nanotubes microstructure electrodes. *J. Phys. D-Appl. Phys.* **2021**, *54*, 155101. [CrossRef]
65. Fu, M.; Zhang, J.M.; Jin, Y.M.; Zhao, Y.; Huang, S.Y.; Guo, C.F. A Highly Sensitive, Reliable, and High-Temperature-Resistant Flexible Pressure Sensor Based on Ceramic Nanofibers. *Adv. Sci.* **2020**, *7*, 8. [CrossRef]
66. Yoon, J.I.; Choi, K.S.; Chang, S.P. A novel means of fabricating microporous structures for the dielectric layers of capacitive pressure sensor. *Microelectron. Eng.* **2017**, *179*, 60–66. [CrossRef]
67. Bijender; Kumar, A. Flexible and wearable capacitive pressure sensor for blood pressure monitoring. *Sens. Bio-Sens. Res.* **2021**, *33*, 100434. [CrossRef]
68. Xia, Y.; Gu, H.; Xu, L.; Chen, X.D.; Kirk, T.V. Extending Porous Silicone Capacitive Pressure Sensor Applications into Athletic and Physiological Monitoring. *Sensors* **2021**, *21*, 1119. [CrossRef]

polymers

MDPI

Article

Flexible Positive Temperature Coefficient Composites (PVAc/EVA/GP-CNF) with Room Temperature Curie Point

Chao Du [1], Yangyang Zhang [2], Jiangmin Lin [1], Guotao Fan [1], Can Zhou [1] and Yan Yu [1,*]

[1] Union Hospital Tongji Medical College and School of Integrated Circuits, Huazhong University of Science and Technology, Wuhan 430074, China
[2] Henan Key Laboratory of Nanocomposites and Applications, Huanghe Science and Technology College, Zhengzhou 450061, China
* Correspondence: yuyan19840218@hust.edu.cn

Abstract: Polymeric positive temperature coefficient (PTC) materials with low switching temperature points are crucial for numerous electronic devices, which typically function within the room temperature range (0–40 °C). Ideal polymeric PTC materials for flexible electronic thermal control should possess a room-temperature switching temperature, low room-temperature resistivity, exceptional mechanical flexibility, and adaptive thermal control properties. In this study, a novel PTC material with a room-temperature switching temperature and superb mechanical properties has been designed. A blend of a semi-crystalline polymer EVA with a low melting temperature (Tm) and an amorphous polymer (PVAc) with a low glass transition temperature (Tg) was prepared. Low-cost graphite was chosen as the conductive filler, while CNF was incorporated as a hybrid filler to enhance the material's heating stability. PVAc0.4/EVA0.6/GP-3wt.% CNF exhibited the lowest room temperature resistivity, and its PTC strength (1.1) was comparable to that without CNF addition, with a Curie temperature of 29.4 °C. Room temperature Joule heating tests revealed that PVAc0.4/EVA0.6/GP-3wt.% CNF achieved an equilibrium temperature of approximately 42 °C at 25 V, with a heating power of 3.04 W and a power density of 3.04 W/cm^2. The Young's modulus of PVAc0.4/EVA0.6/GP-3wt.% CNF was 9.24 MPa, and the toughness value was 1.68 MJ/m^3, indicating that the elasticity and toughness of the composites were enhanced after mixing the fillers, and the mechanical properties of the composites were improved by blending graphite with CNF.

Keywords: polymeric positive temperature coefficient; room temperature Curie Point; EVA and PVAc blends; graphite and CNT co-fillers

Citation: Du, C.; Zhang, Y.; Lin, J.; Fan, G.; Zhou, C.; Yu, Y. Flexible Positive Temperature Coefficient Composites (PVAc/EVA/GP-CNF) with Room Temperature Curie Point. *Polymers* **2024**, *16*, 2028. https://doi.org/10.3390/polym16142028

Academic Editors: Jiangtao Xu and Sihang Zhang

Received: 26 May 2024
Revised: 27 June 2024
Accepted: 2 July 2024
Published: 16 July 2024

1. Introduction

Conductive polymer composites (CPCs) are of widespread interest due to their versatility and dexterity in various engineering applications, such as in overcurrent protectors, electronic devices, electromagnetic shielding, flexible sensors, and self-regulating heaters [1–7]. Specifically, certain CPCs exhibit a thermal resistance behavior, whereby the resistivity increases sharply when the temperature exceeds a certain point, referred to as the switching temperature. This is attributed to the severe thermal expansion near the melting point of the polymer matrix, resulting in an increase in the spacing between conductive particles. This phenomenon is defined as the positive temperature coefficient (PTC) effect and is caused by the mismatch of thermal expansion between the polymer matrix and the filler, as well as by the disruption of the conductive network [8,9]. In general, above the melting point, the resistivity decreases inversely with increasing temperature, exhibiting a negative temperature coefficient effect (NTC) [10]. However, the NTC effect can significantly impact the performance of PTC materials due to the aggregation of conductive particles and the random reconfiguration of the conductive network.

To date, there have been numerous attempts to obtain PTC materials with superior performance, but most of the research has concentrated on the high temperature region, with switching temperatures ranging from 50 to 400 °C [11–15]. However, in numerous regions, the operating temperature range of electronic devices is often required to be within the room temperature range of 0–40 °C [16–18]. To achieve effective temperature control, PTC materials should possess a switching temperature slightly below the control temperature of the electrical device. Therefore, preparing PTC materials with a low switching temperature ranging from 0 to 40 °C has been a crucial issue [19–21].

Ideal polymeric PTC materials for thermal control should possess room-temperature switching temperatures [22–24], low room-temperature resistivity [25–29], good PTC repeatability [14], high PTC strength [19], good mechanical flexibility [30–33], and adaptive thermal control properties [21]. It has been demonstrated that PTC materials with low Curie temperatures can be achieved using low melting point semi-crystalline polymers and low glass transition temperature amorphous polymers as matrix materials. These materials exhibit excellent PTC repeatability and mechanical properties, with PVAc0.4/EVA0.6/GP composites showing the best Joule heating performance, yet their heating stability requires improvement. For polymer-based PTC composites to be practical, heating stability is crucial, directly affecting the material's service life. To enhance heating stability, focus should be on improving PTC repeatability while avoiding exceeding the material's tolerance temperature for extended periods, as this can also shorten its lifespan. Numerous studies have been conducted to improve PTC repeatability, primarily through filler modification [34,35], hybrid fillers [34,36], polymer blending [37–40], and macroscopic assembly of different PTC materials [41–43]. The current study considered polymer blending to enhance PTC reproducibility during the initial design phase, and the results validated this approach. However, further performance enhancement necessitates considering other methods. Modified fillers can improve compatibility between fillers and polymers, preventing large agglomerates and allowing conductive fillers to be reversibly repositioned. The polymer matrix bonds to filler surface-grafted molecules through physical entanglement or chemical bonding. However, due to the complexity of filler modification and potential conductivity losses with increasing grafting modifiers, this strategy is rarely implemented in practical products. Rational device structure design can also enhance PTC performance, but the manufacturing process is intricate, increasing costs and limiting applications in scenarios requiring lighter, simpler materials, such as aircraft wing de-icing or wearable devices. Hybrid fillers leverage the synergy between fillers to hinder cluster formation. The second filler connects conductive filler clusters over long distances, increasing the likelihood of conductive network formation. This approach often reduces room temperature resistivity while enhancing PTC repeatability. High aspect ratio fillers have been reported to restrict other filler movements, inhibiting the NTC effect [44].

In this paper, carbon nanofibers (CNF) with high aspect ratios were chosen as the second conductive filler to be added to the optimal PVAc0.4/EVA0.6/GP composite in order to further enhance the heating stability of the material. Additionally, the heating stability of the material was tested at low temperatures (<0 °C). The material's heating performance at low temperatures (<0 °C) was also examined to explore its potential for de-icing applications.

2. Experimental

2.1. Materials

Graphite (average size about 6.5 µm, density 2.2 g/cm^3) was purchased from Shenzhen Jinda Power Technology Co. (Shenzhan China) PS (density 1.047 g/cm^3, melt flow index 7.9–8.9 g/10 min, 200 °C/5 kg) was provided by Shanghai Aladdin Biochemical Technology Co. EVA (40 w, Mr = 70,000–120,000, density 0.965 g/cm^3, melt flow index 52 g/10 min, 190 °C/2.16 kg) was purchased from DuPont de Nemours & Company (Guangzhou, China). Toluene (AR, Mr = 92.14) was produced by Shanghai Tianteng Technology Co. (Shanghai, China) CNF (purity > 70%, OD 200–600 nm, length 5–50 µm) was provided by Shanghai Aladdin Biochemical Technology Co. (Shanghai, China).

2.2. Sample Preparation

The PVAc/EVA/GP-CNF composites were prepared using the solution blending method. Firstly, a 2/3 mass ratio of PVAc and EVA (total mass 0.4 g) was added to 2.9 mL of toluene at the same time, followed by magnetic stirring at 60 °C for 12 h until the polymer was completely dissolved. After being dissolved, 20 wt.% of graphite as well as mass ratios (1 wt.%, 2 wt.%, 3 wt.%, 4 wt.%, and 5 wt.%) of CNF were weighed and added to the mixed solution, and then the multi-step cyclic dispersion method was performed: magnetic stirring for 1.5 h, sonication for 10 min, oscillation on a vortex mixer for 1 min, and then magnetic stirring for 10 min, sonication for 10 min, oscillation on a vortex mixer for 1 min. Immediately after the dispersion process, the mixed solution in the bottle was sucked up with a rubber-tipped dropper and squeezed out onto a clean glass sheet, which was naturally stretched on the surface using the tension of the liquid. The glass sheet was put into an oven at 60 °C and kept warm for 24 h to allow the solvent to evaporate. After the end of the holding period, it was cooled to room temperature and then annealed: the temperature was slowly increased to 70 °C for 10 min, and then cooled to room temperature. Finally, the sample was obtained by stripping the film, and the thickness of the sample was about 200 µm. The composition of the polymer and conductive filler used in the experiment is shown in Table 1.

Table 1. Surface free energies of the fillers and polymers.

Materials	Total Surface Energy γ (mJ·m^{-2})	Dispersion Component γ^d (mJ·m^{-2})	Polar Component γ^p (mJ·m^{-2})
EVA (40 wt.%VA)	35.9	32.5	3.4
PVAc	36.5	25.1	11.4
Graphite	52.8	41.1	11.7
CNF	94.7	92.1	2.6

2.3. Characterization

The microscopic morphology of the fracture surface and the distribution of graphite in the co-polymer were characterized by field emission scanning electron microscopy. The fracture surfaces of the samples were obtained by breaking the samples after immersion in liquid nitrogen for 5 min. Gold spraying was performed on all studied surfaces before testing.

The thermal properties of PTC materials were determined by differential scanning calorimetry. A sample weighing about 5 mg was heated from 10 °C to 180 °C at a heating rate of 10 °C/min, held at this temperature for 5 min to remove the thermal history, and then cooled from 180 °C to 10 °C at a cooling rate of 10 °C/min, held for 5 min, and then heated from 10 °C to 180 °C at a rate of 10 °C/min.

The DC resistance of the sample along the thickness direction was measured continuously using the PTCR-T characteristic test system with a heating rate of 2 °C/min. The resistivity ρ was calculated as $\rho = RS/\mathrm{d}$, where R, S, and d represent the resistance, area, and thickness of the sample, respectively. The sample was cut to a size of 10 mm × 10 mm before testing, and the upper and lower surfaces of the material were sprayed with gold.

A DC power supply was used to apply a voltage to the samples, and the Joule heating properties of the materials at different voltages were tested by varying the voltage and testing their electrical heating stability. The mechanical properties of the samples were tested using an Instron tensile testing machine to evaluate their mechanical properties.

3. Results and Discussion

3.1. Microscopic Morphological Analysis of PVAc0.4/EVA0.6/GP-CNF Conductive Composites

Figure 1 shows the cross-sectional view of PVAc0.4/EVA0.6/20 wt.% CNF and the corresponding energy dispersive spectroscopy (EDS) image. When the PVAc/EVA mass ratio is 2/3, the morphology of the composite undergoes significant changes when graphite is replaced with the same content of CNF. It has been observed that the region with relatively high O content in the EDS image is primarily PVAc, while the region with low O content is primarily EVA, thus enabling the distinction of regions in Figure 1a based on the EDS (Figure 1b). Initially, when the polymer matrix components were identical, the two phases of the biphasic polymer lost their longitudinal co-continuity after the filler was changed to CNF, resulting in a more distinct division between the EVA and PVAc phases. The EVA phase filled with CNF was more porous and loose, with finer and rougher pores overall. Meanwhile, some pores were observable in the PVAc phase filled with CNF, but they were coarser, and the PVAc phase in the cross-section appeared relatively smoother. Additionally, there were differences in the pattern of CNF filling in the two phases. In the EVA phase, the CNF was smaller, while in the PVAc phase, not only were there some fine CNF interspersed, but also some relatively large CNF tips were visible through the PVAc phase, leading to larger holes in this phase. The reason for this morphological difference may be that EVA is a semi-crystalline polymer, comprising both crystalline and amorphous phases. The introduction of CNF disrupts the crystalline phase structure, resulting in the formation of many loose pores. Conversely, PVAc is a pure amorphous phase with better affinity for CNF, leading to CNF being wrapped in the polymer, resulting in narrower voids and fewer pores.

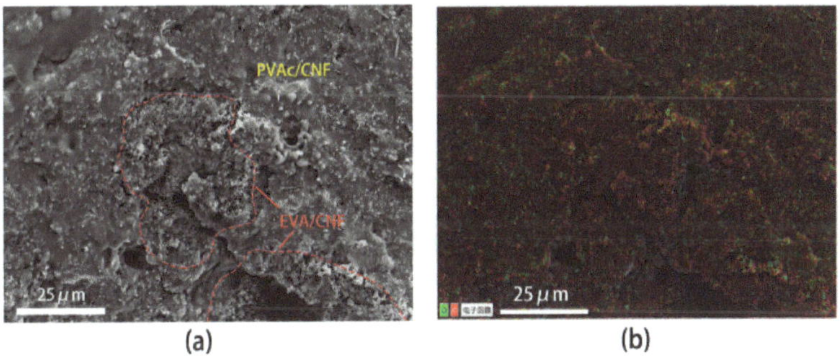

Figure 1. PVAc0.4/EVA0.6/20wt.% CNF: (**a**) SEM image of cross section; (**b**) EDS surface image (the Chinese characters in the figure mean: "Digital Image").

Figure 2 shows SEM cross-sectional images of the PVAc0.4/EVA0.6/GP-CNF composites at different CNF contents. From the figures, it can be seen that the addition of a small amount of second filler CNF on top of 20 wt.% graphite does not significantly affect the morphology of the composites, and the biphasic polymers still exhibit a co-continuous state at these five CNF contents, with no obvious dividing line or partition between the two phases.

According to Figure 3, the distribution of PVAc and EVA phases is similar to that of graphite in the absence of CNF addition. It can be observed that the graphite is well dispersed in the matrix regardless of the CNF content. Graphite plays a significant role in the formation of the conductive network, while CNFs of varying sizes function as a "bridge" connecting graphite particles and forming graphite–CNF–graphite channels. Additionally, some CNFs are directly or indirectly in contact with each other, forming CNF–CNF networks (see Figure S1 in Supporting Information), thus creating a dual conductive network and enhancing the likelihood of forming a conductive pathway. Figure 2f depicts a partial enlargement of the cross-section at a CNF content of 3 wt.%. Notably, CNFs can traverse the polymer phase, serving as a bridge between two-phase regions, which can compensate for the smaller size of graphite particles that may not form long conductive channels alone. This indicates that graphite and CNFs have a synergistic effect on the formation of the conductive network, resulting in a more stable and superior conductive network.

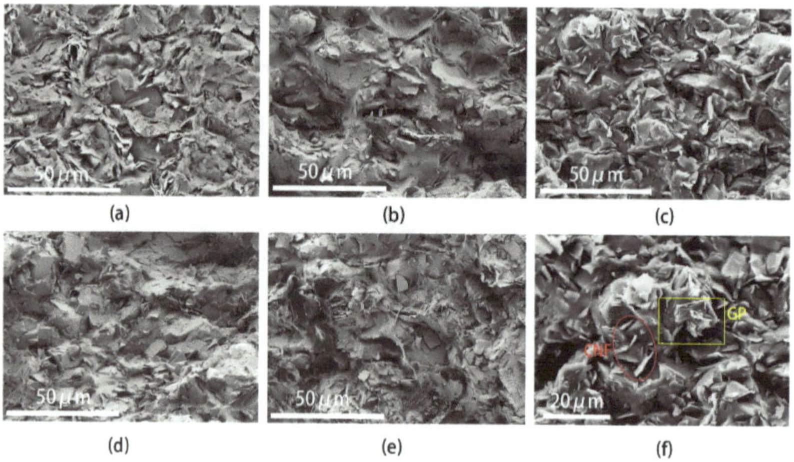

Figure 2. SEM images of PVAc0.4/EVA0.6/GP-CNF composites: SEM images of CNF content of (**a**) 1 wt.%; (**b**) 2 wt.%; (**c**) 3 wt.%; (**d**) 4 wt.%; and (**e**) 5 wt.%. (**f**) Local amplification of (**c**), respectively.

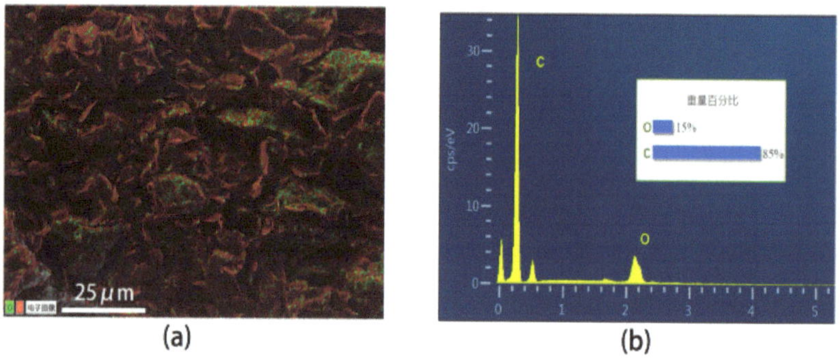

Figure 3. PVAc0.4/EVA0.6/GP-3wt.%CNF composite: (**a**) EDS image: green indicates the distribution of O elements, red indicates the distribution of C elements (the Chinese characters in the figure mean: "Digital Image").; (**b**) EDS energy spectrum (the Chinese characters in the figure mean: "wt.%").

In order to predict the preferential distribution of graphite in the binary-polymer, classical thermodynamics is used. When the equilibrium state is reached, the graphite is either distributed in a specific phase or at the interface of two immiscible polymers, which is dictated by the minimum interfacial energy [45]. The wetting coefficient (ω_a) can assess the equilibrium state of the conductive fillers based on Young's equation, shown as Equation (1) [46]:

$$\omega_a = \frac{\gamma_{CA} - \gamma_{CB}}{\gamma_{AB}} \quad (1)$$

where γ_{CA}, γ_{CB}, and γ_{AB} are the interfacial energies between polymer A and the filler, polymer B and the filler, and between polymer A and polymer B, respectively. The values of $\omega_a > 1$, $\omega_a < -1$, or $-1 < \omega_a < 1$ mean that the fillers would preferentially be localized in polymer B, in polymer A, or at the interface, respectively [47].

The interfacial energies can be calculated from the surface energies of the dispersion and polar parts. According to the type of surfaces, two main approaches are often used to calculate γij, including the harmonic mean equation and the geometric mean equation [48].

Harmonic mean equation:

$$\gamma_{12} = \gamma_1 + \gamma_2 - 4\left(\frac{\gamma_1^d \gamma_2^d}{\gamma_1^d + \gamma_2^d} + \frac{\gamma_1^p \gamma_2^p}{\gamma_1^p + \gamma_2^p}\right) \tag{2}$$

Geometric mean equation:

$$\gamma_{12} = \gamma_1 + \gamma_2 - 2\left(\sqrt{\gamma_1^d \gamma_2^d} + \sqrt{\gamma_1^p \gamma_2^p}\right) \tag{3}$$

where γ_1 and γ_2 are the surface tensions of components 1 and 2; γ_1^d and γ_2^d are the dispersive parts of the surface tensions of components 1 and 2; and γ_1^p and γ_2^p are the polar parts of the surface tension of components 1 and 2.

The results of the interfacial energy obtained using Equations (2) and (3) are presented in Table 2, and the results of the wetting coefficient are shown in Table 3. The wettability coefficients calculated using the harmonic mean equation and the geometric mean equation are 0.3 and 0.4, respectively, suggesting that graphite is primarily situated at the interface between EVA and PVAc, which aligns with our observed findings. However, for CNF, the wettability coefficients determined by the harmonic mean equation and the geometric mean equation are −2.9 and −3.2, respectively, implying that CNF predominantly resides in the EVA phase. Nevertheless, this inference contradicts the observed packing distribution.

Table 2. Interfacial energies as calculated using harmonic mean equation and geometric mean equation.

Materials	Interfacial Energy by Harmonic Mean Equation (mJ·m^{-2})	Interfacial Energy by Geometric Mean Equation (mJ·m^{-2})
EVA/Graphite	5.5	3.0
PVAc/Graphite	3.9	2.0
EVA/CNF	28.6	15.2
PVAc/CNF	43.8	24.1
EVA/PVAc	5.3	2.8

Table 3. Wetting coefficient and predicted location of graphite.

Blends	A	B	ω_a (Harmonic Mean Equation)	ω_a (Geometric Mean Equation)	Predicted Location
Graphite/PVAc/EVA	EVA	PVAc	0.3	0.4	PVAc/EVA interface
CNF/PVAc/EVA	EVA	PVAc	−2.9	−3.2	EVA

3.2. *Influence of Mixed Fillers on Electrical and Thermal Resistance Properties*

Figure 4 shows the room temperature resistivity of PVAc0.4/EVA0.6/GP-CNF with different levels of CNF addition (errors are given in Supporting Information). The room temperature resistivity increased by three orders of magnitude after adding only 1 wt.% CNF or 2 wt.% CNF compared to that without CNF. As the CNF content continued to increase, the room temperature resistivity decreased and then increased, reaching a minimum value of 205.4 Ω·m at 3 wt.% CNF, approximately twice the resistivity without the addition of CNF. The reason for this trend may be that when 1 wt.% CNF or 2 wt.% CNF is added, the amount of CNF is too small not only to form a new conductive chain containing more CNF, but also the insertion of CNF may affect the van der Waals forces between graphite [49] and thus destroy the graphite–graphite conductive network, resulting in a higher resistivity at this time. With the addition of more CNF, CNF starts to participate in the composition of the conducting network, and graphite–CNF–graphite and CNF–CNF conducting channels appear. However, continuing to increase the CNF content after more than 3 wt.% will make it difficult to disperse the CNFs with high aspect ratios by entangling them with each other [40], and the CNF entanglement phenomenon can be observed in Figure 2e. The large amount of CNF agglomeration will reduce the utilization of CNF and make less CNF form the conductive network, and it will affect the composition of the full conductive network, so the continued addition of CNF will instead increase the resistivity.

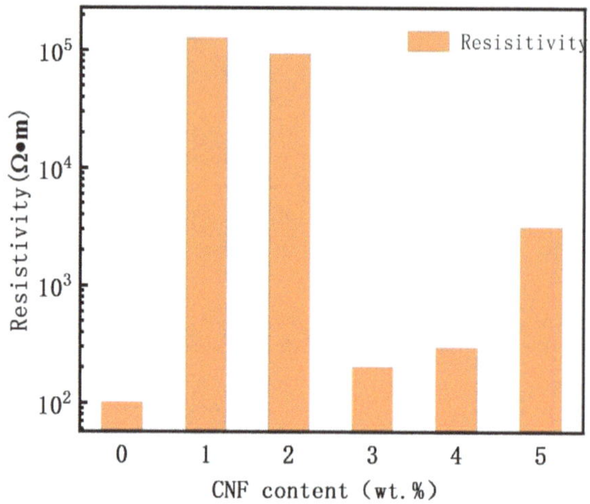

Figure 4. Room temperature resistivity of PVAc0.4/EVA0.6/GP-CNF at different CNF contents.

The temperature resistance characteristic curves of the PVAc0.4/EVA0.6/GP-CNF composites are depicted in Figure 5. When the CNF content is low, the PTC strength is similar to that without CNF addition. The PTC curves at 1 wt.% and 2 wt.% CNF content are similar, suggesting that the conductive network within the material may be comparable at this point and inferior to the graphite conductive network without CNF, resulting in a higher room temperature resistivity. At 3 wt.% CNF content, the PTC strength and room temperature resistivity are closest to those without addition, and there is almost no NTC phenomenon, indicating potential good PTC repeatability. As the CNF content increases, the PTC strength decreases, with 4 wt.% and 5 wt.% CNF exhibiting weaker PTC effects. This is attributed to the rise in conductive particles within the matrix, enhancing the conductive particle percentage and conductive channel formation probability. During warming, conductive fillers may be displaced from their equilibrium positions, moving closer to other conductive particles, enabling the formation of new conductive channels. This second filler addition partially negates the PTC effect. The high CNF content facilitates easier reconfiguration of the conductive network during warming after deconstruction.

Collectively, it seems that the composite with 3 wt.% CNF has the best PTC performance. Therefore, the PTC cycling curve of PVAc0.4/EVA0.6/GP-3wt.%CNF was further tested to investigate its PTC reproducibility. From Figure 6a, it was found that the material exhibits a very weak NTC effect only during the first heating, and the NTC effect disappears during subsequent heating. It can even be observed that the trend at the end of the curve is gradually upward, with only the first curve ending slightly downward. The second curve shows a long "plateau" after 50 °C, and the third to fifth curves are very similar and no longer have a "plateau". However, the end of the curve goes up, indicating that the PTC effect of the material continues from room temperature to the test cutoff temperature. The sixth and seventh cycles also maintain this pattern, with the difference being that the room temperature resistivity remains largely unchanged for the first five cycles but slightly increases for the last two. The gradual elimination of the NTC effect and even the transition to a PTC effect with increasing heating times is an interesting phenomenon, as it is contrary to the results and patterns of many studies [27,50]. This indicates that the

PTC of this material is reproducible and does not deteriorate quickly with increasing use. The significant increase in resistivity in the sixth and seventh cycles compared to previous cycles may be attributed to irrecoverable changes in the structure, volume, aggregation state, and distribution of the conductive fillers in the material after the sixth cycle. This can also be supported by the changes in Curie temperature [51]. From Figure 6b, it can be observed that the Curie temperature is similar for the third, fourth, and fifth cycles, but significantly decreases for the sixth and seventh cycles. Since the Curie temperature is related to the phase transformation process of the composite [29], this change indicates that the material undergoes a different phase transformation in the latter two cycles compared to before.

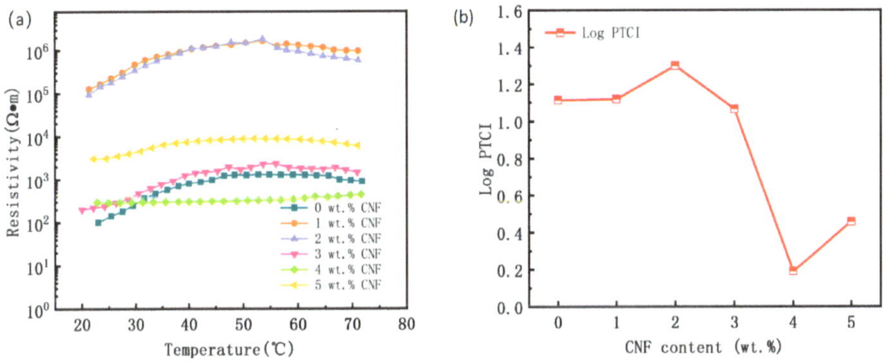

Figure 5. PVAc0.4/EVA0.6/GP-CNF composites with different CNF contents. (**a**) Resistance temperature characteristic curve; (**b**) PTC intensity.

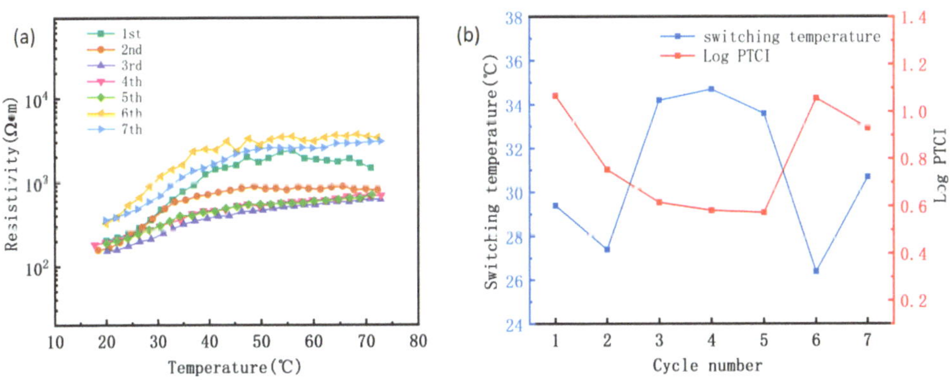

Figure 6. PVAc0.4/EVA0.6/GP-3wt.%CNF composite heating cycle. (**a**) Resistance temperature characteristic curve; (**b**) Curie temperature and PTC intensity.

The Curie temperature during the initial heating of PVAc0.4/EVA0.6/GP-3wt.%CNF increased by approximately 1 °C compared to the material without CNF, reaching 29.4 °C, and remained at approximately 30 °C during cyclic heating. The PTC strength decreased with the number of cycles during the first few cycles, and increased to values close to the initial during the last two cycles. Taken together, it appears that this material has excellent PTC repeatability, which also indicates its potential for long-term use.

To investigate the effect of CNF addition on the melting point and glass transition temperature of the polymer matrix, PVAc0.4/EVA0.6/GP without CNF and three typical concentrations of the material with CNF added were selected for DSC testing. The results are shown in Figure 7. It can be seen that the addition of CNF has essentially no effect on the glass transition temperature and melting temperature of the copolymer matrix, which are around −28 °C for these samples, while the melting peaks at 51 °C and 105 °C also largely overlap. However, with an increase in CNF content, the peak height of the melt peak at 105 °C gradually decreased, indicating that a high content of CNF would affect the formation of crystal structures in EVA and reduce its crystallinity.

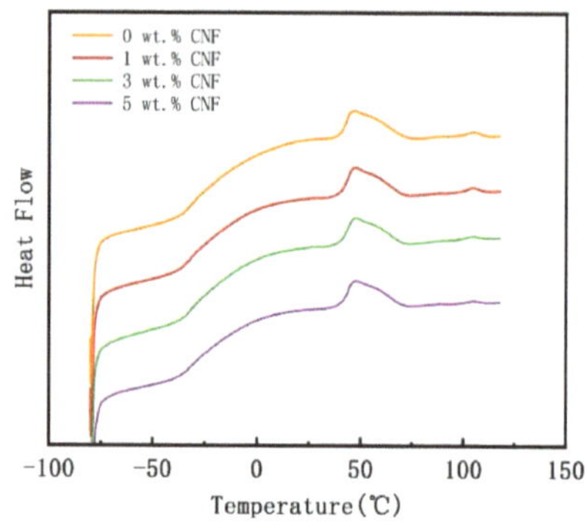

Figure 7. DSC curves of PVAc0.4/EVA0.6/GP-CNF at different CNF contents.

Additional details regarding this Section 3.2 can be found in the Supporting Information (Figures S2–S5 and the related discussions in Supporting Information).

3.3. Impact of Mixed Fillers on Self-Limiting Performance

Based on the results of previous PTC performance tests of the composites, the PVAc0.4/EVA0.6/GP-3wt.%CNF composite was selected as the main object for the next Joule heating performance tests.

(1) Room temperature ambient heating test

From Figure 8a, it can be seen that with an increase in voltage, the material gradually reaches the self-limiting temperature. At 20 V and 25 V, the material shows good self-limiting ability, reaching the self-limiting temperature in approximately 1 min. After this, the temperature increases slowly with time to reach equilibrium temperatures of 42 °C and 40 °C, respectively. However, when the voltage is 30 V, the temperature of the material rises rapidly to 40 °C and then increases to 61 °C due to the large heating power, indicating a lack of good self-limiting ability.

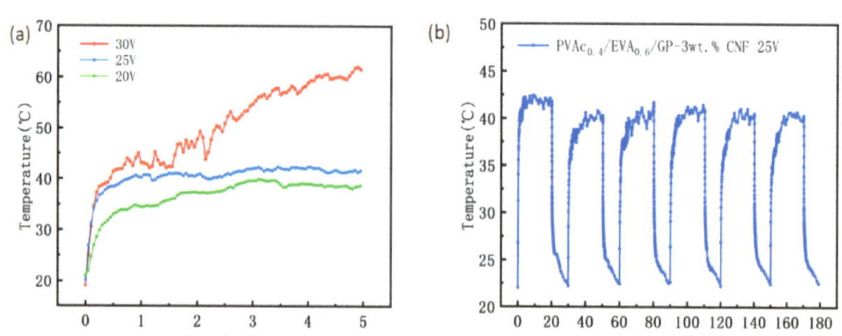

Figure 8. Self-limiting heating of PVAc0.4/EVA0.6/GP-3wt.% CNF composites. (**a**) Temperature-time curves at 20 V, 25 V, and 30 V, (**b**) cyclic heating-cooling test at 25 V cyclic voltage.

To evaluate its electrical heating cycle stability and compare it with the previous material, it was chosen to apply 25 V for six electrical heating tests, and the results are presented in Figure 8b. The initial heating power was 3.04 W, and the power density was 3.04 W/cm^2, approximately six times higher than that of the PS0.4/EVA0.6/GP material and half of the PVAc0.4/EVA0.6/GP material. It is observed that the equilibrium temperature during the first heating was approximately 42 °C. The subsequent five heating processes involved a period of temperature fluctuations within a specific range, not exceeding 2 °C overall, and stabilizing near 40 °C. Although the equilibrium temperature during the last five heating cycles was lower than the first one, the lowest equilibrium temperature was only reduced by approximately 2 °C compared to the first, demonstrating excellent electrical heating stability. This indicates that, although the initial heating power decreases after adding CNF, the electrical heating stability of the composite significantly improves, enhancing its overall self-limiting heating performance. This has a significant impact on the practical application of the material and provides a reference solution to the problem of short service life in low-temperature polymer-based PTC materials, suggesting that this material has great potential for use in low-temperature heating and temperature control applications.

(2) Low-temperature environment heating test

In order to test the electrical heating performance of the material in a low-temperature environment where it can freeze, we cut a PVAc0.4/EVA0.6/GP-3wt.%CNF sample to a size of 20 mm × 20 mm, and coated both sides of the sample with silver paste. Then, the sample was covered with copper foil as the electrode, the DC voltage source used before was connected, and the sample was heated at different voltages, and its temperature variation with time was plotted. The graphs were plotted against time. In this experiment, the heating was started from an ambient temperature of about −10 °C.

As the applied voltage increased and the heating power density increases, the heating rate and the equilibrium temperature of the sample gradually increase (Figure 9). When the applied voltage is less than 5 V, the sample temperature cannot reach the Curie temperature due to the low heating power. When the applied voltage reaches 5 V and above, it takes only about 1 min to reach the equilibrium temperature and keep it around 30 °C, i.e., it remains stable near the Curie temperature. This equilibrium temperature is lower than the equilibrium temperature achieved at room temperature, as the low-temperature environment accelerates the heat dissipation rate of the material, ultimately resulting in a lower equilibrium temperature. The 5 V and 6 V heating curves are closer and show similar heating patterns. In the initial stage, due to the low resistivity, the heating power is high, resulting in a fast heating rate and rapid temperature rise. However, as the Curie temperature is approached, the resistivity of the material rises significantly, leading to a decrease in heating power and a subsequent decrease in the heating rate, thus maintaining the equilibrium temperature near the Curie temperature. In the subsequent heating process, the resistance fluctuates with temperature, but the heating power is automatically adjusted. The material will automatically reduce the power when overheating and increase the power when cooling, ultimately achieving the self-limiting capability of the material. Given this, at these voltages, PVAc0.4/EVA0.6/GP-3wt.%CNF exhibits good automatic temperature control in low-temperature environments, indicating its potential application in the field of de-icing.

(3) Breakage heating test

The heating performance of PVAc0.4/EVA0.6/GP-3wt.%CNF on complex surfaces is shown in Figure 10, with the same upper electrode as in the low temperature environment test. The infrared images of the material in the rectangular, "H", and "back" heating states are shown in the figure.

It is found that the material can be heated successfully in intact, complex, and broken states, and the equilibrium temperature is over 40 °C. Since the surface of the material is covered with a silver paste layer and a copper foil layer, and the metal blocks infrared radiation, the infrared image cannot fully show the internal heat distribution of the material.

However, based on the infrared image, the surface and edge of the material are warmer than the ambient temperature (25 °C), indicating that the damaged sample can still be heated normally. Figure 10d shows the surface temperature of the sample directly tested with a thermocouple, reaching 50 °C, which also confirms the heating function of the damaged sample. This is due to the use of a double-sided electrode, and the partial damage of the sample does not affect the heating performance in other areas, demonstrating the reliability of the material in some actual harsh environments. If the heating material is broken due to external factors, this is a fatal issue for conventional metallic heating materials used in aircraft de-icing. The results also demonstrate that this material can be used in applications requiring irregular or complex shapes, where many heating materials have previously been unable to achieve heating on complex surfaces [52,53].

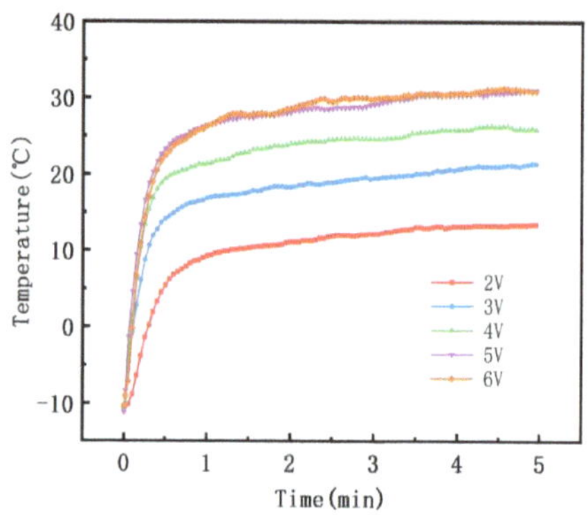

Figure 9. Electrical heating performance test of PVAc0.4/EVA0.6/GP-3wt.%CNF at different voltages at −10 °C.

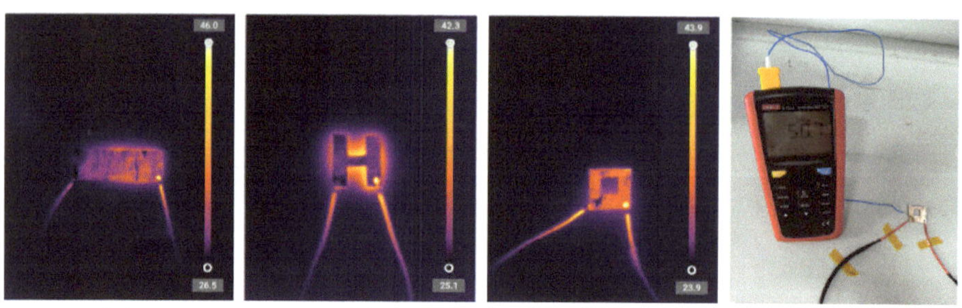

Figure 10. Infrared imaging of PVAc0.4/EVA0.6/GP-3wt.%CNF composite heated in different states: (**a**) rectangle; (**b**) "H" shape; (**c**) "back" shape; (**d**) thermocouple test of the surface temperature of the "H" shaped sample.

Additional details regarding this Section 3.3 can be found in the Supporting Information (Figure S6 and the related discussions in Supporting Information).

3.4. Effect of Mixed Filler on Mechanical Properties

Figure 11a depicts the stress–strain curves of PVAc0.4/EVA0.6/GP and PVAc0.4/EVA0.6/GP-3wt.%CNF. Upon observing the two curves, it is evident that the strain at break of the material increased considerably from 85% to 142% after the addition of 3 wt.% CNF. This indicates a significant enhancement in the ductility of the material following the addition of the second-phase filler, CNF. However, the ultimate tensile strength of the material decreased slightly after the inclusion of CNF, suggesting a reduction in the stress at the onset of necking. The Young's modulus of PVAc0.4/EVA0.6/GP-3wt.%CNF was calculated to be 9.24 MPa, and the toughness value was 1.68 MJ/m^3 (while the Young's modulus of PVAc0.4/EVA0.6/GP was calculated to be 7.92 MPa and the toughness value was 1.23 MJ/m^3). This implies that the stresses required to induce elastic deformation in this material are relatively lower, and the composites with mixed fillers exhibit greater elasticity. Furthermore, the toughness value with the addition of CNF is higher than without it, demonstrating that the hybrid filler also enhances the toughness of the material. In summary, the blending of graphite and CNF fillers appears to have improved the overall mechanical properties of the composite.

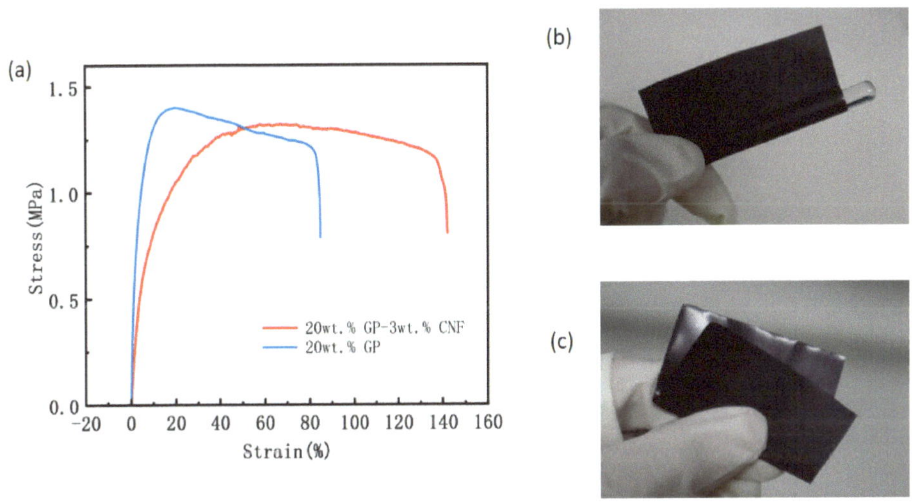

Figure 11. (**a**) Stress–strain curves of PVAc0.4/EVA0.6/GP and PVAc0.4/EVA0.6/GP-3wt.%CNF. PVAc0.4/EVA0.6/GP-3wt.%CNF (**b**) winding around glass rod and (**c**) folded physical diagram.

Figure 11b,c show the physical images of PVAc0.4/EVA0.6/GP-3wt.%CNF wound and folded around a glass rod with a diameter of 5 mm, respectively, as can be seen from the figures. This film has very good flexibility and can be bent and folded at will, which can be applied to complex shapes or curved surfaces.

4. Conclusions

In this paper, PVAc0.4/EVA0.6/GP composites with varying CNF contents were prepared by incorporating CNF as a secondary conductive filler into the PVAc0.4/EVA0.6/GP composite with the best overall performance. This approach enhanced the stability of the conductive network and the thermal stability of the materials. The study also examined the impact of CNF content on the microscopic morphology, electrical conductivity, and PTC properties of the composites. Additionally, the Joule heating properties and mechanical properties of the optimal composition, PVAc0.4/EVA0.6/GP-3wt.%CNF, were tested at room and low temperatures. The PTC material exhibited a low switching temperature point (<30 °C), low room temperature resistivity (205.4 Ω·m), high PTC repeatability and flexibility, high heating stability and reliability, short heating response time, and excellent adaptive thermal control performance.

Supplementary Materials: The following supporting information can be downloaded at: https://www.mdpi.com/article/10.3390/polym16142028/s1, Figure S1. Cross section SEM images of $PVAc_{0.4}/EVA_{0.6}$/GP-CNF composites; Figure S2. Cross section SEM images of four different PVAc/EVA/GP composite; Figure S3. EDS diagrams of the two composite materials; Figure S4. Cross section SEM images of $PVAc_{0.4}/EVA_{0.6}$/GP-CNF composites after heating; Figure S5. FTIR result of $PVAc_{0.4}/EVA_{0.6}$/GP-CNF composites; Figure S6. (a) Influence of humidity on resistance of the composites, (b) application of the composites as temperature sensor on skin.

Author Contributions: Methodology, C.D.; Formal analysis, J.L.; Investigation, Y.Z.; Data curation, C.Z.; Writing—original draft, C.D.; Visualization, G.F.; Supervision, Y.Y. All authors have read and agreed to the published version of the manuscript.

Funding: This research received no external funding.

Institutional Review Board Statement: Not applicable.

Data Availability Statement: The original contributions presented in the study are included in the article/Supplementary Material, further inquiries can be directed to the corresponding author/s.

Conflicts of Interest: The authors declare no conflict of interest.

References

1. Liu, Y.; Zhang, H.; Porwal, H.; Busfield, J.J.; Peijs, T.; Bilotti, E. Pyroresistivity in conductive polymer composites: A perspective on recent advances and new applications. *Polym. Int.* **2018**, *68*, 299–305. [CrossRef]
2. Bautista-Quijano, J.R.; Pötschke, P.; Brunig, H.; Heinrich, G. Strain sensing, electrical and mechanical properties of polycarbonate/multiwall carbon nanotubemonofilament fibers fabricated by melt spinning. *Polymer* **2016**, *82*, 181–189. [CrossRef]
3. Yu, F.; Deng, H.; Zhang, Q.; Wang, K.; Zhang, C.; Chen, F.; Fu, Q. Anisotropic multilayer conductive networks in carbon nanotubes filled polyethylene/polypropylene blends obtained through high speed thin wall injection molding. *Polymer* **2013**, *54*, 6425–6436. [CrossRef]
4. Kim, D.C.; Shim, H.J.; Lee, W.; Koo, J.H.; Kim, D.-H. Material-Based Approaches for the Fabrication of Stretchable Electronics. *Adv. Mater.* **2020**, *32*, e1902743. [CrossRef] [PubMed]
5. Beletskii, E.; Ershov, V.; Danilov, S.; Lukyanov, D.; Alekseeva, E.; Levin, O. Resistivity-Temperature Behavior of Intrinsically Conducting Bis(3-methoxysalicylideniminato)nickel Polymer. *Polymers* **2021**, *12*, 2925. [CrossRef]
6. Liu, Y.; Zhang, H.; Porwal, H.; Tu, W.; Wan, K.; Evans, J.; Newton, M.; Busfield, J.J.C.; Peijs, T.; Bilotti, E. Tailored pyroresistive performance and flexibility by introducing a secondary thermoplastic elastomeric phase into graphene nanoplatelet (GNP) filled polymer composites for self-regulating heating devices. *J. Mater. Chem. C* **2018**, *6*, 2760–2768. [CrossRef]
7. Zhang, L.; Du, W.; Nautiyal, A.; Liu, Z.; Zhang, X. Recent progress on nanostructured conducting polymers and composites: Synthesis, application and future aspects. *Sci. China Mater.* **2018**, *61*, 303–352. [CrossRef]
8. Feng, J.; Chan, C.-M. Positive and negative temperature coefficient effects of an alternating copolymer of tetrafluoroethylene–ethylene containing carbon black-filled HDPE particles. *Polymer* **2000**, *41*, 7279–7282. [CrossRef]
9. Zhang, X.; Zheng, S.; Zheng, X.; Liu, Z.; Yang, W.; Yang, M. Distinct positive temperature coefficient effect of polymer–carbon fiber composites evaluated in terms of polymer absorption on fiber surface. *Phys. Chem. Chem. Phys.* **2016**, *18*, 8081–8087. [CrossRef]
10. Xiang, Z.; Chen, T.; Li, Z.; Bian, X. Negative Temperature Coefficient of Resistivity in Lightweight Conductive Carbon Nanotube/Polymer Composites. *Macromol. Mater. Eng.* **2009**, *294*, 91–95. [CrossRef]
11. Zhang, X.; Zheng, S.; Zou, H.; Zheng, X.; Liu, Z.; Yang, W.; Yang, M. Two-step positive temperature coefficient effect with favorable reproducibility achieved by specific "island-bridge" electrical conductive networks in HDPE/PVDF/CNF composite. *Compos. Part A Appl. Sci. Manuf.* **2017**, *94*, 21–31. [CrossRef]
12. Zhang, R.; Tang, P.; Li, J.; Xu, D.; Bin, Y. Study on filler content dependence of the onset of positive temperature coefficient (PTC) effect of electrical resistivity for UHMWPE/LDPE/CF composites based on their DC and AC electrical behaviors. *Polymer* **2014**, *55*, 2103–2112. [CrossRef]
13. Zha, J.-W.; Li, W.-K.; Liao, R.-J.; Bai, J.; Dang, Z.-M. High performance hybrid carbon fillers/binary–polymer nanocomposites with remarkably enhanced positive temperature coefficient effect of resistance. *J. Mater. Chem. A* **2013**, *1*, 843–851. [CrossRef]
14. Chen, L.; Wu, X.; Zhang, X.-F.; Zhang, J.-M. Enhanced Reproducibility of Positive Temperature Coefficient Effect of CB/HDPE/PVDF Composites with the Addition of Ionic Liquid. *Chin. J. Polym. Sci.* **2021**, *39*, 228–236. [CrossRef]
15. Seo, M.-K.; Rhee, K.-Y.; Park, S.-J. Influence of electro-beam irradiation on PTC/NTC behaviors of carbon blacks/HDPE conducting polymer composites. *Curr. Appl. Phys.* **2011**, *11*, 428–433. [CrossRef]
16. Wang, Y.-H.; Yang, Y.-T. Three-dimensional transient cooling simulations of a portable electronic device using PCM (phase change materials) in multi-fin heat sink. *Energy* **2011**, *36*, 5214–5224. [CrossRef]
17. Etminan, E.; Alimohammadi, M.; Barani, E.; Taheri, A.; Passandideh-Fard, M.; Sardarabadi, M. On the performance of an innovative electronic chipset thermal management module based on energy storage unit concept utilizing nano-additive phase change material (NPCM). *J. Energy Storage* **2022**, *50*, 104244. [CrossRef]
18. Wang, R.-J.; Cheng, W.-L. A novel flexible room temperature positive temperature coefficient material for thermal management. *Adv. Compos. Hybrid Mater.* **2019**, *2*, 83–92. [CrossRef]
19. Jeon, J.; Lee, H.-B.-R.; Bao, Z. Flexible Wireless Temperature Sensors Based on Ni Microparticle-Filled Binary Polymer Composites. *Adv. Mater.* **2013**, *25*, 850–855. [CrossRef]

20. Lai, F.; Wang, B.; Zhang, P. Enhanced positive temperature coefficient in amorphous PS/CSPE-MWCNT composites with low percolation threshold. *J. Appl. Polym. Sci.* **2019**, *136*, 47053. [CrossRef]
21. Yu, A.; Li, Q. Studies on preparation and thermal control behavior of hybrid fillers/binary-polymer composites with positive temperature coefficient characteristic. *Sci. China Technol. Sci.* **2021**, *64*, 1447–1458. [CrossRef]
22. Cheng, W.-L.; Yuan, S.; Song, J.-L. Studies on preparation and adaptive thermal control performance of novel PTC (positive temperature coefficient) materials with controllable Curie temperatures. *Energy* **2014**, *74*, 447–454. [CrossRef]
23. Luo, X.L.; Schubert, D.W. Positive temperature coefficient (PTC) materials based on amorphous poly (methyl methacrylate) with ultrahigh PTC intensity, tunable switching temperature and good reproducibility. *Mater. Today Commun.* **2022**, *30*, 103078. [CrossRef]
24. Chen, L.; Zhang, J. Designs of conductive polymer composites with exceptional reproducibility of positive temperature coefficient effect: A review. *J. Appl. Polym. Sci.* **2021**, *138*, 49677. [CrossRef]
25. Liu, P.; Liu, J.; Zhu, X.; Wu, C.; Liu, Y.; Pan, W.; Zhao, J.; Guo, X.; Liu, C.; Huang, Y.; et al. A highly adhesive flexible strain sensor based on ultra-violet adhesive filled by graphene and carbon black for wearable monitoring. *Compos. Sci. Technol.* **2019**, *192*, 107771. [CrossRef]
26. Duy, L.T.; Trung, T.Q.; Hanif, A.; Siddiqui, S.; Roh, E.; Lee, W.; Lee, N.-E. A stretchable and highly sensitive chemical sensor using multilayered network of polyurethane nanofibres with self-assembled reduced graphene oxide. *2D Mater.* **2017**, *4*, 025062. [CrossRef]
27. Yokota, T.; Inoue, Y.; Terakawa, Y.; Reeder, J.; Kaltenbrunner, M.; Ware, T.; Yang, K.; Mabuchi, K.; Murakawa, T.; Sekino, M.; et al. Ultraflexible, large-area, physiological temperature sensors for multipoint measurements. *Proc. Natl. Acad. Sci. USA* **2015**, *112*, 14533–14538. [CrossRef]
28. Zhang, Q.-H.; Chen, D.-J. Percolation threshold and morphology of composites of conducting carbon black/polypropylene/EVA. *J. Mater. Sci.* **2004**, *39*, 1751–1757. [CrossRef]
29. Li, J.; Chang, C.; Li, X.; Li, Y.; Guan, G. A New Thermal Controlling Material with Positive Temperature Coefficient for Body Warming: Preparation and Characterization. *Materials* **2019**, *12*, 1758. [CrossRef]
30. Yu, Y.; Jiang, S.; Zhou, W.; Miao, X.; Zeng, Y.; Zhang, G.; Zhang, Y.; Zhang, Q.; Zhao, H. Ultrafast room temperature wiping-rubbing fabrication of graphene nanosheets as flexible transparent conductive films with high surface stability. *Appl. Phys. Lett.* **2009**, *101*, 023119. [CrossRef]
31. Jara, A.D.; Betemariam, A.; Woldetinsae, G.; Kim, J.Y. Purification, application and current market trend of natural graphite: A review. *Int. J. Min. Sci. Technol.* **2019**, *29*, 671–689. [CrossRef]
32. Chehreh Chelgani, S.; Rudolph, M.; Kratzsch, R.; Sandmann, D.; Gutzmer, J. A Review of Graphite Beneficiation Techniques. *Miner. Process. Extr. Met. Rev.* **2016**, *37*, 58–68. [CrossRef]
33. Zhao, Z.; Chen, H.; Liu, X.; Wang, Z.; Zhu, Y.; Zhou, Y. The development of electric heating coating with temperature controlling capability for anti-icing/de-icing. *Cold Reg. Sci. Technol.* **2021**, *184*, 103234. [CrossRef]
34. Liu, K.; Xu, Z.; Mei, J.; Han, J.; Zheng, F.; Wang, H.; Huang, Y.; Wu, Q.; Qin, G.; Jiang, J.; et al. Strategies for improving positive temperature effects in conductive polymer composites—A review. *J. Mater. Chem. C* **2023**, *11*, 4966–4992. [CrossRef]
35. Zha, J.-W.; Wu, D.-H.; Yang, Y.; Wu, Y.-H.; Li, R.K.Y.; Dang, Z.-M. Enhanced positive temperature coefficient behavior of the high-density polyethylene composites with multi-dimensional carbon fillers and their use for temperature-sensing resistors. *RSC Adv.* **2017**, *7*, 11338–11344. [CrossRef]
36. Chen, L.; Hou, J.; Chen, Y.; Wang, H.; Duan, Y.; Zhang, J. Synergistic effect of conductive carbon black and silica particles for improving the pyroresistive properties of high density polyethylene composites. *Compos. Part B Eng.* **2019**, *178*, 107465. [CrossRef]
37. Mamunya, Y.; Maruzhenko, O.; Kolisnyk, R.; Iurzhenko, M.; Pylypenko, A.; Masluchok, O.; Godzierz, M.; Krivtsun, I.; Trzebicka, B.; Pruvost, S. Pyroresistive Properties of Composites Based on HDPE and Carbon Fillers. *Polymers* **2023**, *15*, 2105. [CrossRef] [PubMed]
38. Jiang, Y.; Liu, L.; Yan, J.; Wu, Z. Room-to-low temperature thermo-mechanical behavior and corresponding constitutive model of liquid oxygen compatible epoxy composites. *Compos. Sci. Technol.* **2024**, *245*, 110357. [CrossRef]
39. Li, H.; Zhang, Y.; Tai, Y.; Zhu, X.; Qi, X.; Zhou, L.; Li, Z.; Lan, H. Flexible transparent electromagnetic interference shielding films with silver mesh fabricated using electric-field-driven microscale 3D printing. *Opt. Laser Technol.* **2022**, *148*, 107717. [CrossRef]
40. Deng, H.; Skipa, T.; Zhang, R.; Lellinger, D.; Bilotti, E.; Alig, I.; Peijs, T. Effect of melting and crystallization on the conductive network in conductive polymer composites. *Polymer* **2009**, *50*, 3747–3754. [CrossRef]
41. Ma, L.-F.; Bao, R.-Y.; Dou, R.; Liu, Z.-Y.; Yang, W.; Xie, B.-H.; Yang, M.-B.; Fu, Q. A high-performance temperature sensitive TPV/CB elastomeric composite with balanced electrical and mechanical properties via PF-induced dynamic vulcanization. *J. Mater. Chem. A* **2014**, *2*, 16989–16996. [CrossRef]
42. Xu, H.-P.; Dang, Z.-M.; Jiang, M.-J.; Yao, S.-H.; Bai, J. Enhanced dielectric properties and positive temperature coefficient effect in the binary polymer composites with surface modified carbon black. *J. Mater. Chem.* **2008**, *18*, 229–234. [CrossRef]
43. Liu, Y.; Zhang, H.; Porwal, H.; Tu, W.; Evans, J.; Newton, M.; Busfield, J.J.C.; Peijs, T.; Bilotti, E. Universal Control on Pyroresistive Behavior of Flexible Self-Regulating Heating Devices. *Adv. Funct. Mater.* **2017**, *27*, 1702253. [CrossRef]
44. He, X.J.; Du, J.H.; Ying, Z.; Cheng, H.M. Positive temperature coefficient effect in multiwalled carbon nanotube/high-density polyethylene composites. *Appl. Phys. Lett.* **2005**, *86*, 062112. [CrossRef]

45. Göldel, A.; Marmur, A.; Kasaliwal, G.R.; Pötschke, P.; Heinrich, G. Shape-Dependent Localization of Carbon Nanotubes and Carbon Black in an Immiscible Polymer Blend during Melt Mixing. *Macromolecules* **2011**, *44*, 6094–6102. [CrossRef]
46. Wu, S. *Polymer Interface and Adhesion*, 1st ed.; M. Dekker: New York, NY, USA, 1982.
47. Sumita, M.; Sakata, K.; Asai, S.; Miyasaka, K.; Nakagawa, H. Dispersion of fillers and the electrical conductivity of polymer blends filled with carbon black. *Polym. Bull.* **1991**, *25*, 265–271. [CrossRef]
48. Wu, S. Polar and Nonpolar Interactions in Adhesion. *J. Adhes.* **1973**, *5*, 39–55. [CrossRef]
49. Wu, S.; Li, T.; Tong, Z.; Chao, J.; Zhai, T.; Xu, J.; Yan, T.; Wu, M.; Xu, Z.; Bao, H.; et al. High-Performance Thermally Conductive Phase Change Composites by Large-Size Oriented Graphite Sheets for Scalable Thermal Energy Harvesting. *Adv. Mater.* **2019**, *31*, e1905099. [CrossRef]
50. Zhao, S.G.; Li, G.J.; Liu, H.; Dai, K.; Zheng, G.Q.; Yan, X.R.; Liu, C.; Chen, J.; Shen, C.; Guo, Z. Positive temperature coefficient (ptc) evolution of segregated structural conductive polypropylene nanocomposites with visually traceable carbon black conductive network. *Adv. Mater. Interfaces* **2017**, *4*, 10. [CrossRef]
51. Dai, K.; Zhang, Y.C.; Tang, J.H.; Ji, X.; Li, Z.M. Anomalous attenuation and structural origin of positive temperature coefficient (PTC) effect in a carbon black (CB)/poly(ethylene terephthalate) (PET)/polyethylene (PE) electrically conductive microfibrillar polymer composite with a preferential CB distribution. *J. Appl. Polym. Sci.* **2012**, *125*, E561–E570.
52. Yao, X.; Falzon, B.; Hawkins, S.; Tsantzalis, S. Aligned carbon nanotube webs embedded in a composite laminate: A route towards a highly tunable electro-thermal system. *Carbon* **2018**, *129*, 486–494. [CrossRef]
53. Liu, X.; Chen, H.; Zhao, Z.; Yan, Y.; Zhang, D. Slippery liquid-infused porous electric heating coating for anti-icing and de-icing applications. *Surf. Coatings Technol.* **2019**, *374*, 889–896. [CrossRef]

polymers

MDPI

Article

Optimized Design of Material Preparation for Cotton Linters-Based Carbon Black Dispersion Stabilizers Based on Response Surface Methodology

Xiongfei An, Xupeng Yang, Canming Hu and Chengli Ding *

College of Chemical Engineering, Xinjiang University, Urumqi 830046, China; 107552101230@stu.xju.edu.cn (X.A.)
* Correspondence: dingchengli@xju.edu.cn; Tel.: +86-0991-9582887

Abstract: Carbon black particles possess dimensions on the nanometer or sub-nanometer scale. When utilized, these particles have a tendency to aggregate, which compromises their stability under storage conditions. To address this issue, a dispersant was prepared using cotton short fibers as raw materials through etherification and graft polymerization with acrylamide (AM) and 2-acrylamido-2-methylpropane sulfonic acid (AMPS) as raw materials. The dispersant was then used to disperse carbon black to test its dispersing performance. A response surface optimization test was utilized to ascertain the influence of AMPS monomer mass, AM monomer mass, and potassium persulfate (KPS) initiator mass on the dispersibility of carbon black during dispersant preparation, and a set of optimal preparation conditions were obtained. The dispersion stability of carbon black in water was assessed using Fourier transform infrared spectroscopy (FTIR), X-ray diffraction (XRD), elemental analysis (EA), thermogravimetric analysis (TG), zeta potential analysis, high magnification scanning electron microscopy (SEM), and contact angle measurements. Results revealed that the optimum mass ratio of carboxymethyl cellulose (CMC) to AMPS to AM was 1:0.69:1.67, with the KPS initiator comprising 1.56% of the total monomer mass. By incorporating the dispersant at a concentration of 37.50%, the particle size of carbon black particles was observed to decrease from 5.350 μm to 0.255 μm, and no agglomeration of carbon black particles occurred even after 3 weeks of storage.

Keywords: cotton linters; response surface; dispersant; carbon black

Citation: An, X.; Yang, X.; Hu, C.; Ding, C. Optimized Design of Material Preparation for Cotton Linters-Based Carbon Black Dispersion Stabilizers Based on Response Surface Methodology. *Polymers* **2024**, *16*, 1964. https://doi.org/10.3390/polym16141964

Academic Editors: Jiangtao Xu and Sihang Zhang

Received: 30 May 2024
Revised: 24 June 2024
Accepted: 5 July 2024
Published: 9 July 2024

1. Introduction

Carbon black is a dark powder, usually produced from heavy oil in petroleum refineries, coal, natural gas, and biomass under specific conditions. The composition of carbon black consists of discrete, nearly spherical particles, with the size of each particle varying based on the production method, often reaching the nanometer and sub-nanometer range [1]. Carbon black possesses excellent rubber reinforcement properties. Additionally, it is highly conductive and has anti-static properties. It is also used as a coloring agent and exhibits ultraviolet absorption. Due to these versatile properties, it is extensively utilized in various industries including printing and dyeing, and the production of paint, ink, rubber, plastic, foam, ceramics, silicone, leather, and cement building materials [2–7]. Even though carbon black is used in a variety of applications, it possesses a large specific surface area and exudes high surface free energy, which results in a high propensity for cluster formation and poor dispersion, making it challenging to uniformly disperse the particles within a matrix or substrate [8,9]. The technology for dispersing carbon black is pivotal for enabling practical applications of carbon black materials. Techniques such as vigorous agitation, ultrasonic treatment, and other dispersion methods have been proposed, but the dispersion of carbon black often falls short of expectations, leading to particle clumping and sedimentation [10]. Consequently, enhancing the dispersion performance of carbon black has become the focal point in this field.

Cellulose is an abundantly available polysaccharide that is composed of interconnected glucose molecules through β-1,4-glycosidic bonds [11]. Naturally available cellulose has poor utility due to its low solubility in water and vulnerability to degradation by microorganisms and enzymes [12]. Carboxymethyl cellulose (CMC) is another versatile material that has several useful properties such as water retention, along with thickening, stabilizing, and dispersing capabilities [13–15]. These properties arise from its relatively large specific surface area, the presence of hydroxyl functional groups, and the ability to withstand mechanical stress to a certain extent. In recent years, dispersing stabilizers based on carboxymethyl cellulose have garnered significant attention and interest in cellulose chemistry research. This is mainly due to their biodegradability, ease of chemical modification, and capacity to maintain stable dispersions of nanoparticles and emulsions, including nanosized iron tetraoxide (Fe_3O_4), carbon nanotubes (CNTs), graphene oxide (rGO), and others for applications in pigments, coatings, and functional composite materials [16–19].

This study selected Xinjiang's abundant and low-cost cotton linters as raw materials. Through grafting polymerization modification to introduce functional groups such as amino and sulfonic acid groups, a high-performance waterborne dispersant was successfully prepared, enhancing the economic value of cotton linters. Utilizing response surface methodology for multi-factor analysis and response surface modeling, it optimizes multiple factors and their interactions, clarifying the effects of various process parameters on sample performance. This study determined the optimal process conditions for CMC-AMPS-AM and utilized carbon black to test dispersibility, evaluating the effectiveness of the prepared dispersant.

2. Materials and Methods

2.1. Experimental Drugs and Instruments

The cotton short staple is purchased from Aksu Prefecture, Xinjiang, China, at the Tiancheng Cotton Short Fiber Market Development Co., Ltd. (Detailed composition is listed in Table 1). Acetone (C_3H_6O, analytical grade), nitric acid (HNO_3, analytical grade), urea (analytical grade), potassium persulfate ($K_2S_2O_8$, analytical grade) were purchased in Tianjin, China, all provided by Tianjin Xinbo Chemical Co., Ltd. Anhydrous ethanol (C_2H_5OH, analytical grade) and sodium hydroxide (NaOH, analytical grade) were purchased from Tianjin, China, both supplied by Tianjin Zhiyuan Chemical Reagent Co., Ltd. Acrylamide (C_3H_5NO, Analytical grade) was procured in Tianjin, China and supplied by Tianjin Yongsheng Fine Chemical Co., Ltd. Ammonium cerium nitrate ($Ce(NH_4)_2(NO_3)_6$, Analytical grade) was purchased in Shanghai, China, supplied by Shanghai McLean Biochemical Technology Co., Ltd. 2-acrylamide-2-methylpropanesulfonic acid ($C_7H_{13}NO_4S$, Analytical grade) was purchased from Shanghai, China, supplied by Shanghai Beide Pharmaceutical Technology Co., Ltd. Carbon Black (Analytical grade) procured in Shanghai, China, supplied by Shanghai Yien Chemical Technology Co., Ltd. Chloroacetic acid ($ClCH_2COOH$, Analytical grade), procured in Tianjin, China, supplied by Tianjin BASF Chemical Co., Ltd.

Table 1. Cotton linters composition and content.

Component Name	Cellulose	Pectin	Wax	Lignin	Ash Content
content/%	>95	0.5	1	1~2	1

The DF-101S solar collector-type magnetic heating stirrer was purchased from the Medical Instrument Factory in Jintan, Jiangsu Province, China; the SCIENTZ-10N freeze dryer was purchased from Ningbo Xinzhi Biotechnology Co., Ltd. in Ningbo, China; the BT-9300S laser particle size analyzer was purchased from Dandong Bates Instruments Co., Ltd. in Dandong, China; the QM-3SP04 planetary ball mill equipment was purchased from Nanjing ND Instruments Co., Ltd. in Nanjing, China; and the BZY-101 surface tension meter was purchased from Shanghai Jitai Electronic Technology Co., Ltd. in Shanghai, China.

2.2. Synthesis of CMC-AMPS-AM Hyperdispersan

2.2.1. The Mechanism of Cotton Linters Staple Cellulose Modification

Cellulose from cotton linters is converted into CMC via alkalinization followed by derivatization with chloroacetic acid [20–22]. The product is typically a white-to-grayish-white, water-soluble, odorless solid [23]. The preparation reaction scheme for CMC is shown in Figure 1.

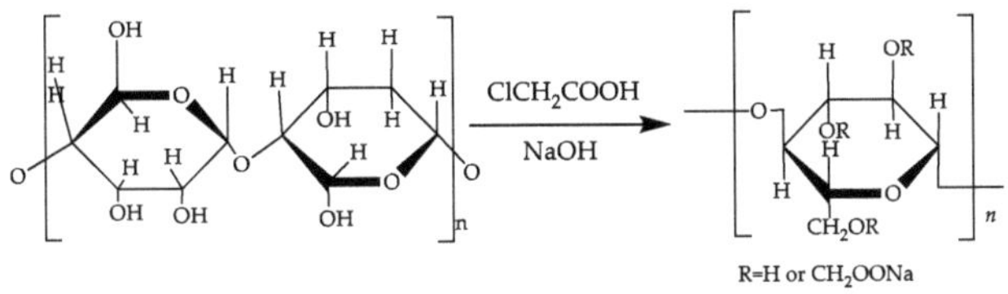

Figure 1. The reaction scheme for preparing CMC from cellulose.

Ceric ammonium nitrate (CAN) grafting mechanism [24]: During the grafting polymerization process, the acidic medium of ceric ammonium nitrate forms chelates with hydroxyl groups of cellulose, particularly between the vicinal hydroxyl groups at C2 and C3 of the glucose units in cellulose. Subsequently, through homolytic cleavage, radicals are generated on the cellulose molecular chain, opening the C2–C3 bond of the glucose ring of cellulose. This allows the cellulose to undergo free radical polymerization with the polymerizable monomers, as shown in Figure 2.

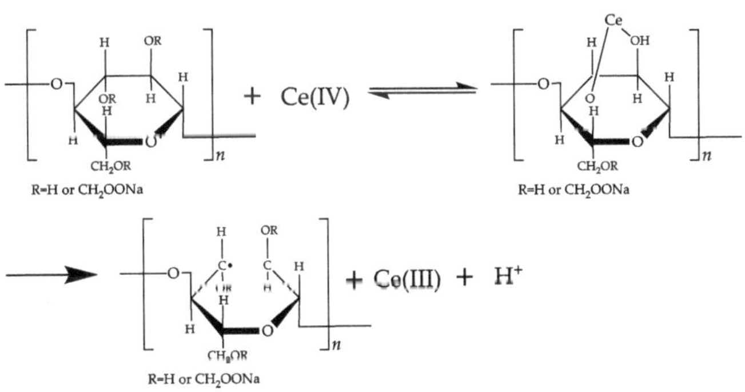

Figure 2. Cerium ions form complex compounds with hydroxyl groups.

Using ceric ammonium nitrate [24] and potassium persulfate [25] to initiate monomer polymerization reactions, the main reactions are depicted in Figure 3.

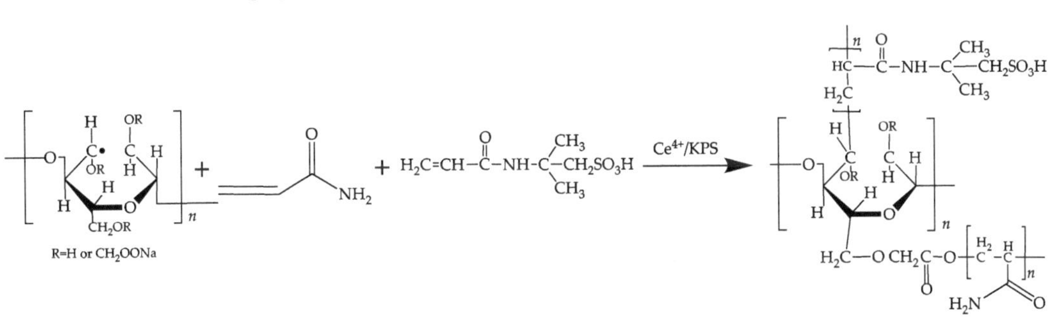

Figure 3. AMPS/AM modified carboxymethyl cellulose mechanism.

2.2.2. Sample Preparation

The cotton linters were shredded, dried, and alkalized to obtain refined cellulose, which was then etherified with chloroacetic acid to produce carboxymethyl cellulose (CMC). Then, 2 g of CMC and 0.125 mmol/L cerium ammonium nitrate solution (relative to a liquid volume of 50 mL of the reactant) were taken and the pH was adjusted to 1.0. The solution was placed in a constant temperature water bath at 50 °C and stirred for 15 min. 2-acrylamido-2-methylpropanesulfonic acid (AMPS) was slowly added to this solution, and a reaction was maintained for 4.5 h. Then, sodium hydroxide and urea were added to form a stable NaOH/Urea/H_2O solution system (with mass fractions of 7% NaOH, 12% urea, and 81% water). The solution was frozen at −12 °C for 12 h. It was then thawed and heated to 50 °C while stirring. Potassium persulfate (KPS) and acrylamide (AM) were slowly added and reacted for 4.5 h. After the reaction, the crude CMC-AMPS-AM copolymer was precipitated and dried in double the amount of ethanol, followed by grinding. Using ethanol and acetone as solvents, pure CMC-AMPS-AM copolymer was obtained after 48 h through extraction with a Soxhlet apparatus. The preparation process is shown in Figure 4.

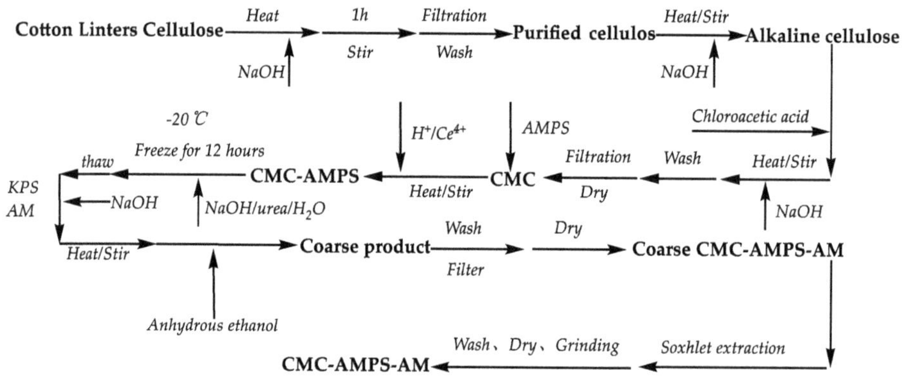

Figure 4. Process flow diagram for the preparation of CMC-AMPS-AM.

2.3. Conversion Rate Calculation

A conversion rate calculation was conducted using the elemental tracing method as shown in the following equations:

$$m_{AMPS} = \frac{m_S \times 207.250}{32} \tag{1}$$

$$\text{Conversion Rate}_{AMPS} = \frac{m_{AMPS}}{1.382} \times 100\% \tag{2}$$

$$m_{N_{AMPS}} = \frac{m_s \times 14.0067}{32} \tag{3}$$

$$m_N = m_{N_{AMPS}} + m_{N_{AM}} \tag{4}$$

$$m_{AM} = \frac{m_{N_{AM}} \times 71.078}{14.007} \tag{5}$$

$$\text{Conversion Rate}_{AM} = \frac{m_{AM}}{3.335} \times 100\% \tag{6}$$

Here, m_S is the mass of sulfur in the polymer (g); 207.25 is the relative molecular mass of AMPS; 32 is the relative atomic mass of element S; m_{AMPS} is the mass of AMPS in the polymer (g); 1.382 is the single cell input for AMPS (g); ConConversion Rate$_{AMPS}$ is the AMPS monomer conversion rate (%); $m_{N_{AMPS}}$ is the nitrogen element mass from AMPS in the polymer (g); 14.007 is the relative atomic mass of element N; $m_{N_{AM}}$ is the mass of element N in the polymer from AM (g); m_N is the total mass of nitrogen in the polymer (g); 71.078 is the relative molecular mass of AM; m_{AM} is the mass of AM in the polymer (g); 3.335 is the single cell input for AM (g); Conversion $Rate_{AM}$ is the AM monomer conversion rate (%).

2.4. Response Surface Optimisation Experimental Design

The influential factors chosen for the response surface method were the inputs of AMPS, AM, and KPS, while the value of surface tension in the dispersant solution was considered as the response value (Y). This was then analyzed using the Box-Behnken design, which is a three-factor, three-level response surface analysis. The outcomes of this design process are documented in Table 2.

Table 2. Response surface experimental design.

Level	Considerations		
	A:AMPS (g)	B:AM (g)	C:KPS (g)
−1	1	2	0
0	3	3	0.075
1	5	4	0.150

2.5. Description

A Fourier transform infrared spectrometer (FTIR, VEETEX-70, BRUKE, Germany) was employed for scanning within the spectral range of 400–4000 cm^{-1}, with a sampling rate of 80 spectra per second. The XRD measurements were performed using a Rigaku Ultima IV diffractometer produced by Rigaku Corporation (Atsugi, Kanagawa, Japan), with a scanning range of $2\theta = 10°\sim80°$. The Bettersize 2600 laser particle size analyzer was used to measure the particle size of carbon black in the slurry. Elemental content analysis of C, H, N, and S was determined using the Vario EL cube type elemental analyzer from Germany. To obtain a uniform fine powder, the samples were subjected to a process of vacuum drying followed by grinding. Subsequently, the content of S and N was measured on the elemental analyzer. The weight loss process of the samples and the extent of heat resistance of the products were analyzed using a thermogravimetric analyzer (TA, TGA550, Milford, MA, USA). Testing conditions involved measuring the samples under a nitrogen atmosphere, with a temperature increase from room temperature to 800 °C at a rate of 10 °C/min. The contact angle between the sample and water was measured utilizing the LSA-100 (LAUDA Scientific GmbH) contact angle tester. The sample platen exerted a pressure of 10 MPa, while the observed volume of the test droplets was 2 µL. Furthermore, the appearance and morphology of the samples were observed using a scanning electron microscope (SEM, ZSISS Gemini 300, Baden-Württemberg, Germany). The Zeta potential of the sample was measured using a Zetasizer Nano ZS90 (manufactured by Malvern, UK). The testing method involved mixing carbon black and dispersant material in a ratio of 10:37.5 in ultrapure water, followed by sonication to prepare a 0.1 wt% dilute solution. Zeta potential measurements were then conducted on the dilute solution samples. The molecular weight and polymer dispersity index of dispersants materials are tested using the American Agilent 1260 Infinity II Gel Permeation Chromatography (GPC). The test is conducted at room temperature with deionized water as the mobile phase.

3. Results

3.1. Response Surface Modelling and Analysis of Variance (ANOVA)

According to the experimental design in Table 2, the independent variables selected are AMPS (A), AM (B), and KPS (C). The experimental response value (Y) is the surface tension in the dispersant solution. The optimal process conditions are determined through response surface methodology (RSM) analysis to explore the effects among these variables, with experimental results shown in Table 3.

Table 3. Response surface experimental design and results.

Std	AMPS	AM	KPS	Surface Tension of Solution (mN/m)
16	0	0	0	48.9
15	0	0	0	52.0
4	1	1	0	55.3
17	0	0	0	49.6
5	−1	0	−1	60.4
7	−1	0	1	49.8
9	0	−1	−1	64.2
3	−1	1	0	51.2
1	−1	−1	0	61.8
14	0	0	0	50.6
2	1	−1	0	64.8
12	0	1	1	57.3
6	1	0	−1	59.5
10	0	1	−1	62.3
8	1	0	1	61.3
13	0	0	0	47.5
11	0	−1	1	63.9

Response surface test design data were designed and optimized using Design-Expert 13 software using Box-Behnken (BBD). The response value (Y) and the influence factor conformed to the quadratic response regression equation (Equation (7)).

$$
\begin{aligned}
Y = {} & 129.30375 - 4.14125 \times A - 41.0025 \times B - 194.23333 \times C + 0.1375 \times A \times B - 20.66667 \times A \times C \\
& -15.66667 \times B \times C + 0.5475 \times A^2 + 6.365 \times B^2 + 1038.2222 \times C^2
\end{aligned}
\tag{7}
$$

The model is capable of forecasting the correlation between the autonomous factors (A, B, and C) and the response variable (Y). The analysis utilized Fisher's statistical test for the purpose of conducting an analysis of variance (ANOVA), with the outcomes of ANOVA and confidence analysis of the equations featured in Table 4. The results of the confidence analysis of the quadratic response surface regression model demonstrate that the correlation coefficient of the model fit, $R^2 = 0.9483$, exceeds 0.8, indicating an excellent fit of the model with the experimental data, along with a low experimental error.

Table 4. Analysis of variance of regression equations and credibility of regression model.

Source	Sum of Squares	df	Mean Square	F-Value	*p*-Value	Significance
Model	576.94	9	64.10	14.26	0.0010	Significant
A-AMPS	39.16	1	39.16	8.71	0.0214	
B-AM	102.25	1	102.25	22.74	0.0020	
C-KPS	24.85	1	24.85	5.53	0.0510	
AB	0.3025	1	0.3025	0.0673	0.8028	
AC	38.44	1	38.44	8.55	0.0222	
BC	5.52	1	5.52	1.23	0.3043	
A^2	20.19	1	20.19	4.49	0.0718	
B^2	170.58	1	170.58	37.94	0.0005	
C^2	143.60	1	143.60	31.94	0.0008	
Residual	31.47	7	4.50			
Lack of Fit	19.88	3	6.63	2.29	0.2205	not Significant
Pure Error	11.59	4	2.90			
Cor Total	608.41	16				

The p value for the lack-of-fit term is 0.2205, which is greater than 0.05, suggesting that the lack-of-fit is not statistically significant [26]. Figure 5 depicts the correlation between the factual surface tension and the anticipated surface tension. The proximity in the correspondence between the projected quantities and the actual quantities signifies an elevated level of concordance between them. In this manner, it showcases the dependability of the quadratic response regression model [27].

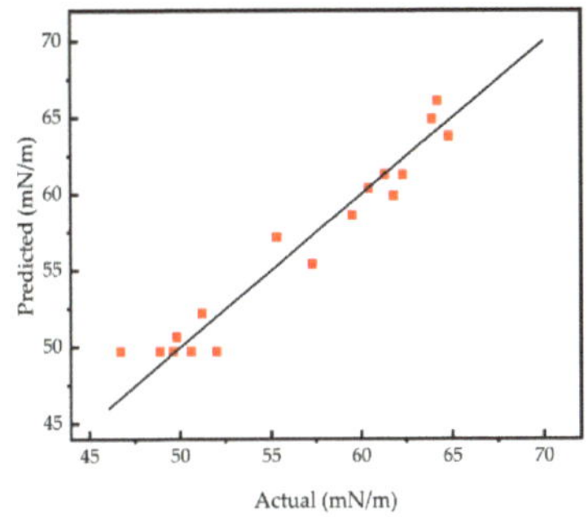

Figure 5. Relationship between the anticipated and actual values of solution surface tension.

3.2. Response Surface 3D Graph

Figure 6 depicts visualizations of the impact of each factor, namely AMPS (Figure 6a), AM (Figure 6b), and KPS (Figure 6c), on the surface tension of the solution. By observing the slope of the surface and the density of contour plots, the magnitude of the influence on the surface tension can be determined. Additionally, Table 4 provides further insight by presenting the F-values of AMPS input (A), AM input (B), and KPS input (C), which are 8.71, 22.74, and 5.53, respectively. Thus, it can be concluded that the influence of each factor on the surface tension of the solution follows this order: AM input > AMPS input > KPS input. Utilizing the proposed model, the optimization of CMC: AMPS: AM and KPS ratios resulted in an ideal composition of 1:0.69:1.67 for CMC, AMPS, and AM, respectively, with KPS being approximately 1.56% of the total monomer amount. Furthermore, the elemental content analysis revealed the N, C, H, and S contents in the samples. By employing the elemental tracing method, the sources of N and S in the CMC-AMPS-AM samples were identified, and the measurements obtained are listed in Table 5.

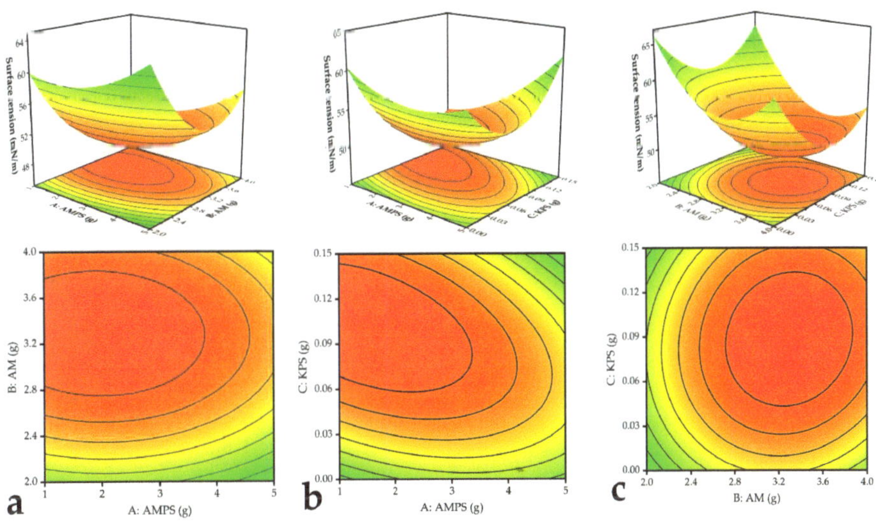

Figure 6. (**a**) Effect of AMPS input and AM input on dispersant surface tension. (**b**) Effect of AMPS input and KPS input on surface tension of dispersant. (**c**) Effect of AMPS input and KPS input on surface tension of dispersant.

Table 5. Elemental analyzer measurement data table.

Specimens	N (%)	C (%)	H (%)	S(%)
CMC-AMPS-AM	4.77	33.67	5.46	2.69

Based on Equations (1)–(6) in Section 2.3, the AMPS and AM grafting rates were calculated. The AMPS grafting rate was calculated to be 70.21%, and the AM grafting rate was 33.80%; the low AM grafting rate was mainly due to the polymerization of AM.

3.3. *Carbon Black Dispersion Stability Performance*

When investigating the dispersing properties of carbon black, it is crucial to take into account several pivotal factors. Firstly, it is imperative to ensure an appropriate amount of dispersant is utilized, as an insufficient quantity would not achieve the desired dispersion of carbon black, while an excessive amount would result in the formation of "bridges" by the dispersant, causing carbon black re-agglomeration. Secondly, the pH value of the dispersant also exerts a certain influence on the dispersion performance. Within an optimal pH range, the surface charge of the particles can be fine-tuned, thus facilitating the homogeneous dispersion of carbon black particles. Furthermore, the stability requirements of the dispersant should also be considered. A stable dispersant not only sustains long-term dispersion efficacy but also enhances the stability and performance of the end product.

3.3.1. Effect of Dispersant Concentration

The dosage of dispersant is dependent upon the type of particles and desired particle size. In the case of pigment dispersion, the dispersant for organic pigment typically ranges from 5% to 30%, as specified by the industry. Similarly, the dispersant for inorganic pigment ranges from 5% to 20%, while for carbon black, it falls between 25% and 50%. For the high-surface-energy carbon black used in automotive coatings, such as FW 200, the recommended amount of dispersant is between 60% and 150%.

A specific quantity of dispersing stabilizer was combined with 1.5 g of carbon black, followed by the addition of 10 mL of deionized water as the solvent. Subsequently, 30 g of zirconia ball mill beads with a particle size of 0.5 mm were added, and the carbon black slurry was acquired through grinding for 5 h in a ball mill rotating at a speed of 300 rpm/min. Simultaneously, the paint without hyperdispersant was prepared as the control group, the results are shown in Figure 7. As the dispersant concentration rose from 0% to 37.5%, the size distribution of the carbon black particles shifted from a wide to narrow distribution, and the median particle size D50 decreased from 5.350 μm to 0.255 μm. Yet, when the dispersant concentration reached 100%, the median particle size D50 of carbon black particles increased to 0.288 μm. In addition, compared with other studies on carbon black dispersants, Table 6 shows that the CMC-AMPS-AM dispersant demonstrates significant advantages in enhancing carbon black dispersibility.

The agglomeration of carbon black particles is a consequence of van der Waals forces coming into play, leading to a gradual elevation in particle density and subsequent settling due to gravitational forces. The function of the dispersant is to counteract these inter-particular forces. The concentration of the dispersant directly influences the balance between dispersant molecules and carbon black particles, thereby affecting the dispersion effect. At low concentrations, insufficient dispersant coverage on the carbon black particles hinders the ability to counteract inter-particle forces. With increasing dispersant concentration, the adsorption between dispersant molecules and carbon black particles gradually rises, effectively diminishing the inter-particle interaction force and thwarting particle agglomeration, and thus resulting in reduced particle size.

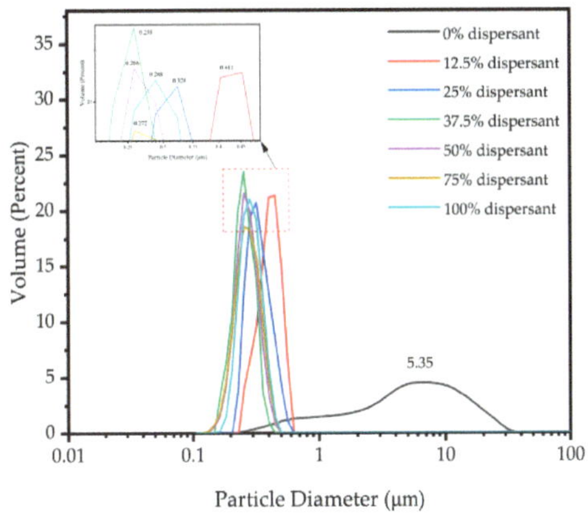

Figure 7. Particle size distribution of dispersed carbon black with different dispersant concentrations.

Table 6. Comparison of the effects of different dispersants on the particle size of carbon black.

Dispersed Materials	D50 Particle Size of Carbon Black (μm)	Reference
DP-3537	1.80	[28]
IMD	0.275	[29]
PIB	0.100	[30]
BYK-192	0.785	[29]

3.3.2. Zeta Potential Analysis

The magnitude of the Zeta potential serves as a crucial factor in assessing the stability of a dispersed system. Greater values of the zeta potential indicate enhanced electrostatic repulsion, thereby leading to superior suspension stability [31–33].

As shown in Figure 8a, the Zeta–pH curve illustrates the dispersion of carbon black by dispersants. According to the double electric layer exclusion mechanism, charged particles adhere to the surface of particles forming a double electric layer, which results in electrostatic repulsion. In an acidic environment, there is an abundance of positive ions H+. Sulfonate ions (SO_3^-) readily react with H+ ions, leading to an increase in potential. Sulfate groups have a lower negative charge density, hence the Zeta potential is relatively lower in acidic environments.

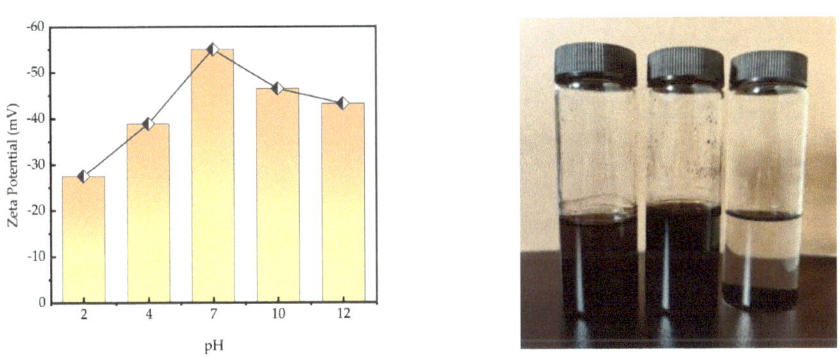

(a) (b)

Figure 8. (**a**) Zeta–pH curves of carbon black and CMC-AMPS-AM. (**b**) Addition of 37.5% dispersant to carbon black at different pH (right pH = 12, middle pH = 7, and left pH = 2). Natural settling effect can be seen for the right vial with high pH.

In alkaline environments, a significant number of negative ions (such as OH-) are present and they interact with the positive charge on the polymer surface, causing the surface to have a negative potential. This ion exchange affects the surface potential and Zeta potential of the polymer. In an alkaline environment, the absolute value of the Zeta potential is mainly due to the presence of a large number of OH- ions in the solution, which leads to the following phenomena: (1) Charge shielding effect: A large number of negative ions, such as OH- ions, adsorb on the surface of the anionic dispersant. When the OH- concentration is sufficiently high, a charge shielding layer is formed on the surface of the dispersant material, which partially shields the negative charges, reduces the absolute value of the potential, and weakens the electrostatic repulsion. (2) Double layer thickness variation: The Zeta potential actually measures the potential difference of the double layer. When a large number of OH- ions adsorb on the surface of the anionic dispersant, the thickness of the double layer increases, which affects the transport speed of the charges, and the absolute value of the Zeta potential may decrease. (3) Solvent polarization effect: A large number of OH- ions increase the degree of solvent polarization, increase the thickness of the double layer, and weaken the electrostatic interaction between the dispersant and the solution. This reduces the absolute value of the Zeta potential. As shown in Figure 8b, the natural sedimentation effect of carbon black particles can be observed with the addition of the same amount of dispersant at different pH levels. In conclusion, CMC-AMPS-AM dispersants tend to induce agglomeration and instability of particles in strongly acidic and strongly alkaline environments.

3.3.3. Stability of Carbon Black Slurry

After undergoing mechanical fracturing and abrasion with the ball mill, the particle size of carbon black decreases while the surface energy increases. This results in an elevated adhesion force between different particles that results in particle agglomeration. Hence, the stability of dispersants holds utmost significance [34,35]. In reference to Section 3.3.1 for preparing the slurry, add 10% dispersant to the carbon black, then ball mill the sample and store at room temperature. The variation in particle size is monitored every seven days. As shown in Figure 9, the particle size of the carbon black exhibits minimal fluctuation, remaining within the range of 0.469 μm before storage to 0.474 μm at the end of the fourth week. Facilitated by the dispersant, the carbon black particles surmount inter-particle adhesion as well as electrostatic forces, displaying no discernible trend of "coarsening". This suggests the dispersant confers commendable dispersion stability.

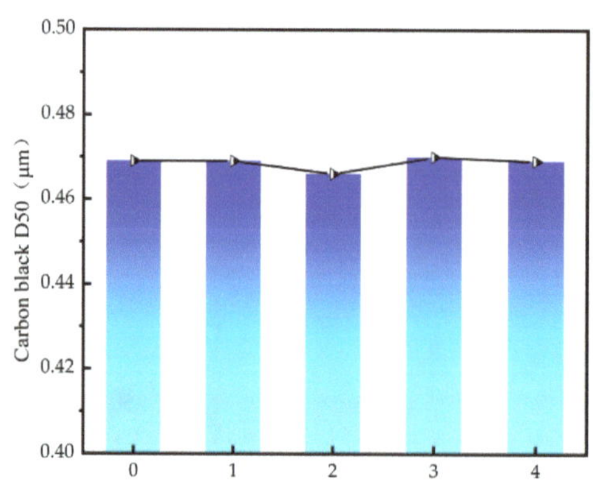

Figure 9. Variation of carbon black particle size with storage time.

As shown in Figure 10, the left bottle contains carbon black slurry without dispersant added, with 0.2 g taken and diluted with distilled water 20 times as a blank group. The middle/right bottles contain carbon black slurry with 37.5% dispersant added, with 0.2 g taken and diluted with distilled water 40/160 times, respectively. The natural settling effects after 0, 1, 2, and 3 weeks are represented by a, b, c, and d. As shown in the figure, carbon black particles without dispersant coating precipitate within 7 days due to the force of attraction between molecules, causing the micro particles to aggregate into larger particles. In contrast, carbon black particles with dispersant effectively overcome the intermolecular forces between carbon black particles caused by the presence of dispersant, resulting in continuous dispersion of particles in water.

Figure 10. Settling of carbon black slurry in water (left bottle diluted 20 times, middle bottle diluted 80 times, and right bottle diluted 160 times): week 0 (**a**); week 1 (**b**); week 2 (**c**); and week 3 (**d**).

3.3.4. Scanning Electron Microscopy (SEM) Analysis

Figure 11 depicts high-magnification scanning electron microscope images of carbon black slurries without dispersant and with 37.5% dispersant added after drying. Upon examining Figure 11a,c, it is evident that the carbon black particles, in the absence of any added dispersant, coalesce together, forming dense clusters with larger-sized particles. From Figure 11b,d, it can be observed that individual carbon black particles typically have smaller diameters, while larger particles are often formed by aggregation of smaller ones. Therefore, after the addition of dispersants, the dispersants effectively inhibit the aggregation between carbon black particles, improving the dispersibility and stability of carbon black.

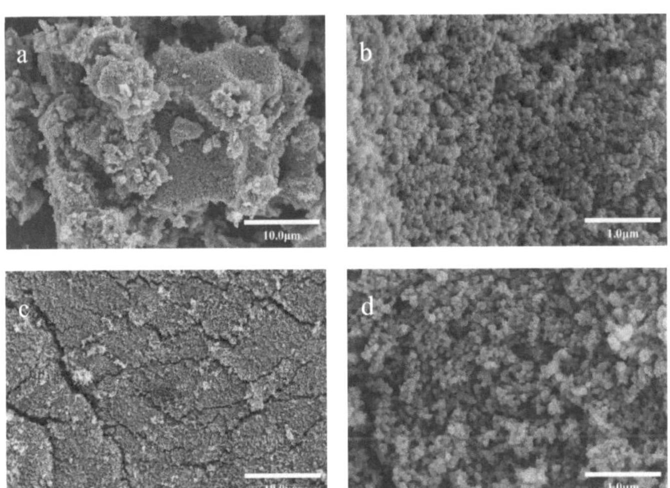

Figure 11. (**a–d**) SEM images of carbon black slurry without dispersant; (**c**,**d**) SEM images of carbon black slurry with 37.5% dispersant added.

3.4. *Characterisation of Dispersant Materials*

3.4.1. Infrared Spectroscopy Test (FTIR) Analysis

From the FTIR spectra shown in Figure 12, it can be observed that the −NH absorption peak of the secondary amide occurred at 2934 cm^{-1}, whereas the C-S stretching vibration peak in AMPS appeared at 628 cm^{-1}. The -SO^{3-} in AMPS exhibited S-O bond asymmetry and S=O bond symmetry vibration absorption peaks near 1045 cm^{-1} and 1219 cm^{-1} [36]. The characteristic absorption peak at 3445 cm^{-1} is attributed to the -NH_2 stretching vibration in AM. Furthermore, the characteristic peaks at 1673 cm^{-1} and 1119 cm^{-1} are related to the stretching vibration of the carbonyl C=O and C-N in the amide group of AM [37]. The deformation of methylene resulted in the generation of the characteristic absorption peak at 1455 cm^{-1}. Through the FTIR analysis, it can be concluded that AMPS and acrylamide have been successfully grafted onto the cellulose skeleton.

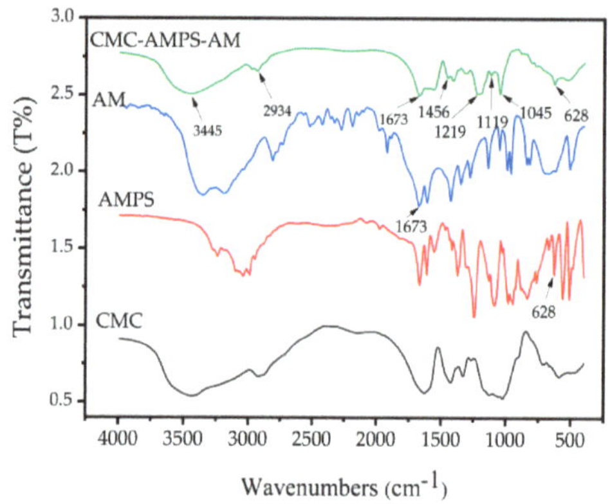

Figure 12. FTIR spectra of CMC/AM/AMPS and CMC-AMPS-AM.

3.4.2. Thermogravimetric (DT) Analysis

TG-DTG curves illustrate the weight loss of CMC and CMC-AMPS-AM, as shown in Figure 13. For CMC (Figure 13a), it is discernible that before the temperature reaches 32 °C–234 °C, there is a relatively insignificant weight loss of approximately 10%, which is attributed to the removal of adsorbed water on the carboxymethyl cellulose surface. Between 234 °C and 325 °C, there is a prominent weight loss of about 37.5%, stemming from the rupture and decay of C=O, C-O-C, and -OH bonds. Notably, the thermal degradation of cellulose attains a maximum when the temperature peaks at 289 °C, primarily due to a substantial number of C–C bond ruptures, leading to a vast mass reduction in cellulose. Throughout the entire process, spanning from 32 °C to 800 °C, the overall weight loss accounts for approximately 66.6% of the mass.

In the case of CMC-AMPS-AM (Figure 13b), a negligible reduction in mass of the CMC-AMPS-AM copolymer is observed within the temperature interval of 36 °C to 100 °C. This phenomenon can be attributed to the low amount of moisture adhered to the CMC-AMPS-AM copolymer, as well as the weight decrease caused by the evaporation of acetone. In contrast, the weight loss percentage of carboxymethyl cellulose is approximately 14.7% within the temperature range of 226 °C to 331 °C and 37.5% within the range of 234 °C to 325 °C. This indicates a considerable decline in weight loss in CMC, likely due to the grafting reaction that disrupts the cyclic structure in the sugar component and imparts greater stability. In the temperature range of 331 °C to 371 °C, the weight loss was approximately 18.9%, possibly due to the thermal decomposition of the amide group in AM and the subsequent imidization reaction. Throughout the temperature range of 32 °C to 800 °C, the weight of the CMC-AMPS-AM copolymer witnessed a total decrease of approximately 69%.

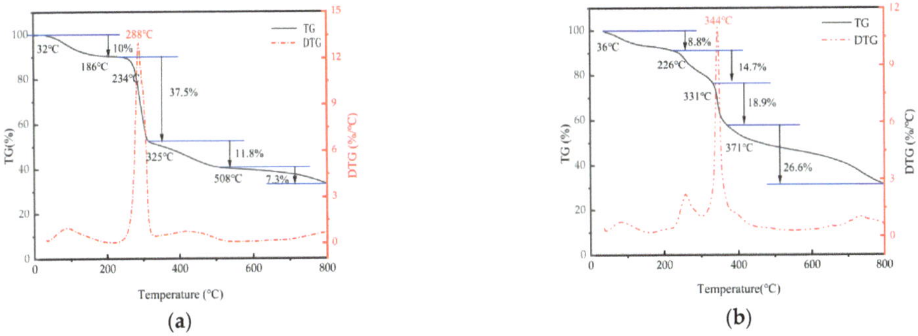

Figure 13. TG-DTG curve of (**a**) CMC and (**b**) CMC-AMPS-AM.

3.4.3. X-ray Diffraction (XRD) Analysis

XRD results for CMC and CMC-AMPS-AM samples are shown in Figure 14, where it can be observed that in the grafting of AMPS, AM could impact the chemical reaction and molecular structure, subsequently causing deviations in the crystal structure, diminished grain size, and defects in the cellulose crystals. As a result of the decreased grain size and crystallinity, the peaks in the XRD pattern became broader, with a rounded shape and lower intensity.

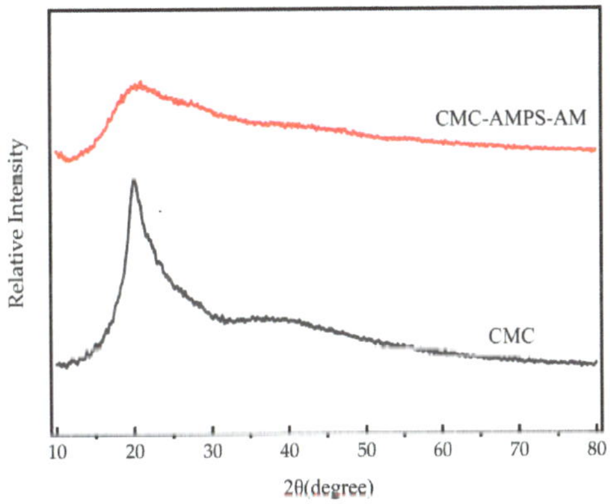

Figure 14. XRD spectra of CMC-AMPS-AM and CMC.

3.4.4. Molecular Weight Testing

According to the dispersion mechanism of dispersants, the molecular weight of dispersants is closely related to their dispersing performance. Gel permeation chromatography (GPC) tests were conducted on CMC-AMPS-AM dispersant materials, and the test results are shown in Table 7.

Table 7. The results of the GPC test for the CMC-AMPS-AM dispersant material.

Specimens	Mn	Mw	PDI
CMC-AMPS-AM	148159	278299	1.878381

3.4.5. Contact Angle Analysis

The dispersant composition not only affects the dispersion of pigment particles but also affects the surface energy of the resulting dry pigment particles. The smaller the contact angle, the higher is the hydrophilicity of the particles [38,39]. After mixing carbon black with CMC-AMPS-AM dispersant material in a 1:1 ratio, measure the contact angle with water. As shown in Figure 15, the untreated carbon black surface is rich in polar functional groups such as hydroxyl carbonyl groups, which attracts hydrogen atoms of water molecules, rendering it hydrophilic. Carboxymethyl (-CH_2COOH) functional groups were prepared through etherification of cellulose. Due to the grafting of AMPS and AM, the concentration of -COOH groups in carboxymethyl cellulose decreased, the contact angle increased, and the hydrophilicity decreased. The contact angle increased after adding CMC-AMPS-AM to carbon black. In the application of carbon black coatings, highly hydrophilic carbon black particles are not compatible with waterborne coatings, and also not conducive to the water resistance of coatings [40,41].

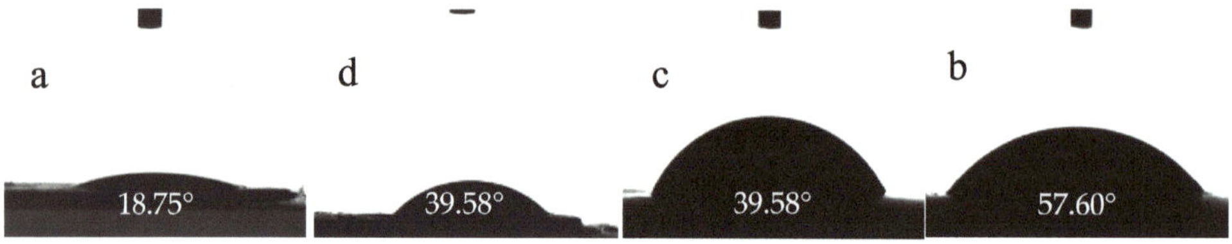

Figure 15. Contact angle between material and water: (**a**) CB; (**b**) CMC; (**c**) CMC-AMPS-AM; and (**d**) CMC-AMPS-AM and CB.

3.5. Dissemination Mechanisms

According to the dispersant dispersing mechanism, dispersants mainly rely on steric effects and electrostatic interactions [42–45]. Steric hindrance: dispersant molecules or ions adsorb on the surface of solid particles, forming physical barriers that hinder direct contact and aggregation of particles. Electrostatic repulsion force refers to the repulsive force generated by electrostatic interaction between particles of the same charge, effectively preventing particle aggregation and settling, promoting particle dispersion, and maintaining suspension stability. In practical dispersion, both mechanisms work synergistically [46]. The CMC-AMPS-AM material disperses carbon black as shown in Figure 16, achieving optimized dispersion through the synergistic effects of steric hindrance and electrostatic repulsion force.

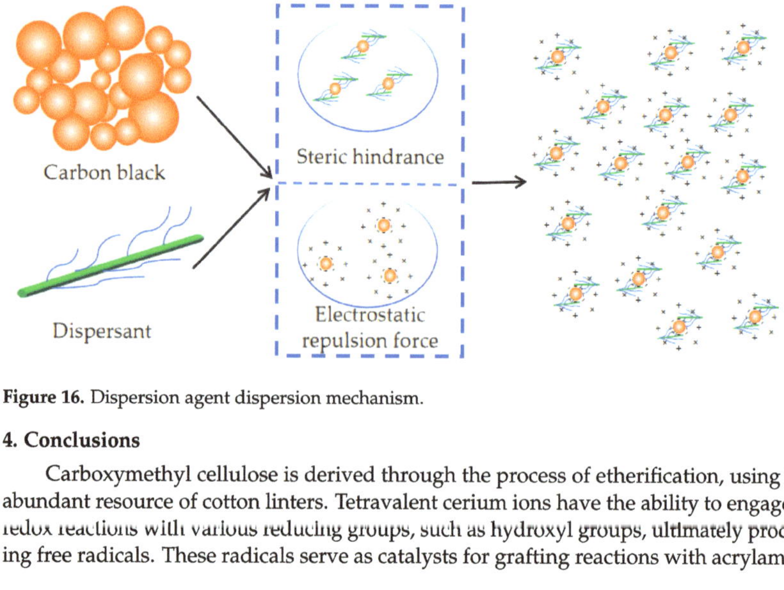

Figure 16. Dispersion agent dispersion mechanism.

4. Conclusions

Carboxymethyl cellulose is derived through the process of etherification, using the abundant resource of cotton linters. Tetravalent cerium ions have the ability to engage in redox reactions with various reducing groups, such as hydroxyl groups, ultimately producing free radicals. These radicals serve as catalysts for grafting reactions with acrylamide monomer and 2-acrylamide-2-methylpropanesulfonic acid monomer. By employing the response surface method, the optimal values for three influential factors were determined. The initiator KPS comprised 1.56% of the total amount of monomers, and the ideal ratio of raw materials CMC:AMPS:AM was established as 1:0.69:1.67. The order of the impact of these factors on the conversion rate was AM > AMPS > KPS. Through elemental tracing, the contents of S and N were determined under the optimal preparation process. Consequently, the elemental tracking method enabled the determination of S and N contents, and the conversion rates of AMPS and AM were determined to be 70.21% and 33.80%, respectively. Upon application of the CMC-AMPS-AM dispersant to carbon black materials, the median particle size (D50) of the carbon black decreased to 0.255 μm. Moreover, the distribution of particle sizes became more concentrated as the amount of dispersant additive reached 37.5%. Notably, even after a storage period of four weeks, the particle size of the carbon black remained stable, without exhibiting the so-called "re-coarsening" phenomenon.

In the optimization of dispersant synthesis process, there are challenges in terms of process complexity and long cycles, which pose a significant challenge for industrial production. The stability assessment of dispersants does not consider performance under extreme conditions, such as high and low temperatures. Additionally, industrial applications often require compatibility with various additives such as ink additives, coating resins, and tire rubber, but this study does not cover their compatibility and synergistic effects. Since cellulose is insoluble in water, etherification treatment is required to prepare water-based dispersants, converting it into carboxymethyl cellulose, which increases costs and industrial difficulty. Cotton short fiber cellulose has a high content and is easy to extract, but most biomass cellulose content in nature is low. The future direction of research is to use low-cellulose biomass for modification and directly prepare water-based dispersants.

Author Contributions: X.A.: investigation, experimental procedures, methodology, validation, writing-original draft, writing–review and editing, and project administration; X.Y.: supervision and writing–review and editing; C.H.: investigation and writing–review and editing; C.D.: experimental design, writing–review and editing, and resources. All authors have read and agreed to the published version of the manuscript.

Funding: This research received funding support from the National Natural Science Foundation of China (Grant No. 22268041) and the Key Research and Development Program of Xinjiang Uygur Autonomous Region (Grant No. 2023LQ01004).

Institutional Review Board Statement: Not applicable.

Data Availability Statement: Data are contained within the article.

Acknowledgments: We would like to express our sincere gratitude to the following individuals and organizations for their support and assistance in the completion of this research: First and foremost, we would like to thank Chengli Ding for his invaluable guidance and insightful advice throughout the research process. His expertise and wisdom have been instrumental in driving our research forward. We are grateful to The College of Chemical Engineering at Xinjiang University for providing the necessary research facilities and technical support. These resources have created a conducive research environment for us to conduct our experiments smoothly. Furthermore, we extend our appreciation to Xupeng Yang and Canming Hu and his team for their collaborative efforts and support in this research. Their expertise and cooperation have enriched our research work with valuable experiences and insights. Lastly, we would like to thank all individuals and organizations who have provided assistance and support in this research. Without your support, our research work would not have been possible. We sincerely thank all the individuals and organizations mentioned above for their support and assistance!

Conflicts of Interest: We declare that we have no financial and personal relationships with other people or organizations that can inappropriately influence our work; there is no professional or other personal interest of any nature or kind in any product, service, and/or company that could be construed as influencing the position presented in, or the review of, this manuscript.

References

1. Song, J.; Yan, W.; Chen, Z.; Liu, Y.; Hong, S. Effect of nano-carbon black content on wetting phenomenon of molten steel and alumina–carbon ceramic filter substrates. *J. Iron Steel Res. Int.* **2024**, 1–14. [CrossRef]
2. Pu, Z.; Fan, X.; Su, J.; Zhu, M.; Jiang, Z. Aqueous dispersing mechanism study of nonionic polymeric dispersant for organic pigments. *Colloid Polym. Sci.* **2022**, *300*, 167–176. [CrossRef]
3. Lin, Q.; Gao, M.; Chang, J.; Ma, H. Adsorption properties of crosslinking carboxymethyl cellulose grafting dimethyldiallylammonium chloride for cationic and anionic dyes. *Carbohydr. Polym.* **2016**, *151*, 283–294. [CrossRef] [PubMed]
4. Nguyen, V.H.; Pazhouhanfar, Y.; Delbari, S.A.; Shaddel, S.; Babapoor, A.; Mohammadpourderakhshi, Y.; Van Le, Q.; Shokouhimehr, M.; Asl, M.S.; Namini, A.S. Beneficial role of carbon black on the properties of TiC ceramics. *Ceram. Int.* **2020**, *46*, 23544–23555. [CrossRef]
5. Goh, Y.; Lauro, S.; Barber, S.T.; Williams, S.A.; Trabold, T.A. Cleaner production of flexographic ink by substituting carbon black with biochar. *J. Clean. Prod.* **2021**, *324*, 129262. [CrossRef]
6. Chollakup, R.; Suethao, S.; Suwanruji, P.; Boonyarit, J.; Smitthipong, W. Mechanical properties and dissipation energy of carbon black/rubber composites. *Compos. Adv. Mater.* **2021**, *30*, 26349833211005476. [CrossRef]
7. Ji, C.; Zhang, Q.; Jing, Z.; Liu, Y.; Han, D.; Wang, J.; Zhang, W.; Sang, S. Highly sensitive wearable flexible pressure sensor based on conductive carbon black/sponge. *IEEE Trans. Electron Devices* **2021**, *68*, 5198–5203. [CrossRef]
8. Du, W.; Liu, J.; Li, Z. High self-dispersibility carbon black particles prepared via hydroxylation and urethane chains encapsulation for enhancing properties of waterborne polyurethane composite films. *Colloids Surf. A Physicochem. Eng. Asp.* **2018**, *543*, 46–55. [CrossRef]
9. Ali, M.; Lin, L.; Cartridge, D. High electrical conductivity waterborne dispersions of carbon black pigment. *Progr. Org. Coat.* **2019**, *129*, 199–208. [CrossRef]
10. Jansen, J.; Kraus, G. New Methods for Estimating Dispersibility of Carbon Blacks in Rubber. *Rubber Chem. Technol.* **1980**, *53*, 48–65. [CrossRef]
11. Ghose, T. Measurement of cellulase activities. *Pure Appl. Chem.* **1987**, *59*, 257–268. [CrossRef]
12. Wei, S.; Wu, X.; Li, X. Solubility analysis of nano particles, cellulose crystalline region and cellulose molecule, and the impact study of crystalline region on properties of cellulose insulating paper. *Mol. Simul.* **2021**, *47*, 1522–1529. [CrossRef]
13. Srivastava, A.; Mandal, P.; Kumar, R. Solid state thermal degradation behaviour of graft copolymers of carboxymethyl cellulose with vinyl monomers. *Int. J. Biol. Macromol.* **2016**, *87*, 357–365. [CrossRef] [PubMed]
14. Xue, Y.; Lu, Y.; Feng, K.; Zhang, C.; Feng, X.; Zhao, Y.; Chen, L. Preparation of the self-accelerating photocatalytic self-cleaning carboxymethyl cellulose sodium-based hydrogel for removing cationic dyes. *Int. J. Biol. Macromol.* **2023**, *250*, 125891. [CrossRef] [PubMed]
15. Li, X.; Lv, J.; Li, D.; Wang, L. Rapid fabrication of TiO_2@ carboxymethyl cellulose coatings capable of shielding UV, antifog and delaying support aging. *Carbohydr. Polym.* **2017**, *169*, 398–405. [CrossRef] [PubMed]
16. Ding, C.; Su, B.; Wang, G.; Meng, Q.; Wang, J.; Zhang, C. Terahertz absorption characteristics of the sodium carboxymethyl cellulose colloid based on microfluidic technology. *Int. J. Opt.* **2021**, *2021*, 5555325. [CrossRef]
17. Zhang, J.; Zhang, G.; Wang, Y.; Qin, X.; Lu, Y.; Wu, Z.; Zhang, C.; Xu, L.; Han, J.; Yang, S. l-Arginine carboxymethyl cellulose hydrogel releasing nitric oxide to improve wound healing. *Eur. Polym. J.* **2023**, *189*, 111940. [CrossRef]
18. Magdassi, S.; Bassa, A.; Vinetsky, Y.; Kamyshny, A. Silver nanoparticles as pigments for water-based ink-jet inks. *Chem. Mater.* **2003**, *15*, 2208–2217. [CrossRef]
19. Zhang, D.; Huang, Y. Dispersion characterizations and adhesion properties of epoxy composites reinforced by carboxymethyl cellulose surface treated carbon nanotubes. *Powder Technol.* **2022**, *404*, 117505. [CrossRef]
20. He, X.; Wu, S.; Fu, D.; Ni, J. Preparation of sodium carboxymethyl cellulose from paper sludge. *J. Chem. Technol. Biotechnol. Int. Res. Process Environ. Clean Technol.* **2009**, *84*, 427–434. [CrossRef]
21. Ismail, N.; Bono, A.; Valintinus, A.; Nilus, S.; Chng, L. Optimization of reaction conditions for preparing carboxymethyl cellulose. *J. Appl. Sci.* **2010**, *10*, 2530–2536. [CrossRef]
22. Saputra, A.H.; Qadhayna, L.; Pitaloka, A.B. Synthesis and characterization of carboxymethyl cellulose (CMC) from water hyacinth using ethanol-isobutyl alcohol mixture as the solvents. *Int. J. Chem. Eng. Appl.* **2014**, *5*, 36. [CrossRef]
23. Batelaan, J.; Van Ginkel, C.; Balk, F. Carboxymethylcellulose (cmc). In *Detergents*; Springer: Berlin/Heidelberg, Germany, 1992; pp. 329–336.
24. Liu, Y.; Wan, X.; Xiang, X.; Li, Y.; Wu, S. Preparation of nano zero-valent iron-microfibrillated cellulose and its application in papermaking wastewater advanced treatment. *China Pulp Pap.* **2015**, *34*, 21–26.
25. Zhang, H.; Lou, T.; Wang, X. Synthesis and properties of xanthan gum-acrylamide-carboxymethyl cellulose ternary anionic flocculants. *J. Water Process Eng.* **2023**, *56*, 104455. [CrossRef]
26. Mao, J.; Guanhua, N.; Yuhang, X.; Hui, W.; Zhao, L.; Zhenyang, W. Modeling and optimization of mechanical properties of drilling sealing materials based on response surface method. *J. Clean. Prod.* **2022**, *377*, 134452. [CrossRef]
27. Ofgea, N.M.; Tura, A.M.; Fanta, G.M. Activated carbon from H3PO4-activated Moringa Stenopetale Seed Husk for removal of methylene blue: Optimization using the response surface method (RSM). *Environ. Sustain. Indic.* **2022**, *16*, 100214. [CrossRef]
28. Subramanian, S.; Øye, G. Aqueous carbon black dispersions stabilized by sodium lignosulfonates. *Colloid Polym. Sci.* **2021**, *299*, 1223–1236. [CrossRef]

29. Jiang, S.; Bai, J.; Ying, X.; Han, J.; Zhu, K.; Mao, L. Synthesis and performance investigation of carbon black hyperdispersant IMD. *J. Coat. Technol. Res.* **2023**, *20*, 1529–1539. [CrossRef]
30. Lee, P.T.; Chiu, C.-W.; Chang, L.-Y.; Chou, P.-Y.; Lee, T.-M.; Chang, T.-Y.; Wu, M.-T.; Cheng, W.-Y.; Kuo, S.-W.; Lin, J.-J. Tailoring pigment dispersants with polyisobutylene twin-tail structures for electrowetting display application. *ACS Appl. Mater. Interfaces* **2014**, *6*, 14345–14352. [CrossRef] [PubMed]
31. Ramaye, Y.; Dabrio, M.; Roebben, G.; Kestens, V. Development and validation of optical methods for zeta potential determination of silica and polystyrene particles in aqueous suspensions. *Materials* **2021**, *14*, 290. [CrossRef]
32. Fathi, A.; Hatami, K.; Grady, B.P. Effect of carbon black structure on low-strain conductivity of polypropylene and low-density polyethylene composites. *Polym. Eng. Sci.* **2012**, *52*, 549–556. [CrossRef]
33. Liu, A.; Tian, H.; Ju, X.; Wang, W.; Han, P.; Li, W. In-situ growth of layered double hydroxides nanosheet arrays on graphite fiber as highly dispersed nanofillers for polymer coating with excellent anticorrosion performances. *J. Taiwan Instit. Chem. Eng.* **2019**, *104*, 330–340. [CrossRef]
34. Muranaka, M.; Yoshizawa, H.; Ono, T. Design of polylactide-grafted copolymeric stabilizer for dispersion polymerization of D, L-lactide. *Colloid Polym. Sci.* **2009**, *287*, 525–532. [CrossRef]
35. Pirrung, F.O.; Quednau, P.H.; Auschra, C. Wetting and dispersing agents. *Chimia* **2002**, *56*, 170. [CrossRef]
36. Huo, J.H.; Peng, Z.G.; Ye, Z.b.; Feng, Q.; Zheng, Y.; Zhang, J.; Liu, X. Preparation, characterization, and investigation of poly (AMPS/AM/SSS) on application performance of water-based drilling fluid. *J. Appl. Polym. Sci.* **2018**, *135*, 46510. [CrossRef]
37. Wang, Z.; Zhang, G.; Li, Y. Preparation of chitosan/polyacrylamide/graphene oxide composite membranes and study of their methylene blue adsorption properties. *Materials* **2020**, *13*, 4407. [CrossRef]
38. Kukobat, R.; Hayashi, T.; Matsuda, T.; Sunaga, M.; Futamura, R.; Sakai, T.; Kaneko, K. Essential role of viscosity of SWCNT inks in homogeneous conducting film formation. *Langmuir* **2016**, *32*, 6909–6916. [CrossRef] [PubMed]
39. Li, W.; Yu, Y.; Xiong, D.; Qi, Z.; Fu, S.; Yu, X. Effects of chemical dispersant on the surface properties of kaolin and aggregation with spilled oil. *Environ. Sci. Pollut. Res.* **2022**, *29*, 30496–30506. [CrossRef]
40. Kang, Y.; Zhu, A. Synthesis of SMA-PEG hyperdispersant and its stabilization mechanism for carbon black. *J. Appl. Polym. Sci.* **2023**, *140*, e54001. [CrossRef]
41. Zhou, Y.; Ma, Y.; Sun, Y.; Xiong, Z.; Qi, C.; Zhang, Y.; Liu, Y. Robust superhydrophobic surface based on multiple hybrid coatings for application in corrosion protection. *ACS Appl. Mater. Interfaces* **2019**, *11*, 6512–6526. [CrossRef]
42. Xiang, L.; Guo, Z.; Yang, L.; Qin, Y.; Chen, Z., Wang, T.; Sun, W.; Wang, C. Modification of crystal growth of NaA zeolite with steric hindrance agents for removing ammonium ion from aqueous solution. *J. Ind. Eng. Chem.* **2024**, *132*, 326–334. [CrossRef]
43. Shuai, W.; Wang, S.; Sun, T.; Yin, H.; Zu, Y.; Yao, G.; Li, Z.; Qi, Z.; Zhong, M. Improving the steric hindrance effect of linear sulfonated acetone–formaldehyde dispersant and its performance in coal–water slurry. *RSC Adv.* **2022**, *12*, 35508–35516. [CrossRef] [PubMed]
44. Ali, M.K.A.; Xianjun, H. Colloidal stability mechanism of copper nanomaterials modified by bis (2-ethylhexyl) phosphate dispersed in polyalphaolefin oil as green nanolubricants. *J. Colloid Interface Sci.* **2020**, *578*, 24–36. [CrossRef] [PubMed]
45. Yoshioka, K.; Sakai, E., Daimon, M.; Kitahara, A. Role of steric hindrance in the performance of superplasticizers for concrete. *J. Am. Ceram. Soc.* **1997**, *80*, 2667–2671. [CrossRef]
46. Zhao, H.; Deng, M.; Tang, M. Synthesis, characterization and dispersing mechanism of aminosulfonate-phenol-formaldehyde superplasticizer for the cement particles. *Polym. Compos.* **2018**, *39*, 2250–2258. [CrossRef]

MDPI AG
Grosspeteranlage 5
4052 Basel
Switzerland
Tel.: +41 61 683 77 34

Polymers Editorial Office
E-mail: polymers@mdpi.com
www.mdpi.com/journal/polymers

www.ingramcontent.com/pod-product-compliance
Lightning Source LLC
Chambersburg PA
CBHW041345210926
51850CB00013B/2998
* 9 7 8 3 7 2 5 8 3 9 9 8 8 *